Compendium of Plant Genomes

Series Editor

Chittaranjan Kole, Raja Ramanna Fellow, Government of India, ICAR-National Research Center on Plant Biotechnology, Pusa, New Delhi, India

Whole-genome sequencing is at the cutting edge of life sciences in the new millennium. Since the first genome sequencing of the model plant Arabidopsis thaliana in 2000, whole genomes of about 100 plant species have been sequenced and genome sequences of several other plants are in the pipeline. Research publications on these genome initiatives are scattered on dedicated web sites and in journals with all too brief descriptions. The individual volumes elucidate the background history of the national and international genome initiatives; public and private partners involved; strategies and genomic resources and tools utilized; enumeration on the sequences and their assembly; repetitive sequences; gene annotation and genome duplication. In addition, synteny with other sequences, comparison of gene families and most importantly potential of the genome sequence information for gene pool characterization and genetic improvement of crop plants are described.

Interested in editing a volume on a crop or model plant?

Please contact Prof. C. Kole, Series Editor, at ckoleorg@gmail.com

More information about this series at http://www.springer.com/series/11805

This book series is dedicated to my wife Phullara and our children Sourav and Devleena

Chittaranjan Kole

Sandra M. Schmöckel
Editor

The Quinoa Genome

Springer

Editor
Sandra M. Schmöckel
Department of Physiology of Yield
Stability, Institute of Crop Science
University of Hohenheim
Stuttgart, Germany

ISSN 2199-4781 ISSN 2199-479X (electronic)
Compendium of Plant Genomes
ISBN 978-3-030-65239-5 ISBN 978-3-030-65237-1 (eBook)
https://doi.org/10.1007/978-3-030-65237-1

This Springer imprint is published by the registered company Springer Nature Switzerland AG
The registered company address is: Gewerbestrasse 11, 6330 Cham, Switzerland

Preface to the Series

Genome sequencing has emerged as the leading discipline in the plant sciences coinciding with the start of the new century. For much of the twentieth century, plant geneticists were only successful in delineating putative chromosomal location, function, and changes in genes indirectly through the use of a number of "markers" physically linked to them. These included visible or morphological, cytological, protein, and molecular or DNA markers. Among them, the first DNA marker, the RFLPs, introduced a revolutionary change in plant genetics and breeding in the mid-1980s, mainly because of their infinite number and thus potential to cover maximum chromosomal regions, phenotypic neutrality, absence of epistasis, and codominant nature. An array of other hybridization-based markers, PCR-based markers, and markers based on both facilitated construction of genetic linkage maps, mapping of genes controlling simply inherited traits, and even gene clusters (QTLs) controlling polygenic traits in a large number of model and crop plants. During this period, a number of new mapping populations beyond F2 were utilized and a number of computer programs were developed for map construction, mapping of genes, and for mapping of polygenic clusters or QTLs. Molecular markers were also used in the studies of evolution and phylogenetic relationship, genetic diversity, DNA fingerprinting, and map-based cloning. Markers tightly linked to the genes were used in crop improvement employing the so-called marker-assisted selection. These strategies of molecular genetic mapping and molecular breeding made a spectacular impact during the last one and a half decades of the twentieth century. But still they remained "indirect" approaches for elucidation and utilization of plant genomes since much of the chromosomes remained unknown and the complete chemical depiction of them was yet to be unraveled.

Physical mapping of genomes was the obvious consequence that facilitated the development of the "genomic resources" including BAC and YAC libraries to develop physical maps in some plant genomes. Subsequently, integrated genetic–physical maps were also developed in many plants. This led to the concept of structural genomics. Later on, emphasis was laid on EST and transcriptome analysis to decipher the function of the active gene sequences leading to another concept defined as functional genomics. The advent of techniques of bacteriophage gene and DNA sequencing in the 1970s was extended to facilitate sequencing of these genomic resources in the last decade of the twentieth century.

As expected, sequencing of chromosomal regions would have led to too much data to store, characterize, and utilize with the-then available computer software could handle. But the development of information technology made the life of biologists easier by leading to a swift and sweet marriage of biology and informatics, and a new subject was born—bioinformatics.

Thus, the evolution of the concepts, strategies, and tools of sequencing and bioinformatics reinforced the subject of genomics—structural and functional. Today, genome sequencing has traveled much beyond biology and involves biophysics, biochemistry, and bioinformatics!

Thanks to the efforts of both public and private agencies, genome sequencing strategies are evolving very fast, leading to cheaper, quicker, and automated techniques right from clone-by-clone and whole-genome shotgun approaches to a succession of second-generation sequencing methods. The development of software of different generations facilitated this genome sequencing. At the same time, newer concepts and strategies were emerging to handle sequencing of the complex genomes, particularly the polyploids.

It became a reality to chemically—and so directly—define plant genomes, popularly called whole-genome sequencing or simply genome sequencing.

The history of plant genome sequencing will always cite the sequencing of the genome of the model plant Arabidopsis thaliana in 2000 that was followed by sequencing the genome of the crop and model plant rice in 2002. Since then, the number of sequenced genomes of higher plants has been increasing exponentially, mainly due to the development of cheaper and quicker genomic techniques and, most importantly, the development of collaborative platforms such as national and international consortia involving partners from public and/or private agencies.

As I write this preface for the first volume of the new series "Compendium of Plant Genomes," a net search tells me that complete or nearly complete whole-genome sequencing of 45 crop plants, eight crop and model plants, eight model plants, 15 crop progenitors and relatives, and three basal plants is accomplished, the majority of which are in the public domain. This means that we nowadays know many of our model and crop plants chemically, i.e., directly, and we may depict them and utilize them precisely better than ever. Genome sequencing has covered all groups of crop plants. Hence, information on the precise depiction of plant genomes and the scope of their utilization are growing rapidly every day. However, the information is scattered in research articles and review papers in journals and dedicated Web pages of the consortia and databases. There is no compilation of plant genomes and the opportunity of using the information in sequence-assisted breeding or further genomic studies. This is the underlying rationale for starting this book series, with each volume dedicated to a particular plant.

Plant genome science has emerged as an important subject in academia, and the present compendium of plant genomes will be highly useful to both students and teaching faculties. Most importantly, research scientists involved in genomics research will have access to systematic deliberations on the plant genomes of their interest. Elucidation of plant genomes is of interest not only for the geneticists and breeders, but also for practitioners of an array of plant science disciplines, such as taxonomy, evolution, cytology, physiology, pathology, entomology, nematology, crop production, biochemistry, and obviously bioinformatics. It must be mentioned that information regarding each plant genome is ever-growing. The contents of the volumes of this compendium are, therefore, focusing on the basic aspects of the genomes and their utility. They include information on the academic and/or economic importance of the plants, description of their genomes from a molecular genetic and cytogenetic point of view, and the genomic resources developed. Detailed deliberations focus on the background history of the national and international genome initiatives, public and private partners involved, strategies and genomic resources and tools utilized, enumeration on the sequences and their assembly, repetitive sequences, gene annotation, and genome duplication. In addition, synteny with other sequences, comparison of gene families, and, most importantly, the potential of the genome sequence information for gene pool characterization through genotyping by sequencing (GBS) and genetic improvement of crop plants have been described. As expected, there is a lot of variation of these topics in the volumes based on the information available on the crop, model, or reference plants.

I must confess that as the series editor, it has been a daunting task for me to work on such a huge and broad knowledge base that spans so many diverse plant species. However, pioneering scientists with lifetime experience and expertise on the particular crops did excellent jobs in editing the respective volumes. I myself have been a small science worker on plant genomes since the mid-1980s and that provided me the opportunity to personally know several stalwarts of plant genomics from all over the globe. Most, if not all, of the volume editors are my longtime friends and colleagues. It has been highly comfortable and enriching for me to work with them on this book series. To be honest, while working on this series I have been and will remain a student first, a science worker second, and a series editor last. And I must express my gratitude to the volume editors and the chapter authors for providing me the opportunity to work with them on this compendium.

I also wish to mention here my thanks and gratitude to the Springer staff, particularly Dr. Christina Eckey and Dr. Jutta Lindenborn for the earlier set of volumes and presently Ing. Zuzana Bernhart for all their timely help and support.

I always had to set aside additional hours to edit books beside my professional and personal commitments—hours I could and should have given to my wife, Phullara, and our kids, Sourav and Devleena. I must mention that they not only allowed me the freedom to take away those hours from them but also offered their support in the editing job itself. I am really not sure whether my dedication of this compendium to them will suffice to do justice to their sacrifices for the interest of science and the science community.

New Delhi, India Chittaranjan Kole

Preface

Compendium of Plant Genomes: Quinoa

Quinoa (*Chenopodium quinoa*, Willd.) is a nutrient-rich pseudo-cereal with many desirable characteristics as a novel crop. Historically, it has been an important grain crop, "the mother grain", in the Andean regions of South and central America for thousands of years. Recently, it has gained much attention in many other parts of the world for its desirable characteristics such as abiotic stress tolerance, its favorable nutrient profile, and absence of gluten—making it often referred to as a "superfood".

The interests and motivation of the authors of the work here presented in Quinoa research are diverse, but we are all joined by the passion for this plant. Quinoa is a dicotyledonous leafy plant with small seeds that come in various colors. It has become a favorite research object for many, studying, for instance, how it performs so well in harsh environments, investigating its domestication and evaluating it for breeding purposes. To address these research questions, it is necessary to unravel the genome information. Quinoa is an allotetraploid plant, making genome sequencing efforts not a trivial undertaking. However, there are now several genome sequences for quinoa available, allowing in depth genetics and genomics studies to commence. This book is being published as these resources only just become available and will therefore not only describe what has been done, but also delve into what might be possible.

The chapters in this book were contributed by experts in their respective fields of research, starting off with an overview of the history of the quinoas in South America and its initial cultivation thousands of years ago. Chapters 2 and 3 then provide the botanical context for domestication in South America and North America. To set the scene for quinoa genomics, Chap. 4 will focus on quinoa cytogenetics. A detailed description of the genome scale quinoa reference genome assembly is then provided in Chap. 5. Chapter 6 provides an overview of what is known so far regarding structural and functional genomics. Chapter 7 discusses the broad diversity of quinoa and how this influences breeding for future goals. Chapters 8–10 look at three examples of current research targets: Saponins, abiotic stress tolerance, and flowering. Finally, Chap. 11 looks at evolution in the quinoa genome and further developments that can be anticipated.

This book provides a comprehensive reference material in Quinoa genomics research for those already in the field, or new members considering joining of the Quinoa research community.

It has been a great privilege to work with colleagues of the quinoa community in the preparation of this manuscript and to see the wide range of genomics researching being done in this still "novel" species. We are grateful to all the authors for their contribution in writing chapters of high quality. Finally, we would also like to thank the traditional owners and custodians of this ancient grain, throughout the Americas for allowing us to study and to develop this crop so that Quinoa can be grown and appreciated worldwide.

Stuttgart, Germany Sandra M. Schmöckel

Contents

1 **History of the Quinuas in South America** 1
Mario E. Tapia

2 **Botanical Context for Domestication in South America** 13
Ramiro Nestor Curti and Hector Daniel Bertero

3 **Botanical Context for Domestication in North America** 33
Eric N. Jellen, David E. Jarvis, Nuri Benet-Pierce, and Peter J. Maughan

4 **Quinoa Cytogenetics** 51
Bożena Kolano and Maja Orzechowska

5 **A Chromosome-Scale Quinoa Reference Genome Assembly** 65
Bo Li and Damien J. Lightfoot

6 **Structural and Functional Genomics of *Chenopodium quinoa*** 81
Elodie Rey and David E. Jarvis

7 **Quinoa Diversity and Its Implications for Breeding** 107
Katharina B. Böndel and Karl J. Schmid

8 **Saponins of Quinoa: Structure, Function and Opportunities** 119
Sophie Otterbach, Gordon Wellman, and Sandra M. Schmöckel

9 **Abiotic Stress Tolerance in Quinoa** 139
Luke Grenfell-Shaw and Mark Tester

10 **Flowering in *Chenopodium* and Related Amaranths** 169
Helena Štorchová

11 **Quinoa—Evolution and Future Perspectives** 179
Gabriela Alandia, Arnesta Odone, Juan Pablo Rodriguez, Didier Bazile, and Bruno Condori

Contributors

Gabriela Alandia University of Copenhagen, Faculty of Science, Department of Plant and Environmental Sciences, Taastrup, Denmark

Didier Bazile UPR GREEN, CIRAD, Univ Montpellier, Montpellier, France

Nuri Benet-Pierce Department of Biology, San Diego State University, San Diego, CA, USA

Hector Daniel Bertero Faculty of Agronomy, University of Buenos Aires and IFEVA-Conicet, Buenos Aires, Argentina

Katharina B. Böndel Institute for Plant Breeding, Seed Sciences and Population Genetics, University of Hohenheim, Stuttgart, Germany

Bruno Condori Inter-American Institute for Cooperation on Agriculture, Bolivian Representation – IICA, La Paz, Bolivia

Ramiro Nestor Curti Universidad Nacional de Salta and Conicet-CCT-Salta, Salta, Argentina

Luke Grenfell-Shaw Division of Biological and Environmental Sciences and Engineering (BESE), King Abdullah University of Science and Technology (KAUST), Thuwal, Saudi Arabia

David E. Jarvis Department of Plant and Wildlife Sciences, College of Life Sciences, Brigham Young University, Provo, UT, USA

Eric N. Jellen Department of Plant and Wildlife Sciences, Brigham Young University, Provo, UT, USA

Bożena Kolano Institute of Biology, Biotechnology and Environmental Protection, Faculty of Natural Sciences, University of Silesia in Katowice, Katowice, Poland

Bo Li School of Life Sciences, Lanzhou University, Lanzhou, China; Ministry of Education Key Laboratory of Cell Activities and Stress Adaptations, Lanzhou, China

Damien J. Lightfoot Biological and Environmental Science and Engineering, King Abdullah University of Science and Technology, Thuwal, Kingdom of Saudi Arabia

Peter J. Maughan Department of Plant and Wildlife Sciences, Brigham Young University, Provo, UT, USA

Arnesta Odone University of Copenhagen, Faculty of Science, Department of Plant and Environmental Sciences, Taastrup, Denmark

Maja Orzechowska Institute of Biology, Biotechnology and Environmental Protection, Faculty of Natural Sciences, University of Silesia in Katowice, Katowice, Poland

Sophie Otterbach Department Physiology of Yield Stability, Institute of Crop Science, Faculty of Agriculture, University of Hohenheim, Stuttgart, Germany

Elodie Rey Biological and Environmental Sciences and Engineering Division (BESE), King Abdullah University of Science and Technology (KAUST), Thuwal, Saudi Arabia

Juan Pablo Rodriguez Crop Diversification and Genetics Program, International Center for Biosaline Agriculture, ICBA, Dubai, United Arab Emirates

Karl J. Schmid Institute for Plant Breeding, Seed Sciences and Population Genetics, University of Hohenheim, Stuttgart, Germany

Sandra M. Schmöckel Department Physiology of Yield Stability, Institute of Crop Science, Faculty of Agriculture, University of Hohenheim, Stuttgart, Germany

Helena Štorchová Institute of Experimental Botany, Czech Academy of Sciences, Prague 6, Lysolaje, Czech Republic

Mario E. Tapia NGO AEDES, Arequipa, Peru

Mark Tester Division of Biological and Environmental Sciences and Engineering (BESE), King Abdullah University of Science and Technology (KAUST), Thuwal, Saudi Arabia

Gordon Wellman Division of Biological and Environmental Sciences and Engineering (BESE), King Abdullah University of Science and Technology (KAUST), Thuwal, Kingdom of Saudi Arabia

Abbreviations

ABA	Abcisic acid
AED	Annotation Edit Distance
AFLP	Amplified Fragment Length Polymorphism
AGP	A Golden Path
AHAS	Acetohydroxyacid synthase
AP1	*APETALA 1*
APX	Ascorbate peroxidase
BAC	Bacterial artificial chromosome
BESs	Bacterial artificial chromosome End Sequences
bHLH	Basic helix–loop–helix
bp	Base-pair
BSA	Bulk segregant analysis
BTC1	*BOLTING TIME CONTROL 1*
BUSCO	Benchmarking Universal Single-Copy Orthologs
BvBBX19	*DOUBLE B-BOX TYPE ZINC FINGER 19*
CAT	Catalase
CDS	Coding regions
cM	Centimorgans
CMA_3	Chromomycin A_3
CNV	Copy number variation
CO	CONSTANS
cp	Chloroplastic
DE	Differential expression
DTI	Drought tolerance index
EBCs	Epidermal bladder cells
EMS	Ethyl methanesulfonate
ER	Endoplasmic reticulum
EST	Expressed sequence tag
FAO	Food and Agriculture Organization of the United Nations
FD	FLOWERING LOCUS D
FID	Flame ionizing detector
FISH	Fluorescence in situ hybridization
FLC	*FLOWERING LOCUS C*
FPP	Farnesyl pyrophosphate
FT	*FLOWERING LOCUS T*
FTL	*Flowering locus T-like*

GBSS1	*Granule bound starch synthase 1*
GC	Gas chromatography
ghy	Green hypocotyl mutant
GISH	Genomic *in situ* hybridization
GO	Gene ontology
GRIN	Germplasm Resources Information Network
GS	Glutamine synthetase's
gSSRs	Genomic screen-derived SSRs
HED	Hederagenin
HPLC	High-pressure LC
HSPs	Heat-shock proteins
IAA	Indole-3-acetic acid
IGS	Intergenic spacer
Indels	Insertions-deletions
IRs	Inverted repeat regions
Iso-Seq	Isoform sequencing
IYQ2013	The International Year of Quinoa
Kañiwa	*Chenopodium paliidicaule* Aellen
Kiwicha	*Amaranthus caudatus*
Ks	Rate of synonymous substitutions per synonymous site
LC	Liquid chromatography
LC-ESI-MS	Liquid chromatography-electrospray mass-spectrometry
LGs	Linkage groups
LINE	Long Interspersed Nuclear Element
LSC	Long single copy
LTR	Long terminal repeat
MAF	Minor allele frequency
masl	METERS above sea level
MeJA	Methyl jasmonate
miRNAs	Micro RNAs
MS	Mass spectrometry
MSBP	Membrane steroid binding protein
MVA	Mevalonate pathway
MYA	Million years ago
NIRS	Near-infrared spectroscopy
NJ	Neighbor joining
NMR	Nuclear magnetic resonance
NOR	Nuclear organizing region
NR	Non-redundant
NTS	Nuclear 5S rDNA spacer region
NUE	Nutrient use efficiency
OA	Oleanolic acid
OSC	2,3-oxidosqualene cyclase
PA	Phytolaccagenic acid
PacBio	Pacific Biosciences
PCA	Principal component analysis
quinoa	*Chenopodium quinoa* Willd.

R genes	Disease resistance genes
RAPD	Random Amplified Polymorphic DNA
RFT1	*RICE FLOWERING LOCUS T 1*
RIL	Recombinant inbred line
rRNAs	Ribosomal RNAs
SA	Serjanic acid
SAAT	Sonication-assisted *Agrobacterium*-mediated transformation
SI	Self-incompatibility
SMTA	Standard Material Transfer Agreement
SNP	Single nucleotide polymorphism
SOD	Superoxide dismutase
SOS1	*Salt Overly Sensitive 1*
SSC	Short single copy
SSR	Simple sequence repeat
TALENs	Transcription activator-like effector nucleases
TE	Transposable element
TIR	Terminal Inverted Repeat
TLC	Thin-layer chromatography
TMS	Trimethylsilyl
TR	Tandem repeats
tRNAs	Transfer RNAs
TSAR	*Triterpene saponin biosynthesis activating regulator*
TSARL	*TSAR-like*
UGTs	Uridine diphosphate UDP-dependent glycosyltransferases
UPGMA	Unweighted pair-group method with arithmetic averages
USDA	United States Department of Agriculture
VIGS	Virus-induced gene silencing
WGS	Whole genome sequencing
ZFNs	Zinc finger nucleases

History of the Quinuas in South America

1

Mario E. Tapia

Abstract

Quinua (*Chenpopodium quinoa Willd*) is an Andean food grain domesticated manly in the highlands of the central part of the Andes in South America by the quechua and aimara people since at least 3,000 years. The geographical distribution of the crop covers territories from south of Colombia to north of Argentina and central part of Chile. The actual technological conditions and production areas in each country are discussed and presented a differentiation of five main types of quinua including, the geographical distribution: the Andean valleys; the high plateau around the lake Titicaca; the salares in Bolivia , the sea level and the yunga or subtropical quinua types. Suggestions for a future need of research as well as agro industrial process with references to the different traditional uses from the roots, grains , leaves as stems is describe.

Keywords

Origin · Quinua distribution · Main types · Research needed

M. E. Tapia (✉)
NGO AEDES, Arequipa, Peru
e-mail: mario.tapia.n@gmail.com

The Andean settlers cultivated, besides potato and corn, the Andean Grains such as quinoa (*Chenopodium quinoa* Willd), kiwicha (*Amaranthus caudatus*) and Amaranthus mantegazzianus and later kañiwa (*Chenopodium paliidicaule* Aellen) for thousands of years (Hunsiker 1952; Cardenas 1969; Nuñez 1970; Pulgar Vidal 1954). The Andrean Grains are known specially for their high nutritional value in quality proteins to replace the lack of milk.

1.1 Origin

The age of the domestication of quinoa and the beginning of its use as a food can be dated at least 2,000 to 3,000 years BC because of its presence in archeological remains. Towle (1961) mentions several archeological findings of quinoa, with fruitful panicles, branches and loose grains, found in different regions of Peru and in the coastal zone of Arica, Chile.

As Nuñez (1970) indicates, it is not well known how quinuas were domesticated. However, findings in northern Chile (Chinchorro complex), suggests that quinoa was used before 3000 BC. In the area of Ayacucho (Peru), Uhle 1919 describes remains of quinua grains and gives an even earlier date, 5,000 years BC, as the beginning of domestication.

In a Khipu (a traditional accounting system, based on alpaca fiber with different size knots

S. M. Schmöckel (ed.), *The Quinoa Genome*, Compendium of Plant Genomes,
https://doi.org/10.1007/978-3-030-65237-1_1

and different colors, which allows recording the production of food crops in the Inca Empire) of the sixteenth century, studied by Murra (1975), the importance of quinoa production is noted in the central highlands of Peru. The production of quinoa precedes the potato production in the khipu.

Ulloa Mogollon refers in 1586 to the use of quinoa in the province of Collaguas (Bolivia). As already mentioned, there is evidence that quinoa was widely cultivated in the valleys of northern Chile. In 1558, Cortes Hogea, first to visit the island of Chiloe (Chile), found quinoa sowings. In the North Argentinian territory, Pedro Sotelo (1583) mentions these crops in the valley of Calchaquies and in the vicinity of Cordoba.

However, it should be noted that there is little information about religious rites with the use of quinoa. Cárdenas (1969) does not believe that quinoa has completely replaced corn in the highlands, and it is not being used for religious ceremonies. It points out the tradition that the natives of the mountain range still have, of traveling to the valleys to exchange quinoa for corn or salt.

It is still not understood when and from which wild species the native domesticated quinoa varieties or land races were obtained and more research is necessary to provide evidence on domestacation; however, there are important hypotheses.

For Andean researchers, the center of origin and domestication is the altiplano, around Lake Titicaca, shared by countries such as Peru and Bolivia (Gandarillas 1968; Mujica 1977). Other scholars consider the existence of different centers of origin in the inter-Andean valleys and that the quinoa has been taken to the altiplano of Lake Titicaca, which is considered the great center of diversification.

Other researchers suggest that cultivated quinoa would have an ancestor in *Chenopodium berlandieri,* a wild species from North America; however, there is little evidence of its use as a crop or food in that region (Maughan 2013).

The Andean origin of quinoa is supported by the existence of the different regional names of quinoa. There are many regions within the native language. Robledo, quoted by Pulgar Vidal (1954), specifies that the Chibchas (Colombia) called it "pasca" and with great surprise it has been defined that "pasca" etymologically means the pot or food of the father.

The name "suba or supha" (Chibcha language) is indicated by Pulgar Vidal (1954). As the primitive name of quinoa in the Bogota area, and the author relates it always with the Aymara term "hupha", which is used in some regions of Bolivia. In the rest of the territory that is now Colombia, the Quechua Quinoa name was generally used, but in Cundinamarca the indigenous name was "pasca".

In the Aymara language, quinoa has different names, depending on the variety: the wild quinoa is called "cami", while the cultivated white and most appreciated is called "ppfique"; the red one is called "kana llap"; the yellow one, "cchusllunca"; another yellowish variety is called "ccachu yusi"; and the wild one "isualla", according to Latchman (1936). However, this author confuses the quinoa with the kañihua (*Chenopodium pallidicaule* Aellen) and includes it within the quinoa varieties, calling it "cinnamon or cañagua". The same author adds that in the North of Chile quinoa was cultivated and that it is called in atacameño "dahue". Bertonio (1879) added Aymara names for the varieties, such as "aara", "callapi" and "vocali". It also mentions a variety between colorada and negra, la "cami hupa".

An additional source of information to recognize the origin and distribution of this Andean grain is the tradition that exists in the diverse ways of consuming this grain in Colombia, Ecuador, Peru, Bolivia, Northern Chile and Argentina, both in the preparation of different dishes, and drinks and processed foods. Traditional dishes are known as "lawa", thick quinoa soup; with fat and katawi as the preparation of white chicha, etc. Quinoa's tender leaves, known as "liccha", are used extensively in salads, and to prepare the ashes of the roots and stem for the preparation of the "llipta", alkali, used to chew coca leaves (Beyerdorf and Blanco 1984); the names of the surplus residue of the leaves and stems and the name of "jipi" to designate the residues of the grains and small seeds.

Samples of Amaranthus with erect inflorescence (very similar to those of quinoa) have been found in the area of Tarija, Bolivia. This receives the name of "coimi"; however, indistinctly it is also called quinua (Tapia 1979).

Quinoa has a close relative cultivated in the highlands of Mexico as *Chenopodium nuttaliae,* called "huauzontle", which resembles Andean quinoa (Wilson 1976). Apparently, it had high importance in times of the Aztec empire as a food, according to the codex of Antonio Mendoza, first Viceroy of Mexico between the years 1535 and 1550, in which the tributes are indicated in grains like "huauzontle", from each of the 363 vassal peoples of the Aztec Empire paid annually, to the central government, with this grain (quoted by Hunkizer 1952) (Photo 1.1).

The first Spanish to mention the cultivation of quinoa in the new world was Pedro de Valdivia, who informed the Emperor Carlos I in 1591 about the crops in the surroundings of Concepción (Chile) and indicated that "the region is abundant in all the maintenance that the Indians plant for their sustenance as well as corn, potatoes, quinoa." (Hunsiker 1952).

Photo 1.1 Ceremonial vessel of the Wari culture, Ayacucho, Peru, ninth century, with a drawing of the quinoa plant on the right

However, there was a strong confusion in the seventeenth century when quinoa was not identified in all cases. The first Spaniards, for example, always related quinoa to the pigweed that grows as a weed in the Iberian Peninsula. BERNABE Cobo (1653) says "quinoa is a plant very similar to the 'bledos'". Bledos were described in Europe as annual plants with creeping stems of the family of the Chenopodiaceae and Amaranthaceae, like *Amaranthus blitum L.* Confusion develops as the eminent botanist Carolus Clusius, in his Historia Rariorum Plantorum of 1601, presents the first illustration of a species that he called quinoa, but which is actually a plant of *Amaranthus caudatus* L.

1.2 Geographical Distribution

Quinoa nowdays has been widely distributed in the Andean countries. Cieza de León (1560) reports that in the South of Colombia, quinoa was also cultivated in the highlands between the cities of Pasto and Quito, and he writes: "In all these towns, it gives little corn or almost none, because of frost, but other species as quinua does".

The Inca Garcilaso de la Vega, in his famous Royal Comments, says about quinoa "in the second place of the crops that are raised on the face of the earth give what they call quinoa and, in Spanish, 'millet', or small rice, because in the grain and the taste it resembles something". This historian refers to his trip to Spain in the sevententh century with the first export of quinoa grains to the old world, which fatally failed to spread because "the seeds arrived dead" (Photo 1.2).

The Spanish chroniclers did not always treat these Andean Grains with the just evaluation. The fact that quinoa resembled the "Spanish bledo", evoked that quinoa was considered a plant without value ("worth a damn"). It is also true that there were numerous confusions between quinuas, cañihua and amaranth, because of the similar architecture, morhology of the plant and the shape of the grain, and it was called, for example, cinchona quinoa or cañihua.

In the North of Peru the cultivation of quinoa was common, but in most fields quinoa was

Photo 1.2 Inca Gracilazo de la Vega, the first to introduce quinua in Europe

grown together with corn. Further to the South, quinoa gained importance both in the "Callejon de Huaylas" and in the Mantaro Valley, where it was widely cultivated by the Huancas tribe. Numerous sources report that the Spanish were surprised finding in each region, the "ccolcas" or food deposits with large amounts of quinoa grain.

Relatives of quinoa are very common, especially in the agricultural area around lake Titicaca and neighboring areas, as a group of Chenopodium species called "ayaras or ajaras". In the Arequipa countryside, several species from the Chenopodium genus can be found, known locally as "llicchas", which is the same name given to the tender quinoa leaves in the Puno highlands. These species are considered as weeds, although their tender leaves are consumed in green rolls called "loritos" (Tapia et al. 2014) (Photo 1.3).

Quinoa was initially studied by archeologists, Towle, 19. Uhle (1919) botanics as Hunsiker (1952), Cardenas (1969), Leon (1964) followed by nutricionists, White (1955), geographers, Pulgar Vidal (1954) finally researched by agronomists Gandarillas (1967), Rea (1944), Mujica (1977), Canahua (2012), Rojas (2013), Gomez (2013) and ultimately specialists in molecular studies. Maughan (2013), Jiménez (2014).

1.3 Current Production Centers

From the North to South of the South American continent, the following current centers of increased production of quinoa can be identified.

Photo 1.3 Cultivars of quinoa from the high plateau of Puno, Peru

1.3.1 Colombia

According to Acosta (1948), the cultivation of quinoa was abundant in the past, but now almost abandoned in the Colombian savannas, because most of the agricultural areas of the cold lands of Cundinamarca and Boyaca became pastures for grazing.

Pulgar Vidal (1954) believes that quinoa can and should be sown in the Cundinamarca area, a name that etymologically means "the country of frost", and that this plant could have a safe production.

At present, the region with the greatest cultivation is the province of Nariño, with the towns of Ipiales, Pueres, Contadero, Cordova, San Juan, Mocandino and Pasto (Pulgar Vidal 1954).

Since 1947 Professor Braulio Montenegro of the University of Nariño has dedicated his effort to the promotion of the cultivation of quinoa in Colombia. In 1958, he obtained an improved variety, the Dulce de Quitopamba that gives yields of 1500 to 2000 kg/ha, with fertilizations of 30 at 50 kg of nitrogen per hectare.

Romero (1976) made experimental plantings at the Marengo Agricultural Center of the National University in Bogota to test the production of new varieties. He found yields of 400 to 500 g per plant with the Bolivian Sajama variety.

1.3.2 Ecuador

In Ecuador, quinoa has persisted among peasants in the area of Carchi, Imbabura, Pichincha, Cotopaxi, Chimborazo and Loja (Cardozo 1976; Tapia 1976; Romero 1976; Peralta et al. 2012).

Morales (1975) compared some 18 ecotypes from the areas of Imbabura, Cayambe, Cotopaxi and Chimborazo with material from Bolivia, and highlighted the Ecuadorian ecotypes Chaucha, Punin, Grande and Staquinua with very good yields.

The quinoas of Latacunga, Ambato, Carchi, Riobamba and Cuenca are of small grain, in general of high size and quite bitter grain. It is estimated that the total cultivated area is about 1,200 ha in all the country (Freere et al. 1975). However, this area has increased remarkably in the last ten years.

1.3.3 Peru

At present, Peru is the country where most farmers cultivate and consume quinoa and where a large number of varieties have been selected (Tapia 1979; Mujica 1977).

In the region of the inter-Andean valleys, quinoa is cultivated within the fields of corn and beans, or as the border of potato crops. But it is in the highlands, where corn does not grow, where quinoa's cultivation becomes more important.

In Cajamarca it is customary to plant 6 to 10 rows of corn followed by one of quinoa, in a system known as Chaihua. Only in the highlands near the jalca you can see small quinoa fields in monoculture.

Other important areas are the Callejon de Huaylas in Ancash, the Mantaro valley and the Jauja highlands in Junin, Andahuaylas in Ayacucho, as well as the highlands of the department of Cusco.

In the Mantaro valley and the upper part of Jauja, the varieties Blanca and Rosada de Junín are planted, with very uniform grains and low saponin content.

In the Vilcanota valley between Cusco and Sicuani, at altitudes of 3000 and 3600 m with precipitation of more than 500 mm, the Yellow of Marangani variety is cultivated, whose yields can exceed 2000 kg/ha.

The Quinua White, variety of Junín has been fully adapted to the conditions of Anta in Cusco at 3700 m, there you can find the most extensive cash crops, up to 150 ha, with yields above 2000 kg/ha. As in the Marangani area, in Cusco, it is recognized as the Marangani variety.

Quinoa is really important in the Altiplano del Collao, department of Puno, above 3800 m, where corn cannot be produced. The quinoa cultivation plots appear in the small ravines or lands near lagoons or Lake Titicaca. The Cheweca variety has been selected around the Orurillo lagoon, which produces a small, almost sweet, very soft and special grain to make flours. Lescano from the region of Cabanillas comes the variety Kanccolla (del Collao), of almost sweet grains that has high yields. The variety called "jiura" rice of small grains, called Blanca de Ayaviri, has white and semi sweet grains.

Finally, on the Peruvian side of the lake quinoa ecotype "Blanca de Juli" is grown. Other local ecotypes include "Chullpi", with transparent grains.

According to the General Directorate of Statistics of Peru, the cultivation of quinoa covered more than 42,000 ha per year in 1951. This area initially decreased, but in the last decades it has been recovered to more than 65,000 ha. Of this area, more than 55% is concentrated in the department of Puno, in the South-west of the country, in the highlands bordering Bolivia.

1.3.4 Bolivia

In the Altiplano, salty plains, and in the inter-Andean valleys, quinoa has been maintained above all as a self-consumption crop for the thousands of farmers who appreciate its nutritional value (Gonzales 1917; Gandarlllas 1986 Rea).

Here we must also distinguish between the quinoa of the Altiplano with a plant of smaller size (up to 1.60 m) and the quinuas of the valleys that can reach 2 m or more. Around Lake Titicaca, the crops are concentrated in the peninsula of Copacabana and, with higher density, between the area of Desaguadero and Guaqui. Quinoa cultivation decreases in frequency and extension from the South until Oruro.

The main quinoa producing areas with some 3000–4000 ha are the provinces of Quijarro, Nor

Lipez and Daniel Campos, of the department of Potosi, and Ladislao Cabrera, of the department of Oruro, where the "quinua real" produced. It has a bigger seed size and a high content of saponins (Bautista 1976).

The region of the salt plains of Coipasa and Uyuni is characterized by the more xerophytic conditions in which quinoa is cultivated. This environment determined an adaptation of the methods of cultivation for about two centuries. This is the region from where quinoa is mostly exported to Europe and the United States.

1.3.5 Chile

In this country, quinoa is grown in two very different ecological and geographical zones. On the one hand, in the Chilean highlands (for example, Isluga, Iquique) in the North of the country, the conditions and varieties are very similar to the Bolivian altiplano (Lanino 1076). On the other hand, in the fields of the Concepción area, in the South, at sea level, with longer photoperiod, very different ecotypes of small, flattened, somewhat transparent (as cooked) grain are found. As an example, we mentioned the Catentoa variety (Junge et al. 1973).

Bazile Didier article

1.3.6 Argentina

Quinoa is grown mostly in the North-western region of Argentinia, in small areas of about 100 m in the highlands of Jujuy and Salta in the North. Vorano and Garcia (1977) believe that quinoa, despite having a series of difficulties in its use (mainly self-consumption), is an irreplaceable species for the conditions of the Argentine highlands (Arraguez 2017).

The cultivation of quinoa is spread throughout the Andes, however, in the case of Peru, it can be mentioned that there are at least 6 assembly centers in Cajamarca, the Callejon de Huaylas, in Junin, Ayacucho, Cusco and mostly in the highlands of Puno.

1.4 Main Types of Quinuas

The quinuas, according to their agro ecological adaptation, can be grouped into five major types (Tapia 1997):

- Quinuas including dry valleys (Junín, Ayacucho, Cochabamba) and humid valleys (Cajamarca in Peru and valleys of Ecuador and Colombia).
- Quinuas from the altiplano (around Lake Titicaca). White and colored are more frost resistant from the Suni agro-ecological zone.
- Quinuas from salt plains "salares" (South plains of Bolivia).
- Sea level from South latitude quinoa (Chile, Concepcion,).
- Quinuas from the Yunga agroecological zone (Bolivia at 1,500 m).

In the book Razas de Quinuas del Peru (Quinua Races in Peru), 24 major different groups are distinguished according to their morphology, grain characteristics, agroecological zones of cultivation and different ways of consumption (Tapia 2013). With the same approach of classification, Quinua races of Bolivia have been published by Gandarillas (1968) and of Ecuador by Gandarillas et al. (1989).

Quinuas from the valley environments could be differentiated between cultivars from dry areas grown on irrigated land, or under rainfed viroments as in Cajamarca, Cusco, Huaraz, valley Mantaro, Ayacucho, Abancay and Cochabamba (Bolivia). There is also the influence of increased precipitation North of Peru, which extends into Ecuador, and into Southern Colombia. In the area of Nariño, Colombia and Northern Ecuador there is a tall ecotype, highly branched, light green leaves and very white and sweet grain that gave rise to the variety Nariño, cultivated in Peru (Tapia 1982).

Quinoa from the altiplano is also produced under variable conditions: low rainfall and favorable weather conditions such as around Lake Titicaca; in lands close to lagoons or ravines near rivers like Kancolla cultivar is grown,

Blanca de Juli, and Tahuaco are native of South of Puno. Some cultivars are adapted to the high plains, at 3 900 m.a.s.l., for instance, Cheweca, Ccoitu, Wariponcho, Chullpi and Witulla, with colored panicles and that also withstand lower temperatures.

Cultivars in the region of Puno, Canahua (2012)

1. Janko or yurac. White
2. Chullpi or hyalines. White/transparent Good Clado, pure.
3. witullas. Colored Red/red, purple. High. Kispiño Flours, torrejas.
4. Wariponcho. Yellow/yellow for soups, flours.
5. Kcoito. White or lead/Good Turrets Flours. Plumb, brown.
6. Pasancallas. Red, white/red. High. Manna. Flours
7. Cuchi wila. Red/black High. Chicha Kispiño

The quinuas of the salares group in the south of Bolivia support extreme xerophytic conditions and their initial development is possible because they take advantage of the humidity of the holes dug at the time of sowing. The cultivation of quinoa in this area follows a very special production system: after the harvest, the soil takes a rest for 4 to 8 years, in recent times this period is shortened, producing negative effects due to depletion of the fertility of the soils.

Quinoa at sea level is adapted to humid conditions and with more moderate temperatures; they are located mostly in the latitudes South of 30° S (Concepción and Valdivia, Chile). Didier (2014).

Finally, there is a very small group of quinuas that have adapted to the conditions of yunga agro ecological zone of Bolivia, at altitudes between 1,500 to 2,000 m, with the characteristic of having the stem and perigonium orange to mature state (Tapia 1979). Their adaptation to subtropical climates allows them to tolerate higher levels of precipitation and heat. There is only one collection made in Bolivia and the samples of this group grew adequately in K'ayra, Cusco at 3,300 m above sea level, presenting a long vegetative period of more than 200 days.

Quinua is also cultivated in different production systems depending on the agroecolgical zones. Sometimes it is grown in pure or single crop fields, and sometimes it is grown together with maize or potatoes, or sometimes it is grown as a border of different crops like maize or potatoes (Photos 1.4 and 1.5).

Photo 1.4 Quinoa grown in suka collos or elevated land, Puno, Peru

Photo 1.5 Cultivation of quinoa in the aynokas system, Puno, Peru

1.5 Research Advances

The initial technical meeting related to quinoa was held in Puno, Peru 1968, called the First International Quinoa and Kañiwa Convention with the assistance of researchers from Peru and Bolivia. The second Convention was held in Potosi, Bolivia 1976.

From there, 12 International Andean crops Congress have been held in the six Andean countries (1977–2012). Also, six international congresses dedicated specifically to Andean grains have been organized (2001–2017). All the proceedings of these meetings have been published in Spanish.

Specific research done on this crop includes the floral biology (Rea 1944), nutritional evaluation (White 1944), botanical studies (Cardenes 1969 Leon), agricultural and plant breeding (Gandarillas 1986; Mujica 2013; Gomez 2013; Lescano 1994; Tapia 2012; Canahua 2012; Rojas 2013; Bonifacio et al. 2012).

The future development of this crop requires to propose an integrated plan for a greater diffusion, consumption and cultivation of quinoa, coordinated by the Andean countries, as to know in greater detail the agronomic and climatic adaptation characteristics of the different varieties of this species, that is, the agroecological zoning of the quinoa in south America and the potential for adaptation that has the varied genetic material, in the germplasm banks of this crop for other agro ecosystems.

In the altiplano conditions, the yields can be affected by frequent frosts, especially at the beginning of flowering. The valley varieties are adapted to temperatures that fluctuate between 10 and 18 °C and are not tolerant to frost. The varieties in the salt flats in Bolivia support temperatures of −8 °C, alkaline soils up to pH 8.0 and salinity up to 52 mS/cm (Mujica et al. 2001).

Quinoa had been grown from 2012 in the Peruvian coast, yields obtained are over 4,000 kg/ha under conditions of conventional agriculture with high level of fertilization and irrigation systems (Tapia and Lozada 2017).

1.6 Traditional Quinoa Utilization

Quinoa has been used for centuries by the Andean peasants who have used it in their usual diet, even considering it as an appropriate food for food security, that is, to keep it for the years of low harvests that occur periodically in the plateau.

In recent years, quinoa, especially quinuas of colors, have received special attention by the most renowned chefs and with this, its value and acceptance is spreading in urban areas.

In addition, the farmers keep the tradition of the integral use of the plant with different purposes that it has tried to recover in different events and field visits during the last 30 years.

Special mention should be given to the work of the NGO Pratec, which in its work of recovery of traditional knowledge has produced more than 1,500 primers, several of them dedicated to cultural traditions in the cultivation of quinoa.

Currently, quinoa is used in the preparation of soups, stews, grains, pures, desserts and drinks; transformation into flour: in baking biscuits, desserts, sweets, beverages, making noodles mixed with other flours, ingredient of sausages and meatballs, flakes; drinks, soups, sweets; (popeadas) or Pipocas: for direct use alone or with yogurt, ice cream, desserts and chocolate.

Quinoa, kañihua and all edible amaranthus species together, constituted an important component of the diet of prehispanic people in the highlands of the Andes. Its use was common in the Andean regions of Peru and Bolivia until the first third of the last century (1940), when it began its decline, when the reception of food aid from the United States began and the massive importation of subsidized wheat increased.

1.6.1 Preparation of the Llipt'a O (Llujt'a)

The stems of quinoa and, sometimes, the main root have an important traditional use, which is maintained until today: they are used in llipt'a (alcali) together with "chacchado", chewing of the coca leaf.

To obtain the llipt'a the stems and roots of quinoa are burnt, the ash is collected and mixed with water. Optionally milk or anise is added, which gives a more pleasant taste. With this, a type of circular muffin is formed, "biscuits", about 6 cm in diameter that are put to dry in the sun.

The ashes, which are alkaline, facilitate the extraction of cocaine, which is an alkaloid contained in coca leaves, during the chewing process and, thereby, has a stimulating effect.

In the different parts of the Sierra del Peru, there are several ways to prepare the llipt'a according to the available ingredients. For example, it can be made with banana peel ash. But in the Southern Sierra it is considered that the llipt'a of quinoa and kañiwa (*Chenopodium pallidicaule* Aellen) is the most pleasant and smooth.

The stems of quinoa have a high cellulose content, and have therefore been used in the production of cartons or used as firewood.

In the Aymara area of Puno "ch'iwa" soup is prepared with dry meat "chalona", potato, peeled barley, chuño, green beans and processed quinoa. After boiling these ingredients, quinoa leaves are added and boiled for a few minutes, with fresh oregano and aji to taste (Pratec 2001).

Quinoa leaves also used for children, boiled and milled, they are mixed with potato pap. Quinoa is used preferentially to spinach as it contains less oxaltes.

The rural population recommends black quinoa seeds to convalescent people, because of its high protein content (19%).

An important aspect is the digestion and absorption of the quinoa protein by the human body. It has been found that the digestion of whole grains is very difficult for children under two years of age, even when quinoa has been cooked. The utilization improves notably when quinoa is consumed as flour. Therefore, it is recommended to prepare quinoa and other Andean grains for young children in porridges, cream soups or beverages based on flours.

Quinoa is the most versatile of the Andean foods in terms of preparation possibilities. You can prepare all kinds of foods, such as soups, breads, savory dishes, cakes, drinks, cookies and much more.

The taste of quinoa can be bitter, if the saponin has not been properly washed away. This fact is often taken as a reason for not wanting to consume quinoa. However, it is not about washing or scarifying the seeds until they are mutilated in their structure, without nutritional value due to loss of the germ and without flavor. It is desirable to find the ideal point of undamaged grain, where valuable nutrients are preserved, and the bitterness is not unpleasant or a nuisance, but allows to appreciate the own flavor of freshly cooked quinoa.

In the traditional kitchen, the type of food that is going to be prepared according to the ingredients available at the time and place is decided. Quinoa is a dry food, which can be kept for a long time if the conditions are adequate, therefore, it is ideal for periods of scarcity and, consequently, it is always necessary to have a sufficient reserve.

- Steam cooking is used to prepare the kispiño; also for tamales and humitas made from quinoa or quinoa flour.
- To make toasted flours the grain is roasted fat-free in a clay pot.
- Fried tortillas are much appreciated, they require fat or oil. Until about thirty years ago, it was a sporadic form of food, nowadays it is becoming more frequent.
- Finally, drinks are prepared such as soft drinks and white quinua chicha. For beverages, fermented or not, small varieties of colored quinoa are used as the quinoa variety witulla.

1.6.2 Traditional Use of Leaves

Quinoa leaves are called "llipcha" or lliqch'a in Quechua and "ch'iwa" in Aymara. "Ch'iwa jauch'a" refers to the tender leaves of quinoa, which are cooked and used to prepare stews and salads. Quinoa is a close relative to spinach.

The consumption of tender leaves is frequent, especially in the month of December until Carnival, (February) and is part of the traditional Christmas dishes. When the plants have reached about 30 cm in size, they proceed to thinning in the field. The plants removed are consumed fresh, preferably on the same day. The most frequent preparations with soups, scrambled eggs, cheese and spices. They can also be an ingredient of the Quinoa Kispiño or an accompaniment of the p'esqe (Quinoa type mazamorra).

References

Acosta J (1948) Compendio histórico del desubrimiento y colonización de la Nueva Granada, en el siglo XVI. Paris

Arraguez GA et al (2017) La importancia d ela quinua en la alimentación de los grupos prehispanicos del noroeste Argentino. En. VI Congreso Mundial de la quiua y granos andinos, Puno, Peru

Baptista S (1976) Breve estudiosobre el cltivo de la quinua. En: I Convencion Internacional de Quenopodiaceas, Potosi, Bolivia

Beyerdorf and Blanco (1984) Diccionario agrícola. Proyecto Pisca, Cusco, Peru

Bertonio L (1879) Vocabulario dela legua aymara. Alemania, Leipzig

Bonifacio A, Aroni G, Vilca YM (2012) Catalogo Etnobotánica de la Quinua Real, Proinpa. McKnight Foundation, Cochabamba, Bolivia

Canahua A (2012) Los tipos de quinuas en el Altiplano de Puno Proyecto SIPAM. FAO, Puno, Perú

Cardenas M (1969) Manua de Plantas Economicas de Bolivia. Editorial Icthus, Cocabamba, Bolivia

Cárdenas M (1965) Manual de Plantas Economicas. Segunda Edicion. Editorial los amigos del Libro., La Paz, Bolivia

Cardozo A (1976) El cultivo d ela quinua en Colombia y Ecuador. En. II Convencion Internacional de Quenopodiaceas, Potosi, Bolivia

Cobo B (1945) Historia del Nuevo Mundo. Bibliteca de autores españoles, Edtorial Atlas, Madid, España

Freere M, Rijks JQ, Rea J (1975) Estudio agroclimático de la zona andina. FAO, UNESCO, OMM, Roma

Gandarillas H (1967) Distribución geográfica de quinuas sin saponina y granos grandes. Sayana 2: 6–7. La Paz, Bolivia

Gandarillas H (1968) Razas de Quinua. Boletín Experimental # 34. Ministerio de Agricultura, La Paz, Bolivia

Gandarillas H (1986) Aspectos Relativos a la Producción, comercialización e industrialización de la quinua. En: Simposio sobre Políticas de Seguridad Alimentaria, Min.isterio de Planificación, UNICEF. La Paz, Bolivia

Gandarillas H, Nieto C, Castillo R (1989) Razas de quinua en el Ecuador. INIAP Boletín Técnico # 67. Quito, Ecuador

Gomez Luz (2013) Mejoramiento Genético de la Quinua. En Congreso Científico Internacional de Quinua y Granos Andinos, Resúmenes, Lima, Perú

Gonzales RS (1917) Investigaciones del Chenopodium quinoa Willd. Boletín de la Sociedad Geográfica de La Paz 44:1–44

Hunsiker A (1952) Los pseudo cereales de South America. Editorial ACME. U. Cordoba, Argentina
Jiménez J (2014) Informe al Proyecto Razas de quinuas. Concytec, ANPE, Lima, Perú
Junge I et al (1973) Lupine and Quinua. Research and development in Chile. Anales de la Escuela de Ingenieria. Universidad de Concepcion, Chile
Latchman RR (1936) Agricultura preolombina en Chile y los países vecinos. Universidad de Chile
Lescano JL (1994) Genetica y Mejoramiento de cultivos Alto Andinos. Puno, Peru, PIWA, INADE, PELT-COTESU
Leon J (1964) Plantas almenticias andinas, Boletin N° 6 IICA, Lima Peru
Maughan J (2013) Avances en la genetica de la quinua y sus parientes silvestres. In: Congreso Cientifico Internacional de la Qunua y Granos andinos-UNALM, La Molina, Lima, Peru
Mujica A (1977) Cultivo de Quinua. Manual 1. INIA, Lima, Peru
Mujica Á, Mar S, Jacobsen S (2001) Producción actual y potencial de la quinua en el Perú. En: Memorias Primer Taller Internacional sobre Quinua UNL, CIP-UNA, Lima, Perú
Mujica Á et al (2013) Producción orgánica de Quinua (Chenopodium quinoa Willd.). Universidad Nacional del Altiplano, Fincagro, Puno, Perú
Murra J (1975) Formaciones económicas y políticas del mundo andino. Instituto de Estudos Peruanos, Lima, Peru
Nuñez L (1970) Agricultura prehistórica en los Andes meridionales. Edi. Universidad del Norte, Chile
Peralta E et al (2012) Manual agrícola de Granos Andnos, Chocho, quinua, amaranto y ataco INIAP. Quito, Ecuador
Pulgar Vidal J (1954) La quinua o suba, alimento básico de los Chipchas. Economia Colombiana
Rea Julio (1944) Observaciones sobre la Biología Floral de la Quinua. Facultad de Agronomia de la Universidad de la Plata, Argentina
Rojas W (2013) La diversidad genética de la quinua en Bolivia. En: congresocientifico de la Quinua, IICA, INIAF, La Paz, Bolivia
Romero A (1976) El cultivo de la quinua en la sabana de Bogotá. En: 1ra mesa redonda sobre la investigación de quinua en Colombia. Comité Interinstitucional Colombiano de la quinua, Bogota, Colombia
Tapia M (1976) Cultivo de la Quinua en los Andes. Anales de la II Convención de Quenopodiaceas. Potosí, Bolivia
Tapia M et al (1979) Quinua y kañiwa, Granos Andinos. Serie: Libros y manuales educativos N° 40. IICA, CIID, Bogota, Colombia
Tapia M (1982) Las quinuas de valle. En. Anales III Congresos Internacional sobre cultivos Andnos, La Paz, Bolivia
Tapia M (2012) La quinua, Hitoria, distribución geográfica, actual producción y usos. En: Ambienta, N°99. Madrid, España
Tapia M, Canahua A, Ignacio YS (2014) Razas de Quinuas del Peru. ANPE, CONCYTEC, Lima, Peru
Tapia M, Lozada M (2017) La Quinua a la conquista de nuevos Ecosistmas. En: Memorias delVI Congreso Internacional dela Qinua, Puno, Peru
Towle M (1961) The Etnobotany of precolombian civilizations, Chicago, USA
Uhle M (1919) La arqelogia de Arica y Tocta. Sociedad Ecuatorianade estudios históricos, Quito, Ecuador
Vorano A, Garcia YR (1977) La quinua en la provincia de Jujuy. Situación actual y perspectivas. In: II Convención Internacional sobre Quenopociaceas. Potosí, Bolivia
White PL et al (1955) Nutrient content and protein quality of quinua and kañihua edibleseed producto of Andes Mountains. J Food Chem
Wilson HD (1976) A biosistematic study of the cultivated Chenopods and related especies. Ph.D. thesis, Blominton University of Indiana

Botanical Context for Domestication in South America

2

Ramiro Nestor Curti and Hector Daniel Bertero

Abstract

Quinoa domestication studies based on seed's morphological traits and conducted in the Central Andes region concluded that it occurred somewhere around Lake Titicaca before 3000 BC. Recent genetic studies showed quinoa (an allotetraploid) resulting from the fusion of two diploid species (carrying the A and B genomes), one Eurasian and one American (probably in North America), from where a tetraploid ancestor migrated to South America. Extant wild relatives are found from the U.S. to South America, and quinoa is part of a complex of domesticates including Chenopodium berlandieri spp. jonesianum and nuttalliae. Quinoa domestication in the Andes appears as a diffuse process occurring in a wide area within the Bolivian Highlands. Here, we pose the hypothesis and provide evidence that quinoa was domesticated twice: in the Andes and Central Chile. The domestication syndrome in quinoa included bigger seeds with a reduced testa width and a range of colours, plus a wide array of plant architectures, panicle morphologies and reproductive partitioning. We widen those studies including root traits and phenological adaptations to a wide climatic range. Finally, the hypothesis that reduced testa width can be related more to reduced restrictions to seed growth than to a reduced dormancy is presented.

2.1 Introduction

Scientific interest in quinoa has grown over the last decades. This may be related to the diffusion of information regarding its positive nutritional properties which lead to increased consumption outside of the Andean region from where it originates and where consumption was restricted until the 80s (Jacobsen 2003; Bazile and Baudron 2015). A recent synthesis of current knowledge showed the state of the art about the crop covering several areas of current research (Bazile et al. 2015). Most of it is focused on nutritional (e.g. Valencia-Chamorro 2004; Jankurova et al. 2009; Wu 2015) and physiological aspects, particularly those related to responses to stress (e.g. Jensen et al. 2000; Geerts et al. 2008; Jacobsen et al. 2009; González et al. 2010; Hariadi et al. 2011; Orsini et al. 2011; Ruiz Carrazco et al. 2011; Adolf et al. 2012; Cocozza et al. 2013; Razzaghi et al. 2011, 2012; Biondi et al. 2015; Zurita Silva et al. 2015; Alandia et al. 2016; Ruiz et al. 2016). Studies related to the molecular and archaeological aspects of domestication have

R. N. Curti
Universidad Nacional de Salta and Conicet-CCT-Salta, Salta, Argentina

H. D. Bertero (✉)
Faculty of Agronomy, University of Buenos Aires and IFEVA-Conicet, Buenos Aires, Argentina
e-mail: bertero@agro.uba.ar

S. M. Schmöckel (ed.), *The Quinoa Genome*, Compendium of Plant Genomes,
https://doi.org/10.1007/978-3-030-65237-1_2

received less attention, partly because questions as to when and where quinoa was domesticated are considered as answered by many authors, who assumed that quinoa was domesticated just once in the Central Andes from where it became distributed to the North and South of the region (Gandarillas 1979; Bazile et al. 2013; Martinez et al. 2015). However, recent available evidence suggests a more complex picture. In this chapter we will show what is known about the genetic relations between quinoa and related species, its genetic structure and relationships with geographic origin, the antiquity of quinoa cultivation and the evolution of the domestication syndrome, what traits were affected (but not fixed) by domestication and recent hypotheses about where quinoa was domesticated.

2.2 Quinoa Wild Ancestors and Relationships with Related Species

Quinoa (2n = 4x = 36) is an allotetraploid and there are two processes associated to its origin: the polyploidization which gave origin to the tetraploid ancestor of the cultivated form and related species and domestication. The main hypotheses about the origin of quinoa were posed by Hugh Wilson more than 20 years ago (Wilson 1990): a common tetraploid ancestor of quinoa and related species originated in North America from where one of them migrated to South America. The polyploidization event gave origin to at least two wild species: *Chenopodium berlandieri* Moq. (distributed in North and Central America) and *Chenopodium hircinum* Schrad. (distributed in the lowlands of South America), all belonging to subsection Favosa (former *Cellulata*) of the *Chenopodium* genus, section *Chenopodium* (Aellen and Just 1943; Mosyakin and Clements 1996). After that, there were at least three independent domestication events: one in North America, giving origin to *Chenopodium berlandieri* spp. *jonesianum*, currently extint (Smith 2006; Kistler and Shapiro 2011), another in Mexico giving origin to *C. berlandieri* spp. *nuttalliae* (Wilson and Heiser 1979) and a third one in South America giving origin to *Chenopodium quinoa* (Wilson 1990). This hypothesis arose from morphological studies combined with interspecific crosses and allozyme analyses, conforming the first molecular studies conducted on the species (Wilson 1981, 1988a, b, c, 1990; Wilson and Manhart 1993). An alternative hypothesis was posed by the bolivian researcher Antonio Gandarillas (Gandarillas 1979), suggesting quinoa arose in the Andes after a polyploidization event in South America with *Chenopodium pallidicaule* (canihua, a diploid domesticate from the high Andes) as the source of one of the diploid genomes.

There are two sub-genomes in quinoa and related species: A, originated in the Western Hemisphere and the B genome present in Eurasian diploids and shared with the hexaploid species *Chenopodium album* (Jellen et al. 2015). Based on the DNA sequence of the single-copy nuclear locus *Salt Overly Sensitive 1 (SOS1)* Walsh et al. (2015) constructed a phylogeny of 34 *Chenopodium* species, with the aim of exploring the origin of the genomes of polyploid members of the group and to identify the potential wild ancestor of domesticated species. Four clades were identified by the study. These clades, named A to D include diploid (A), tetraploid (A & B) and hexaploid (B, C & D) species. American tetraploids are represented in clades A and B, while Eastern Hemisphere hexaploids in clades B, C and D. There was high sequence similarity within these clades, which lead to difficulties on the resolution of the relationships among the taxa involved. This was consistent with a recent divergence, but it also called into consideration species delimitation, as these limits are poorly supported by available data. *Chenopodium berlandieri* and *C. hircinum* were genetically indistinguishable as they also are from the morphological point of view (Standley 1917; Aellen 1929). A genetic distinction between the domesticated *C. berlandieri* spp. *nuttalliae* and *C. quinoa* was found as a 10 bp indel in SOS1 intron 17 of *C. berlandieri* spp. *nuttalliae*. It was not possible to elucidate from available data whether or not that means an independent origin of both species. Before that, the main support for an independent origin of both cultivated species *C. quinoa* and *C. berlandieri* spp. *nuttalliae* was provided by Heiser and Nelson (1974), who found that the white seed trait is determined by different genes; hybridization between homozygous white seeded parents of both species produced black seeded F1.

C. standleyanum Aellen, a North American diploid, was closely associated to a clade composed of tetraploids, posing it as the best A genome donor candidate for that lineage. *C pallidicaule*, the South American cultivated diploid, is clearly distinct from quinoa and the other tetraploid species, rejecting Gandarillas (1979) hypothesis of that species being the donor of one of the diploid genomes. Kolano et al. (2016) conducted a phylogenetic analysis of four plastid regions, nrITS and nuclear 5S rDNA spacer region (NTS) using *C. quinoa, C. berlandieri* and several of its diploid relatives. It was found that the A (American) genome pool involved in these tetraploids belonged to the maternal species, while the B genome (Eurasian) to the paternal line. The studied species with the closest relationship to the putative paternal species is *C. ficifolium* Sm. (B genome), while the maternal line remains more elusive due to the lower resolution of the analysis for the American diploids. *C. standleyanum, C. watsoni* A. Nelson and *C. nevadense* Standl. are the more likely relatives, and it was not possible to establish whether *C. berlandieri* and *C. quinoa* have an independent or shared origin. *C. hircinum* was represented by a single accession from Chile in Walsh et al. (2015) and not included in Kolano's et al. (2016), it is probably that more coverage of this species in future analyses will help solve some of these unanswered questions. An advance on this was made by Jarvis et al. (2017) who sequenced five *C. berlandieri* and two *C. hircinum* samples together with 15 quinoa samples, showing *C. berlandieri* placed as a basal member of the group. The *C. hircinum* samples located at different positions and the potential meaning of these results are discussed later as part of the hypotheses about where was quinoa domesticated.

2.3 Quinoa Genetic Structure

Quinoa is described as a crop-weed-wild species complex (Wilson 1990) in which cultivated forms are found together with weedy ones (usually called *ajaras* in the Andes), with partial geographic overlap with wild forms covering a much wider geographic range as described for the species *C. hircinum sensu stricto* (Wilson 1990). There have been several attempts to organize its variability into categories and to study the relationships between them. One of the first classifications was done by Hunziker (1943a) who classified cultivated quinoa into three varieties: *rubescens, viridiscens* and *lutescens*, based on plant colour and a black seeded weedy one, *melanospermum*. Later, Wilson (1990) included all cultivated quinoa as part of *C. quinoa* spp. *quinoa* and the black seeded type as *C. quinoa* spp. *melanospermum*. Gandarillas (1968) distinguished 17 quinoa races but did not analyse the relationships between them. Tapia et al. (1980), after studying germplasm from Colombia, Ecuador, Peru, Bolivia and Chile, proposed variants according to loosely defined morphological types linked to different areas of cultivation and named them ecotypes. These ecotypes were called Valley, Highland, Salares, Subtropical and Sea Level. The Valley type is found in the Interandean Valleys between Colombia and Bolivia, the Highland type in the Andean plateau between Peru and Bolivia, the Salares type around the salt lakes in south Bolivia and the Subtropical type in some humid valleys in Bolivia. Finally, the Sea Level group is found at lower altitudes in Central and Southern Chile (south of its capital Santiago up to 47° S (Tapia et al. 1980; Wilson 1990)). The allozyme and morphological analyses conducted by Wilson (1988a) found a sharp distinction between Sea Level quinoa and its Andean counterpart, and a less clear one within the Andes between Northern and Southern types, but this classification only partially matched the one established by Tapia. All Northern Andes quinoas belong to the Valley ecotype, but it was not possible to distinguish between the Valley, Highland and Salares ecotypes within the Southern Andes type. Bertero et al. (2004) grouped 24 quinoa cultivars into four genetic groups according to yield variation patterns across environments: Valley, Northern Highlands, Southern Highland and Sea Level, and confirmed Wilson's (1988a) observation of the sea level type as the most differentiated one. The Northern and Southern highland groups can be linked to Tapia's Highland and Salares ecotypes. Rojas (2003) studied the phenotypic variability of a $\sim$1500 accessions collection belonging to a Bolivian collection curated by Fundacion Proinpa and distinguished seven groups originated in three environments: mid-altitude valleys, high-altitude valleys and highlands.

More recent DNA-based molecular studies conducted at a national or regional level detected a strong genetic structure. del Castillo et al. (2007), using RAPD markers, detected three genetic groups from over an eight population collection which included cultivated and weedy quinoa forms in the Bolivian highlands analysed at the individual plant level. There was no separation between weedy and cultivated forms and genetic groups were associated with Northern and Central Highland, Southern Highlands and High Altitude Valleys. Individuals clustered according to their population within each group. Anabalon Rodriguez and Thomet Isla (2009) evaluated quinoa from North and Central Chile using AFLP and morphological descriptors, distinguishing between these two origins and differentiating those from Central Chile into two sub-groups: from the coast (Costa) and from between 330 and 600 m of altitude (Cordillera). Costa Tártara et al. (2012) studied 35 accessions from North West Argentina and found variability structured into 4 genetic groups with a strong correspondence with eco-geographic regions. These groups were linked to Dry Highlands, Dry Valleys, Humid Valleys and a High altitude Transition Zone between these last two. Costa Tártara (2014) compared these accessions with non-Argentinian ones from Colombia to Chile, detecting genetic and geographical continuity between her local groups and genotypes coming from similar environments in the Andes. Thus, the Highland group was related to quinoas from South West Bolivia and Northern Chile, the Dry Valley one to accessions from similar environments in the Department of Sucre in Bolivia, Humid Valleys to the Valley quinoas from Tarija to Peru, Ecuador and Colombia, while the Transition group was related to accessions from around Lake Titicaca in Bolivia and Peru. Against previous hypotheses connecting NW Argentina germplasm to Sea Level quinoa (Christensen et al. 2007), central Chile accessions were not related to any Argentinian genetic group. A map showing the geographic distribution of quinoa main genetic groups is shown in Fig. 2.1. It is based on the ecotypes proposed by Tapia et al. (1980), is consistent with the genetic groups proposed by Bertero et al. (2004) and considers more recent knowledge regarding North West Argentina (Costa Tártara el al. 2012; Curti et al. 2012, see inset). The maps also expand the distribution of Sea Level quinoa to the Patagonia region in South West Argentina. The Subtropical quinoa ecotype described by Tapia is not included in the map, it described one accession from the Cochabamba region in Bolivia and is currently reported as lost (M. Tapia, personal communication).

2.4 Where and When Domestication Occurred

2.4.1 Geographic Patterns of Variation in Genetic Variability

"By consensus, all scientist who have conducted morphological, genetic and systematic studies consider that quinoa originates in the Andes of Peru and Bolivia" (Rojas and Pinto 2015). Narrowing the geographic range, Gandarillas (1979) proposed the area of quinoa domestication to be located in the region between Cuzco (Peru) and Oruro (Bolivia). What was the basis for this assertion? One is that the Central Andes is the main center of origin of domesticated plants in South America (Vavilov 1992). A second factor was the assumption that this area is the one with the highest diversity of wild and cultivated quinoa populations (Gandarillas 1979).

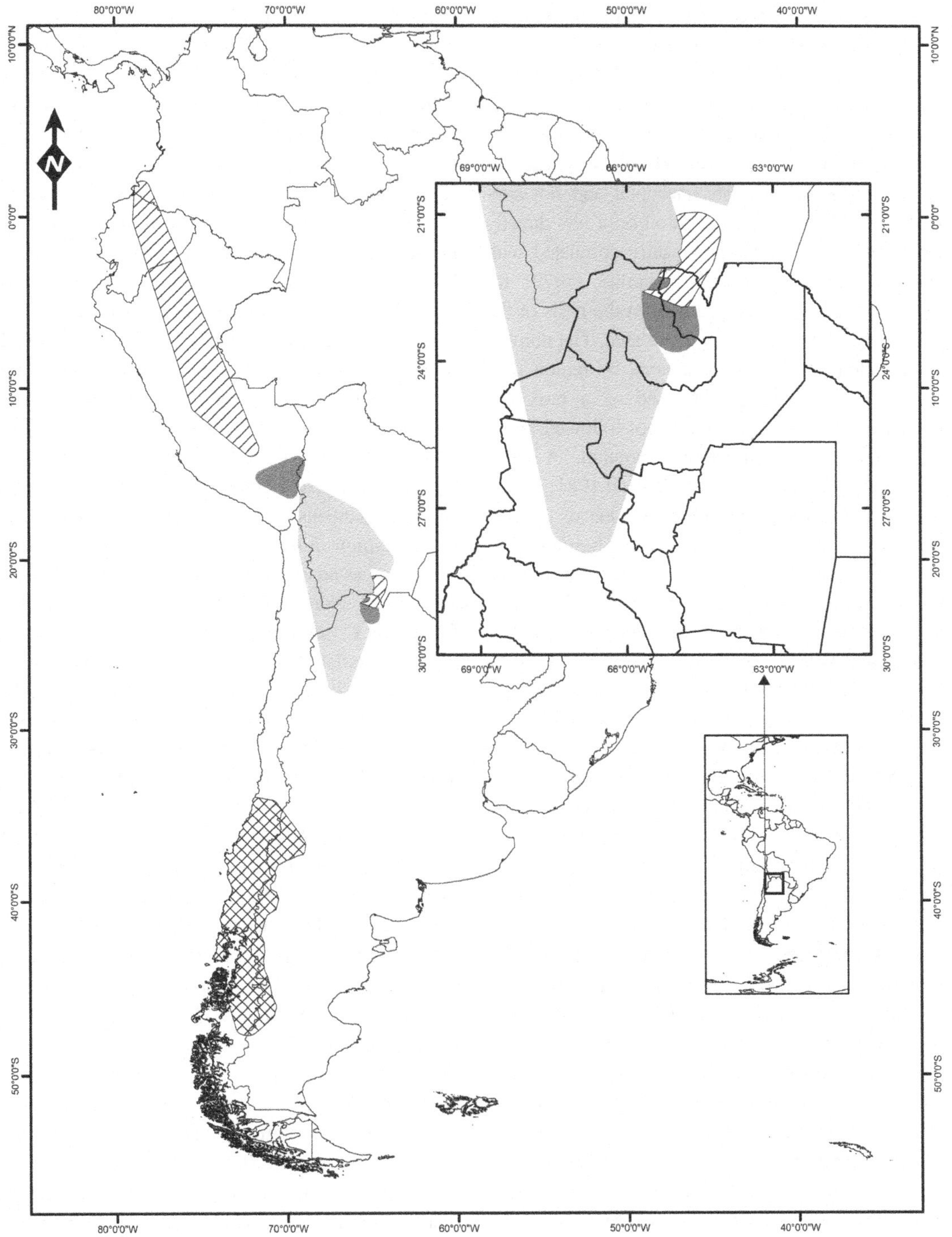

Fig. 2.1 (Map). Geographical distribution of extant quinoa genetic groups. From North to South these are: Valley (diagonal lines), Northern Highlands (dark grey), Southern Highlands (light grey) and Sea Level (crossed lines). The geographical distribution of three of these genetic groups in Northwest Argentina is shown in the inset. A difference with previous maps is that only Southern Highlands quinoa were supposed to be found in North West Argentina, and that quinoa is also reported for the first time in Argentinean Patagonia. The denomination of Northern and Southern Highland quinoas is used instead of Highland and Salares (Tapia et al. 1980) because the Southern Highland group is distributed in an area bigger than that of the Salares of South Bolivia

Considering the geographic patterns of variation in diversity the earliest references speak of it in terms of phenotypic diversity and more specifically that related to the more conspicuous traits like branching, plant and panicle colour, panicle type and compactness, seed size and colour. These references were many times anecdotal and not supported by quantitative data, based on a few directly observed traits and not necessarily correlated with similar genetic variation. Gandarillas (1979) for example, supported his assertion on the basis of the number of races found in a given area. The number of collected germplasm accessions by region (e.g. Rojas et al. 2015) can be used as a proxy of diversity but it is a poor one if not followed by an evaluation of the underlying variation. A more systematic analysis was that of Wilson 1981, 1988a, b, c) who used allozyme markers. One of the first observations done by Wilson was that besides seeds colour, it is not possible to genetically discriminate between cultivated quinoa and ajaras (weedy forms), in fact a higher similarity was found between cultivated and weedy populations of the same region (evidence of recurrent introgression between both types) than between cultivated forms of different regions (Wilson 1990 and references therein). Wilson detected a higher degree of allozyme electrophoretic variation within the Southern Andes complex consistent with the notion of this area as the one with the highest diversity. Diversity was lower in the Northern Andes group (Interandean Valleys from Peru to Colombia) and also in the Sea Level (Chilean) one. This supported the hypothesis of diffusion from the Central Andes. A high degree of homogeneity (allozyme monomorphism) was detected in comparison with *C. hircinum* from the Argentinian lowlands (Wilson 1990). The analysis by del Castillo et al. (2007) of eight Bolivian populations using RAPD markers detected a higher degree of polymorphism within populations than Wilson's studies, but there were no differences in diversity between the sampled ecoregions. Christensen et al. (2007) exploring a wide range of geographic origins using microsatellite markers also detected an important degree of heterozygosity (all microsatellite markers utilized were polymorphic and 69% highly polymorphic when using heterozygosity (H) values higher than 0.7 as a threshold for high polymorphism). Variation was structured into two big genetic groups (Andean and Lowland), with higher levels of diversity in the Andean group. Within the Andean group, diversity (expressed as the average number of alleles per locus) was higher for Bolivian and Peruvian accessions and lower for Equatorian and Argentinian accessions, supporting the notion of higher diversity in the Central Andes. However, these values could be biased by the lower number of accessions used from the last two countries. Fuentes et al. (2009) compared accessions from the North (Andean) and South (Sea Level) of Chile using SSR markers, and found a higher diversity in Sea Level accessions. Similar number of accessions were used for each origin. Their results question the notion of the Sea Level quinoa genetic group as one of lower diversity. Spontaneous crossings between quinoa and *Chenopodium hircinum* or *Chenopodium album* was posed as an explanation for that diversity. In the analysis of Northwest Argentina germplasm performed by Costa Tártara et al. (2012) a geographic pattern of variation in diversity was also found, with higher diversity in accessions from the Dry Highlands and Dry Valleys than in the Eastern Valleys and Humid Highlands. It was hypothesized that the lower level of diversity in these latter groups was caused by a founder effect related to geographic isolation. In her comparison of Argentinian and non-Argentinian germplasm (Costa Tártara 2014) it was found that around 2/3rd of total genetic variation was found in the Argentinian accessions, and the proportion was even higher when Sea Level accessions were excluded from the analysis.

All these analyses, including either national or regional samples, suggest that a more systematic (including a higher number of accession but, more important, higher geographic coverage) evaluation of the patterns of genetic variation in quinoa is needed before reaching a conclusion about how it is distributed. SSR and ISSR markers were used to characterize ~86% of the Bolivian National quinoa germplasm collection (Rojas and Pinto 2015). This one is the biggest quinoa collection including ~3200 accessions from all the countries where quinoa is traditionally cultivated. Their report provides information about the range of variation in the number of alleles per locus, but nothing about the geographic distribution of variation.

2.4.2 Archaeological Evidence

Evidence of the use of *Chenopodium* seeds by hunter-gatherer's societies since thousands of years ago is abundant in the Andes from Peru to Central Chile and North West Argentina, as is shown in a recent review of archaeological findings in quinoa (Planella et al. 2015). Seeds showing traits related to domestication appear on the archaeological records ~3000 BC in Junin, Peru (Pearsall 2008). For Bolivia, the oldest findings are from Chiripa near Lake Titicaca dated at 1500 BC (Bruno 2008; Langlie 2008; Whitehead 2007) and 1300 BC at La Barca site in the Central highlands (Langlie et al. 2011). In Argentina, findings from Pampa Grande, in the Northwestern Salta province, are dated at 500–700 AD (Hunziker 1943b) and the oldest findings originated at the central provinces of Mendoza and San Juan close to the Chilean border and dated ~390 BC for Mendoza (Lagiglia 2001) and 300 BC for San Juan (Burrieza et al. 2016). Indirect evidence of quinoa use was provided from quinoa-like stems, phytoliths and starch grains associated to grinding stones found in domestic contexts in the highlands of Catamarca, North West Argentina (Babot 2011; Planella et al. 2015). Finally, seeds from the northern coast of Chile are dated between 1600 and 600 BC at the Chomache 1 site (Nuñez 1986), while for Central Chile seeds showing some degree of human manipulation have been dated between ~3400 BP (Planella et al. 2011) and 2960 BP (Planella et al. 2005). These ancient seeds from Central Chile are very small ones and were found in places where agriculture was unlikely, and bigger ones with clear domesticate type traits appear in agricultural contexts later, between 1000 and 1500 BP (Planella and Tagle 2004).

Although there are reports of quinoa seeds being found in contexts dated 5500–6000 BC in Northern Peru (Dillehay et al. 2007), available uncontested evidence suggests that the domestication process was ongoing before 3000 BC and that the crop had already reached most of its current distribution around the beginning of the Common Era (Planella et al. 2015). All this evidence supports the notion that domestication occurred somewhere in the central Andes but does not clarify whether it was a single event or a more diffuse process within a larger area. Available evidence suggests the second. In a study conducted by Maria Bruno (2008) changes in seed size and morphology were followed at the Chiripa archaeological site near Lake Titicaca in Bolivia, starting 1500 years BC. A few hundred km south, half way between Lake Titicaca and Lake Poopo, Langlie et al. (2011) found evidence of an ongoing domestication process dated around 1300 BC. It involved a morphological type which was clearly different from that studied by Bruno. Some of the samples were similar to the other Andean *Chenopodium* domesticate canihua (*Chenopodium pallidicaule*) but it seems that different sets of *Chenopodium* populations were being domesticated at that time.

Further away, south of the Central Andes, the team of Planella (Planella et al. 2005, 2011) found *Chenopodium* seeds, determined as belonging to *Chenopodium quinoa*, which were slightly bigger (1 to 1,5 mm width) than the wild species *C. hircinum* and with truncate margins, a trait related to quinoa domestication. This suggests that some populations could have been under selection at the time of the transition to agricultural societies in the area. Because the area of the location of the finding was not suited for cultivation the authors speculated that they could be connected by exchanges with the current province of Mendoza, Argentina, where fully domesticated (albeit small) quinoa seeds were found a few centuries later (Lagiglia 2001). This is interesting, as it suggests that quinoa could have been independently domesticated in Chile, though the taxonomic identity awaits confirmation and these are the only findings in the region. Sea level quinoa lacks some traits (floury perisperm, amaranthiform panicles) found in Andean populations leading Wilson (1990) to speculate that they represent an archaic form of quinoa which migrated to the area early after domestication. It is also possible however that they are the result of an independent domestication process. Sea Level and Andean populations appear as genetically differentiated in all genetic analyses starting with Wilson's (1981, 1988a, b, c) use of allozyme markers and confirmed by more recent analysis (Mason et al. 2005; Christensen et al. 2007; Fuentes et al. 2009). Findings of synapomorphic polymorphism at the NOR intergenic spacer (IGS) separating Andean and Chilean germplasm pools indicate an early differentiation (Maughan et al. 2006). This kind of analysis requires the inclusion of *Chenopodium hircinum* samples in order to distinguish whether that differentiation occurred before or after domestication. This was done by Jarvis et al. (2017) who found a *C. hircinum* accession from Chile which is basal to Chilean samples but not to Andean ones, supporting the notion of (at least) two independent domestication events.

2.5 Traits Affected by Domestication

The domestication syndrome is the set of traits shared by many different cultivated species which have been modified as a result of selection under cultivation (Gepts 2010). The list includes lack of seed shattering and dormancy reduction as the most important components, plus a more compact growth habit (reduced branching), higher harvest indices (% biomass partitioning to grains) and bigger harvested organs (tubers, roots, seeds or fruits). These traits have been evaluated partially in quinoa by comparing domesticated with wild forms or weedy ones. Root growth was compared between quinoa and *Chenopodium hircinum* and some of the differences could be linked to domestication. "Aesthetic selection" is analysed as the high degree of variation in colour and plant form observed in quinoa seems to reflect concerns beyond those required to grow a crop. Finally, genetic variation for duration of development and its plasticity facilitated the geographic expansion of quinoa and its adaptation to contrasting environments (Bertero 2003). Section 2.5.6 is devoted to describe the type of environmental responses which allowed this adaptation.

2.5.1 Seed Size and Morphology

Changes in seed size and morphology were studied by Maria Bruno (2005a, b; Bruno and Whitehead 2003) at the Chiripa archaeological site on the Southern border of the Titicaca lake in Bolivia and encompassing the 1500 BC–100 AD period. Her comparisons included both cultivated and weedy (*C. quinoa* spp. *melanospermum*) forms of quinoa, plus other local *Chenopodium* species. Changes in seed size (as maximum seed diameter), testa thickness (the episperm, also named coat), coat texture, pericarp patterning and seed margin configuration were analysed. Quinoa seeds (strictly speaking a fruit, as it includes the pericarp, a maternal tissue) are made of a series of layers from the outside to the inside: pericarp, episperm and the embryo plus perisperm (where starch reserves are accumulated). The pericarp is made of two cell layers, the outer one being large and papillose. The episperm or seed coat has two cell layers from the outside to the inside: the exotesta and the endotegmen. The endosperm is present only in the micropylar region and surrounds the hypocotyl-radicle axis (Prego et al. 1998). The range of colours varies from cream to purple for the pericarp and from translucent to black for the episperm (Bioversity International 2013). The perisperm can be either vitreous (semi-translucent) or opaque (floury). Seed coats have smooth to slightly undulating surfaces and pericarps have a reticulate morphology in both quinoa forms which helps to distinguish them from other *Chenopodium* species found in the studied area. Four types of seed margin configurations were found: biconvex, equatorially banded, rounded and truncate (Bruno and Whitehead 2003). All modern quinoa have truncate margins. The flat cotyledons of the embryo create a truncate margin and in seeds with reduced episperm thickness the fruit morphology is influenced by structural features of the embryo (Bruno 2005b).

Temporal patterns of change in seed size were not clear and did not allow to distinguish between cultivated and weedy forms, but changes in the relative testa (episperm) thickness, the ratio of testa width to maximum seed diameter, distinguished both ancient from modern quinoa

Fig. 2.2 Wild (left photo) versus domesticated (right photo) quinoa seeds. Wild seeds (*Chenopodium quinoa* spp. *melanospermum*) are black or dark with reticulate-alveolate coats (pericarp) while domesticated seed can show a range of colours with smooth coats. Photo by R. Curti

and wild from domesticated ones, with lower values in modern and domesticated seeds. Seed morphology was also different; all modern domesticated seeds had truncate margins, while weedy seeds were either biconvex, equatorially banded or rounded. Other changes distinguish domesticated from weedy forms; seed coat is smooth in domesticated and reticulate alveolate in weedy ones (Fig. 2.2). Changes in absolute testa width values were significant besides those in its relative value, it ranged from 1 to 20 microns in cultivated forms and from 22 to 51 microns in weedy ones (Bruno and Whitehead 2003). Changes in testa width have been attributed to selection for a more even germination (arising from reduced seed dormancy), but this interpretation was somehow challenged by recent findings. An analysis of the impact of changes in testa width conducted on two potential sources of pre-harvest sprouting resistance, which requires some degree of dormancy at maturity, showed dormancy to be related to testa width in one of the genotypes and not in the second, which originated from the humid island of Chiloe in southern Chile (Ceccato et al. 2015). That genotype exhibited a high dormancy at any sowing date tested with a reduced range (17 to 20 microns) of variation in testa width and changes in dormancy were related to sensitivity to the hormone abscisic acid (ABA) (Ceccato 2011).

In summary, domestication produced a range of variants in quinoa seed traits which distinguishes it from its weedy counterparts. All modern seed have truncate margins, and overlapping sizes with weedy forms. Seed coats are smooth and thinner (some Bolivian highland seeds have very thin coats, within a few microns range). Seed colour can also vary as detailed in the species descriptors (Bioversity International 2013), most are white or with pale while only few are black. One limitation of Bruno's analysis is that both cultivated and weedy forms studied are co-evolving units (with some degree of mutual crossing) arising from natural or human selection in an agricultural context. *Chenopodium hircinum* was not included in that analysis and a similar detailed study of variation in seed traits in this putative quinoa ancestor is still needed.

2.5.2 Starch Forms

Perisperm (the reserve tissue in *Chenopodium*) is translucent (vitreous) in wild quinoa and many domesticated forms have floury perisperm. The floury perisperm is found only in Andean forms and absent in Sea Level ones. The main component of quinoa perisperm and hence determinant of perisperm appearance is starch, and a particular trait of it is its high proportion of amylopectin (Li and Zhu 2018). Differences between vitreous and floury perisperm could arise from the degree or starch crystallinity, which is related to starch chemical structure and composition (Li and Zhu 2018).

2.5.3 Seed Shattering

Seed shattering is a common trait of all *C. hircinum* populations studied so far (unpublished data). On the other hand, domesticated quinoa is able to retain their seeds until harvest although some populations seem to lose their seeds easily, for example, populations of quinoa Real grown in the Southern Bolivia highlands (R. Joffre, personal communication).

2.5.4 Plant Architecture

Wild *C. hircinum* plants have a highly branched growth habit, loose and non-terminal panicles. Plants with reduced branching, terminal and compact inflorescences are found in domesticated forms, but this is far from being a universal trait, and that is expressed in the variants included as part of the quinoa descriptors (Bioversity International 2013) (Fig. 2.3). Branching varies from non-branched to basal branches resembling cereal's tillers and panicles can be compact or lax, and terminal to diffuse. One important panicle trait differentiating wild from domesticated quinoa is panicle type. Wild quinoa, those from the Sea Level type and many from the Central Andes have glomerulate panicles, where glomeruli (short branches bearing a group of flowers or grains) are supported by third-order branch axes, while amaranthiform inflorescences having glomeruli supported by second-order branches (Bertero et al. 1996) are found only in domesticated types from the Andes (Wilson 1990).

2.5.5 Roots

Root morphology and growth under contrasting conditions of water availability were studied for quinoa and *Chenopodium hircinum* grown in rhizotrons by Alvarez-Flores et al. (2014a, b, 2018). Quinoa samples consisted of two populations from contrasting habitats, one from a low altitude, humid environment in Southern Chile (38° 56′ S) and a second from a very dry, high altitude environment in Southern Bolivia (19° 51′ S), the main quinoa production area in that country. The agricultural system in southern Bolivia relies for initial plant growth on water stored on deeper soil layers during a previous fallow period which encompasses a whole rainy season in a biennial cycle (Joffre and Acho 2008). In this system fast growth after germination through a dry layer facilitates access to water stored deeper in the soil. The *Chenopodium hircinum* sample came from a less dry environment in the northern Bolivian highlands. The quinoa population from southern Bolivia differed from both the Chilean and *C. hircinum* population by a faster root growth (Alvarez-Flores et al. 2014a) and a larger root length per unit soil layer at the deepest depth explored (Alvarez Flores et al. 2014b) and with a more even root length density distribution between layers. Root profiles were similar for both quinoa populations in the upper layers and with higher proliferation than that of *C. hircinum* (Alvarez-Flores et al. 2014b). However differences between the Chilean quinoa and *C. hircinum* were smaller than between both of them and the population from southern Bolivia. When these three populations were compared under contrasting conditions of water availability, both quinoa lines exhibited accelerated taproot growth in dry soils compared with an irrigated control and a higher plasticity in root development (and the lowest growth reduction when experienced) in the Bolivian accession (Alvarez-Flores et al. 2018). The fact that some traits (i.e. root proliferation) are higher in quinoa than in *C. hircinum* even when comparing a humid origin quinoa with a semiarid origin *C. hircinum* suggest that domestication was linked to changes in root traits, but the traits found in quinoa from south Bolivia could be related to adaptations to a drier environment not necessarily connected with domestication, and generalizations from these data are restricted by the low number of populations involved. A relationship

Fig. 2.3 Contrast in plant morphology between wild (*Chenopodium hircinum*, left photo) and domesticated (*Chenopodium quinoa*, right photo) plants. *C. hircinum* plants can be highly branched with a diffuse inflorescence while *C. quinoa* plants can show compact and terminal inflorescences, but this is not an universal trait in quinoa plants. Photos by D. Jarvis and R. Curti

between seed size and initial growth was found, similar to that found in other species (Richards et al. 2007) which led the authors to speculate that selection for bigger seeds by farmers could have been involved in the adaptation of quinoa to these dry environments (Alvarez-Flores et al. 2014b).

2.5.6 Climate Adaptation

Contemporary quinoa is adapted to a very wide range of environments and this required changes in sensitivity of phenological development to environmental factors in order to fit the crop cycle within the restrictions imposed by the growing season. These environments imposed gradients of aridity, for example, North to South in the Bolivian highlands from lake Titicaca to the Lipez region near the border with Argentina and Chile (Geerts et al. 2006), or South to North gradients in Sea level environments of Chile from Chiloe to the VI region near Santiago de Chile. Gradients of altitude are also evident in the transition from valleys to highlands in Peru, Bolivia and Northwest Argentina (Curti et al. 2016). All quinoa cultivars behave as short day plants at least for some phenological phases, accelerating development when exposed to short days, and duration of development is also significantly affected by temperature (Bertero et al. 1999a). There is no evidence of a vernalization requirement. An additional factor, at least in one Bolivian variety is water deficit. Time to anthesis can be significantly delayed by water deficits before that stage, while maturity can be accelerated by a post-anthesis water deficit (Geerts et al. 2008).

The association between both photoperiod sensitivity of time to flowering or the minimal duration to flowering (measured under short days) with the latitude of origin of different cultivars was evaluated for the Andean region. Within that region, an increase in latitude is linked to lower temperatures and precipitation during the growing season. Both photoperiod sensitivity and minimal duration to flowering decreased with increasing latitudes (Bertero

2003). In a narrower geographic range, environmental patterns of variation are associated to changes in altitude of origin in Northwest Argentina, where higher altitudes are associated with lower temperatures and precipitation. Quinoa adaptation to these environments is expressed as a strong negative relationship between time to flowering and altitude (Curti et al. 2016), and this variation was related to a reduction in photoperiod sensitivity the higher the altitude of origin. How this variation arose, either as natural of human selection from an original genepool or by crossing with adapted wild populations in each new environment, is still unknown.

One particularity of quinoa, shared with other species like soybean, is its prolonged photoperiod sensitivity (Sivori 1945; Bertero et al. 1999b). All Andean varieties studied so far have a short day requirement for seed filling (Bertero et al. 1999b; Christiansen et al. 2010). This can be adaptive in the tropical Andean highlands short days, accelerating development at the end of the summer, but results in a complete inhibition of seed growth when varieties from this region are grown at higher latitudes (Bertero et al. 2004), particularly when combined with high temperatures (Bertero et al. 1999b). Sea Level quinoa differs in this trait. These populations are traditionally grown under longer photoperiods during summer linked to higher latitudes ($\sim$32 to 47°S) and they can exhibit some photoperiod sensitivity during seed filling (Christiansen et al. 2010) but that is never followed by the inhibition of seed growth which is a key difference with Andean accessions, allowing these plants to be grown in temperate climates.

2.5.7 Aesthetic Selection

It can be thought that selection for beauty is part of the work of those dealing with ornamental plants while edible plants are selected for practical (functional) traits. In fact, several crops, like rice, wheat or barley are very similar in terms of plant, seed or spike/panicle colour. The highlands in Peru and Bolivia, and many of the towns within it, can look monotonous to the foreign observer. But a pretty different landscape appears when faced with quinoa fields. Viewed from the distance, a patchwork of green, yellow, red or purple fields face the visitor (Winkel 2013). At a closer look, these differences are expressed in terms of leaves, stems, panicles and seed colours. That variation is found in Andean quinoa while sea level types are more homogeneous, with a few red plants and the majority of them plain green. *Chenopodium hircinum* leaves are homogeneously green, plants can have green or red stems and very little variation is observed in panicle colour. Seeds are systematically black. When going to markets, this variation in colour is also found for corn, potatoes, oca (*Oxalis tuberosa*), ulluco (*Ullucus tuberosus*) and other Andean tuber crops.

What does this variation mean? For a start, it is clearly a domesticated trait not found (or not expressed) in wild plants. Though there can be some influence of natural selection (some researchers claim to see a link between red stems and cold tolerance) there is clearly a deliberate human selection. Quinoa colour is determined by pigments called betaxanthins and betacyanins and the co-existence of both pigments generates different shades that could be used to fit specific colour requirements of products for the pharmaceutical, cosmetic or food industries (Escribano et al. 2017). Anthropologist can look for connections between this selection and other symbolic aspects of the Andean civilization but, if asked for an opinion, we prefer to think that this was done for the mere search and enjoyment of beauty.

2.5.8 Saponin Content

Saponins are triterpenoid glucoside compounds found in many species and present in the pericarp of quinoa seeds which confer a bitter taste to them, requiring seeds to be washed thoroughly before cooking or its removal as part of post-harvest processing (Koziol 1991; Mizui et al. 1988, 1990). Varieties are classified as bitter or sweet according to their taste and taste is determined by variation in saponin content. Saponin content is a quantitative trait, although a locus for a major gene has also been identified, where genotypes with a recessive allele of the Sp gene have no detectable saponin content (Ward 2000). Small seeded varieties with low saponin content were identified in the Lake Titicaca region in Bolivia and Peru and included in a breeding program to produce a big seed, low saponin content cultivar baptized Sajama decades ago (Gandarillas and Tapia 1976). This variety was aimed at eliminating or significantly reducing the cost of quinoa processing. When released to farmers, acceptance was low as they found these crops to be highly susceptible to bird attack. Breeding for low saponin content seeds was stopped in the highlands, and some of the cultivars released more recently are bitter ones (Proinpa 2003). As a consequence, all quinoa grown in Bolivia is subject to post-harvest processing for saponin removal. A different strategy was followed by some European based breeding programs, in which some low saponin content cultivars were released (https://www.wur.nl/en/article/Quinoa-cultivation-in-the-Netherlands.htm).

Although there are no available comparisons for saponin content between domesticated and wild quinoa there is no evidence that it was one of the traits affected by domestication, in fact farmers appear to prefer to bear the burden of manual saponin removal (a tedious work before its automatization) to the risk of crop loss to birds (Lopez et al. 2011). The Mexican domesticate *Chenopodium berlandieri* spp nuttalliae has a much lower saponin content than quinoa and is not processed to remove saponins before cooking (Barron-Yañez et al. 2009) but we cannot assume that this is the result of human selection or of the lack of variation for the trait in the original gene pool of that crop.

2.6 Geographical Distribution of the Domestication Syndrome. An Example in North West Argentina

The domestication syndrome does not seem to be evenly distributed among regions where quinoa is cultivated. For example, genotypes from the Central Highlands in Bolivia tend to have more compact inflorescences, bigger grains and shorter plants with reduced branching. But this could be the reflection of recent selection by modern breeders and not the result of long term human selection before that. An analysis of seeds collected from farmers in a region not exposed to modern breeding techniques could shed light on whether there are, or not, differences in which traits have been selected by farmers. We have been able to do this on a recently characterized germplasm collection from North West Argentina. Systematic collections were done in 2006 and 2007, and only after that, the Argentinian National Institute of Agriculture Technology (INTA) began a breeding program using these resources. As described earlier in this chapter, four genetic groups were detected based on phenotypic and genotypic characterizations (Costa Tártara et al. 2012; Curti et al. 2012) reflecting landscape eco-geographic structure. These groups were named from West to East as: Highland, Dry Valleys, Transition Zone and Humid Valleys. This reflects gradients of precipitation (higher to the East) and altitude (higher to the West, save for the Transition Zone separating Dry East from humid West Valleys and where some accessions were collected at $\sim$4000 masl). The biggest contrast appears when comparing accessions from the geographic extremes of the distribution. Those from the highlands have shorter lifecycles, compact glomerulate or amaranthiform panicles clearly differentiated from the rest of the plant, reduced branching and big grains. Those from the Eastern Valleys show longer cycle lengths and taller plants, with longer panicles not clearly differentiated from the rest of the plant, almost all of the glomerulate type, plus strongly branched growth habit and smaller seeds (Curti et al. 2012). Another important difference is found in perisperm type, floury in the highlands and vitreous (ancestral) in the Eastern Valleys. Although some of the differentiating traits can be explained by climate adaptation like cycle length (Curti et al. 2016) and to some point the degree of ramification, others suggest a differentiated human selection trajectory. These traits are panicle type and compactness, perisperm type and seed size, being the Highlands the environment where most derived traits (amaranthiform inflorescence and floury perisperm) are found. Another significant aspect relates to harvest index and seed colour; some accessions from the Transition zone have very low harvest indices, but that was also found in some from the Highlands, which differed from the general pattern for that group. Even more, those from the Highlands with low harvest indices also exhibited a mix of black and white seeds. When asked about the use of these materials, farmers answered that they were used to make "Llipta". Llipta is a mass formed by quinoa stems ash, which is said to enhance the extraction of alkaloids of the coca (*Erithroxilum coca*) leaves. Consumption of quinoa seeds was very low in Northwest Argentina up until few years ago and there is a chance that its use in coca consumption contributed to its conservation there (Costa Tártara et al. 2015), but at least in some cases selection for white seeds and higher yields was abandoned. There is abundant evidence of genetic exchanges between quinoa and its companion weed (Wilson 1990; del Castillo et al. 2007; González Marín 2009) and selection against black seeds was part of the domestication process (Bruno and Whitehead 2003), a common practice by commercial farmers today.

All this shows that the domestication syndrome is far from being an evenly distributed trait and not at all the result of a directional trajectory from wild to fully domesticated forms. The lack of genetic barriers between cultivated, weed and wild forms is the main limitation to the fixation of domestication alleles. A more complicated question to answer is: if the domestication syndrome is unevenly distributed, how did this difference originate? The Humid Valleys from Argentina are separated by the Eastern Andean Mountain Range from the Highlands and Dry Valleys and they lack some of the more "advanced traits" but they are far from isolated. In fact, a branch of the main Inca Road (the Capac Ñam), goes through the Eastern Argentinian Andes, confirming its ancient connection to the main hubs of the Andean civilization and there is evidence of the presence of *mitmakquna* (populations relocated by the Incas for political and/or economic purposes) at this border of the Inca Empire (Oliveto and Ventura 2009). Network exchanges between valleys are still active or were active until recently (Hilgert 1999). This contrasts with the high degree of differentiation of the genetic groups detected (Costa Tártara et al. 2012). Some climatic factors limit the dispersal of some genotypes beyond their ecoregion. As an example, fungal diseases make it impossible to grow Highland origin quinoa in the Humid Valleys (Curti et al. 2014). If traits like the floury perisperm or amaranthiform inflorescence evolved in the highlands, its diffusion to more humid environments would have been strongly restricted. Tracking the geographic dispersal of allelic variants for agronomic factors can be a fascinating task but, for a more thorough understanding of its causes, the help of archaeologists and anthropologists would be mandatory.

2.7 Concluding Remarks and Some Topics for Research

Quinoa can be considered a semi-domesticated species based on the traits analysed so far. Populations with some advanced traits like reduced branching, big seeds with reduced dormancy and high harvest index coexist with populations with a high degree of branching, seed shattering and low harvest indices. This is further complicated by the lack of a genetic barrier between cultivated, wild and weedy forms, demanding a continuous effort to remove black seeded plants from agriculture fields. On the other hand introgression of genes from wild plants could have contributed to quinoa adaptation to new environments, increasing its genetic diversity during crop expansion. As for the geography of domestication, the hypothesis of domestication in the Central Andes finds support in available data, but recent findings suggest at least one additional domestication could have taken place in central Chile. Studies including a wide geographic range are needed in order to confirm where quinoa domestication took place.

Chenopodium domestication archaeologists used relative testa thickness as a demarcation criteria to distinguish domesticate from wild forms, and interpreted changes in testa thickness as the consequence of selection for reduced dormancy (Gremillion 1993; Bruno 2005a, b). Most modern quinoa seeds have thin testas and experiments confirmed an association between testa width and dormancy (Ceccato et al. 2015) but this association is not universal as the same authors detected an accession for which changes in seed dormancy were not related to environmentally induced changes in testa width but to sensitivity to hormones (Ceccato 2011). Although this does not reject the hypothesis of reduced testas as a consequence of selection for more even germination, it shows that other factors (in this case sensitivity to the germination inhibitor abscisic acid) can affect dormancy in quinoa. An alternative hypothesis could be considered. Thick testas pose a restriction to seed expansion and increases in seed size could have been linked to reduction in testa thickness. Some preliminary observations suggest that quinoas with thick testas also have small seeds (e.g. accession Chadmo, Ceccato et al. 2015) and this hypothesis could be tested on modern varieties.

Some traits haven't been evaluated yet for their difference between cultivated and wild populations. This includes plant leaf area, leaf photosynthetic rate and partitioning of vegetative growth. Alvarez et al. (2014a) found no changes in the relative proportions of leaf, stem and root biomass. Only the proportion of secondary branches biomass was higher in wild plants. Some preliminary observations on *C. hircinum* accessions indicate that growth before flowering is very similar between both quinoa and *hircinum* species. Branch growth starts after the beginning of anthesis in *C. hircinum*.

In conclusion, research on quinoa domestication is still incipient and focused mostly on seed traits in the case of archaeological studies. Approaches to the subject from the fields of crop (ej. Alvarez-Flores et al. 2018) or seed (Ceccato et al. 2015 and see discussion of their results in Langlie 2019) physiology are enriching the discussion about quinoa domestication. Molecular studies (e.g. Maughan et al. 2006; Jarvis et al. 2017) are providing some new hypothesis about the way domestication took place and molecular tools have been applied to the study of ancient genetic variability in quinoa (Babot et al. 2015). It is expected that these transdisciplinary approaches will help answer some of the questions raised in this chapter.

References

Aellen P (1929) Beitrag zur Systematik der *Chenopodium*-Arten Amerikas, vorwiegend auf Grund der Sammlung des United States National Museum in Washington, D.C. II. Repert Nov Specierum Regni Veg 26:119–160

Aellen P, Just T (1943) Key and synopsis of the American species of the genus *Chenopodium* L. Am Midl Nat 30:47–76

Adolf VI, Shabala S, Andersen MN, Razzaghi F, Jacobsen S-E (2012) Varietal differences of quinoa's tolerance to saline conditions. Plant Soil 357:117–129

Alandia G, Jacobsen S-E, Kyvsgaard NC, Condori B, Liu F (2016) Nitrogen sustains seed yield of quinoa under intermediate drought. J Agron Crop Sci 202:281–291

Alvarez-Flores R, Winkel T, Degueldre D, Del Castillo C, Joffre R (2014a) Plant growth dynamics and root morphology of little-known species of *Chenopodium* from contrasted Andean habitats. Botany 92:101–108

Alvarez-Flores R, Winkel T, Nguyen-Thi-Truc A, Joffre R (2014b) Root foraging capacity depends on root system architecture and ontogeny in seedlings of three Andean *Chenopodium* species. Plant Soil 380:415–428

Alvarez-Flores R, Nguyen-Thi-Truc A, Peredo-Parada S, Joffre R, Winkel T (2018) Rooting plasticity in wild and cultivated Andean *Chenopodium* species under soil water deficit. Plant Soil 425:479–492. https://doi.org/10.1007/s11104-018-3588-7

Anabalon Rodriguez L, Thomet Isla M (2009) Comparative analysis of genetic and morphologic diversity among quinoa accessions (*Chenopodium quinoa* Willd.) of the south of Chile and Highland accessions. J Plant Breed Crop Sci 1:210–216

Babot MP (2011) South-Central Andes hunter-gatherers and plant processing. A discussion from Southern Argentinian Puna (CA. 7000–3200 years BP). Chungara 43:413–432

Babot MP, Aguirre MG, Arizio CM, Aschero CA, Bertero HD, Costa-Tártara S, Hocsman S, Joffre R, López Campeny SML, Manifesto MM, Winkel T (2015) Diversidad genética de quinoa en los últimos dos milenios: primer caso de estudio en Antofagasta de la Sierra (puna de Catamarca, Argentina). In: Abstracts of the V world quinoa congress, San Salvador de Jujuy, 27–30 May 2015

Barrón-Yánez MR, Villanueva-Verduzco C, García-Mateos MR, Colinas-León MT (2009) Valor nutricio y contenido de saponinas en germinados de huauzontle (*Chenopodium nuttalliae* Saff.), calabacita (*Cucurbita pepo* L.), canola (*Brassica napus* L.) y amaranto (*Amaranthus leucocarpus* S. Watson syn. hypoch. Rev Chapingo Ser Hortic 15:237–243

Bazile D, Baudron F (2015) The dynamics of the global expansion of quinoa growing in view of its high biodiversity. In: Bazile D, Bertero, D, Nieto C (eds) State of the art report on Quinoa around the World 2013. FAO (Santiago de Chile) and CIRAD (Montpellier, Francia), pp 42–55

Bazile D, Fuentes F, Mujica A (2013) Historical perspectives and domestication. In: Barghava A, Srivastava S (eds) Quinoa: botany, production and uses. CABI, Boston, pp 16–36

Bazile D, Bertero D, Nieto C (eds) (2015) State of the art report on quinoa around the world in 2013. FAO (Santiago de Chile) and CIRAD (Montpellier, Francia)

Bertero D, Medan D, Hall AJ (1996) Changes in apical morphology during floral initiation and reproductive development in quinoa (*Chenopodium quinoa* Willd.). Ann Bot 78:317–324

Bertero HD, King RW, Hall AJ (1999a) Modelling photoperiod and temperature responses of flowering in quinoa (*Chenopodium quinoa* Willd.). Field Crops Res 63:19–34

Bertero HD, King RW, Hall AJ (1999b) Photoperiod-sensitive development phases in quinoa (*Chenopodium quinoa* Willd.). Field Crops Res 60:231–243

Bertero HD (2003) Response of developmental processes to temperature and photoperiod in quinoa (*Chenopodium quinoa* Willd.). Food Rev Int 19:87–97

Bertero HD, de la Vega AJ, Correa G, Jacobsen SE, Mujica A (2004) Genotype and genotype-by-environment interaction effects for grain yield and grain size of quinoa (*Chenopodium quinoa* Willd.) as revealed by pattern analysis of international multi-environment trials. Field Crops Res 89:299–318

Biondi S, Ruiz KB, Martínez EA, Zurita-Silva A, Orsini F, Antognini F, Dinelli G, Marotti I, Gianquinto G, Maldonado S, Burrieza H, Bazile D, Adolf VI,Jacobsen SE (2015) Tolerance to saline conditions In: Bazile D, Bertero D, Nieto C (eds) State of the Art Report on Quinoa Around the World 2013. FAO (Santiago de Chile) and CIRAD (Montpellier, Francia), pp 143–155

Bioversity International (2013) Descriptors for quinoa (*Chenopodium quinoa* Willd.) and wild relatives. Bioversity International, Food and Agriculture Organization of the United Nations, Rome, Italy

Bruno MC (2005a) Domesticated or wild? Results of the investigation of *Chenopodium* seeds form Chiripa, Bolivia (1500 BC-100 BC). Textos Antropológicos 2:39–50

Bruno MC (2005b) A morphological approach to documenting *Chenopodium* domestication in the Andes. In: Zeder M, Bradley D, Emhswiller E, Smith D (eds) Documenting domestication: new genetic and archaeological paradigms. University of California Press, Berkeley

Bruno MC (2008) Waranq Waranqa: ethnobotanical perspectives on agricultural intensification in the Lake Titicaca basin (Taraco peninsula, Bolivia). Ph D thesis Dissertation, Washington University of Saint Louis

Bruno MC, Whitehead WT (2003) *Chenopodium* cultivation and formative period agriculture at Chiripa, Bolivia. Lat Am Antiq 14:339–355

Burrieza HP, Sanguinetti A, Michieli CT, Bertero HD, Maldonado S (2016) Death of embryos from 2300-year-old quinoa seeds found in an archaeological site: Plant Sci 253:107–117

Ceccato D (2011) Efecto de las condiciones ambientales durante el llenado de granos sobre la dormición en semillas de quinoa (*Chenopodium quinoa* Willd.), y su relación con la susceptibilidad al brotado pre-cosecha. MsSci thesis Dissertation, University of Buenos Aires, Argentina

Ceccato D, Bertero D, Batlla D, Galati B (2015) Structural aspects of dormancy in quinoa (*Chenopodium quinoa*): importance and possible action mechanisms of the seed coat. Seed Sci Res 25:267–275

Cocozza C, Pulvento C, Lavini A, Riccardi M, d'Andria R, Tognetti R (2013) Effects of increasing salinity stress and decreasing water availability on ecophysiological traits of quinoa (*Chenopodium quinoa* Willd.) grown in a mediterranean-type agroecosystem. J Agron Crop Sci 199:229–240

Christensen SA, Pratt DB, Pratt C, Nelson PT, Stevens MR, Jellen EN, Coleman CE, Fairbanks DJ, Bonifacio A, Maughan PJ (2007) Assessment of genetic diversity in the USDA and CIP-FAO international nursery collections of quinoa (*Chenopodium quinoa* Willd.) using microsatellite markers. Plant Genet Resour 5:82–95

Christiansen JL, Jacobsen S-E, Jørgensen ST (2010) Photoperiodic effect on flowering and seed development in quinoa (*Chenopodium quinoa* Willd.). Acta Agric Scand Sect B — Soil Plant Sci 60:539–544

Costa Tártara SM (2014) Caracterizacion molecular del germoplasma nativo de quinoa (*Chenopodium quinoa* Willd.) del Noroeste Argentino mediante microsatelites. PhD thesis Dissertation, University of La Plata, Argentina

Costa Tártara SM, Manifesto MM, Bramardi SJ, Bertero HD (2012) Genetic structure in cultivated quinoa (*Chenopodium quinoa* Willd.), a reflection of landscape structure in Northwest Argentina. Conserv Genet 13:1027–1038

Costa Tártara SM, Manifesto MM, Curti RN, Bertero HD (2015) Origen, prácticas de cultivo, usos y diversidad genética de quinua del Noroeste Argentino (NOA) en el contexto del conocimiento actual del germoplasma de America del Sur. In: Cruz P, Joffre R, Winkel T (eds) Racionalidades campesinas en los Andes del Sur: reflexiones en torno al cultivo de la quinua y otros cultivos andinos. Editorial de la Universidad Nacional de Jujuy, Jujuy, pp 201–230

Curti RN, Andrade AJ, Bramardi S, Velásquez B, Daniel Bertero H (2012) Ecogeographic structure of phenotypic diversity in cultivated populations of quinoa from Northwest Argentina. Ann Appl Biol 160:114–125

Curti RN, de la Vega AJ, Andrade AJ, Bramardi SJ, Bertero HD (2014) Multi-environmental evaluation for grain yield and its physiological determinants of quinoa genotypes across Northwest Argentina. Field Crops Res 166:46–57

Curti RN, de la Vega AJ, Andrade AJ, Bramardi SJ, Bertero HD (2016) Adaptive responses of quinoa to diverse agro-ecological environments along an altitudinal gradient in North West Argentina. Field Crops Res 189:10–18

del Castillo C, Winkel T, Mahy G, Bizoux J-P (2007) Genetic structure of quinoa (*Chenopodium quinoa* Willd.) from the Bolivian altiplano as revealed by RAPD markers. Genet Resour Crop Evol 54:897–905

Dillehay TD, Rossen J, Andres TC, Williams DE (2007) Preceramic adoption of peanut, squash and cotton in Northern Peru. Science 316:1890–1893

Escribano J, Cabanes J, Jiménez-Atiénzar M, Ibañez-Tremolada M, Gómez-Pando LR, García-Carmona F, Gandía-Herrero F (2017) Characterization of betalains, saponins and antioxidant power in differently colored quinoa (*Chenopodium quinoa*) varieties. Food Chem 234:285–294

Fuentes FF, Martinez EA, Hinrichsen PV, Jellen EN, Maughan PJ (2009) Assessment of genetic diversity patterns in chilean quinoa (*Chenopodium quinoa* Willd.) germplasm using multiplex fluorescent microsatellite markers. Conserv Genet 10:369–377

Gandarillas H (1968) Razas de quinua. Boletin Experimental No 34. Ministerio de Agricultura. Instituto Boliviano de Cultivos Andinos

Gandarillas H (1979) Genética y origen. In: Tapia ME, Gandarillas H, Alandia S, Cardozo A, Mujica A (eds) Quinua y kaniwa: Cultivos Andinos. Instituto Interamericano de Ciencias Agrícolas, Bogotá, pp 45–64

Gandarillas H, Tapia M (1976) La variedad de quinua dulce Sajama. In Convencion Internacional de Quenopodiaceas 2. Potosi, Bolivia. IICA, Informes de Conferencias, Cursos y Reuniones No. 96

Geerts S, Raes D, Garcia M, del Castillo C, Buytaert W (2006) Agro-climatic suitability mapping for crop production in the Bolivian highlands: a case study for quinoa. Agric Forest Meteorol 139:399–412

Geerts S, Raes D, Garcia M (2008) Indicators to quantify the flexible phenology of quinoa (*Chenopodium quinoa* Willd.) in response to drought stress. Field Crops Res 108:150–156

Gepts P (2010) Crop domestication as a long-term selection experiment. Plant Breed Rev 24:1–44

González JA, Bruno M, Valoy M, Prado FE (2010) Genotypic variation of gas exchange parameters and leaf stable carbon and nitrogen isotopes in ten quinoa cultivars grown under drought. J Agron Crop Sci 197:81–93

González Marín SP (2009) Estudio del flujo de genes en quinua (*Chenopodium quinoa* Willd.) en campos de agricultores mediante el uso de marcadores microsatelitales. Undergraduate thesis dissertation Escuela Politecnica del Ejercito, Sangolqui, Ecuador

Gremillion KJ (1993) Crop and weed in Prehistoric Eastern North America: the *Chenopodium* example. Amer Ant 58:496–509

Hariadi Y, Marandon K, Tian Y, Jacobsen S-E, Shabala S (2011) Ionic and osmotic relations in quinoa (*Chenopodium quinoa* Willd.) plants grown at various salinity levels. J Exp Bot 62:185–193

Heiser CB, Nelson DC (1974) On the origin of the cultivated Chenopods (*Chenopodium*). Genetics 78:503–505

Hilgert NI (1999) Las plantas comestibles en un sector de las Yungas meridionales (Argentina). An Jardín Botánico Madrid 57:117–138

Hunziker AT (1943a) Las especies alimenticias de *Amaranthus* y *Chenopodium* cultivadas por los indios de América. Rev Arg Agron 10:297–354

Hunziker AT (1943b) Granos hallados en el yacimiento arqueológico de Pampa Grande (Salta, Argentina). Rev Arg Agron 10:146–154

Jacobsen S-E (2003) The worldwide potential for quinoa (*Chenopodium quinoa* Willd.). Food Rev Int 19:167–177

Jacobsen S-E, Liu F, Jensen CR (2009) Does root-sourced ABA play a role for regulation of stomata under drought in quinoa (*Chenopodium quinoa* Willd.). Sci Hortic 122:281–287

Jankurova M, Minariviciva L, Dandar A (2009) Quinoa-a review. Czech. J Food Sci 27:71–79

Jarvis DE, Ho YS, Lightfoot DJ, Schmöckel SM, Li B, Borm TJA, Ohyanagi H, Mineta K, Michell CT, Saber N, Kharbatia NM, Rupper RR, Sharp AR, Dally N, Boughton BA, Woo YH, Gao G, Schijlen EGWM, Guo X, Momin AA, Negrão S, Al-Babili S, Gehring C, Roessner U, Jung C, Murphy K, Arold ST, Gojobori T, van der Linden CG, van Loo EN, Jellen EN, Maughan PJ, Tester M (2017) The genome of *Chenopodium quinoa*. Nature 542:307–312

Jellen EN, Maughan PJ, Fuentes F, Kolano BA (2015) Botany, phylogeny and evolution. In: Bazile, Bertero D, Nieto D (eds) State of the Art Report on Quinoa Around the World 2013. FAO (Santiago de Chile) and CIRAD (Montpellier, Francia), pp 12–23

Jensen CR, Jacobsen S-E, Andersen MN, Núñez N, Andersen SD, Rasmussen L, Mogensen VO (2000) Leaf gas exchange and water relation characteristics of field quinoa (*Chenopodium quinoa* Willd.) during soil drying. Eur J Agron 13:11–25

Joffre R, Acho J (2008) Quinoa, descanso y tholares en el sur del altiplano Boliviano. Habitat 75:38–48

Kistler L, Shapiro B (2011) Ancient DNA confirms a local origin of domesticated Chenopod in eastern North America. J Archaeol Sci 38:3549–3554

Kolano B, McCann J, Orzechowska M, Siwinska D, Temsch E, Weiss-Schneeweiss H (2016) Molecular and cytogenetic evidence for an allotetraploid origin of *Chenopodium quinoa* and *C. berlandieri* (Amaranthaceae). Mol Phylogenet Evol 100:109–123

Koziol MJ (1991) Afrosymetric estimation of threshold saponin concentration for bitterness in quinoa (*Chenopodium quinoa* Willd.). J Sci Food Agric 54:211–219

Lagiglia H (2001) Los origenes de la agricultura en Argentina. In: Berberian EE, Nielsen AE (eds) Historia

Argentina Prehispanica, vol 1. ed. Brujas, Cordoba, Argentina, pp 41–81

Langlie B (2008) Paleoethnobotanical analysis of Formative Chiripa, Bolivia. Ph D Thesis dissertation, University of California

Langlie B (2019) Morphological analysis of late Prehispanic Peruvian *Chenopodium* spp. Veg Hist Archaeobot. https://doi.org/10.1007/s0033401806778

Langlie BS, Hastorf CA, Bruno MC, Bermann M, Bonzani RM, Condarco WC (2011) Diversity in andean *Chenopodium* domestication: describing a new morphological type from La Barca, Bolivia 1300-1250 B.C. J Ethnobiol 31:72–88

Li G, Zhu F (2018) Quinoa starch: structure, properties, and applications. Carbohydr Polym 181:851–861

López LM, Capparelli A, Nielsen AE (2011) Traditional post-harvest processing to make quinoa grains (*Chenopodium quinoa* var. quinoa) apt for consumption in Northern Lipez (Potosí, Bolivia): ethnoarchaeological and archaeobotanical analyses. Archaeol Anthropol Sci 3:49–70

Martínez EA, Fuentes FF, Bazile D (2015) History of quinoa: its origin, domestication, diversification, and cultivation with particular reference to the Chilean context. In: Murphy K, Matanguihan J (eds) Quinoa: improvement and sustainable production. Wiley, Hoboken, NJ, USA, pp 19–24

Mason SL, Stevens MR, Jellen EN, Bonifacio A, Fairbanks DJ, Coleman CE, McCarty RR, Rasmussen AG, Maughan PJ (2005) Development and use of microsatellite markers for germplasm characterization in quinoa (*Chenopodium quinoa* Willd.). Crop Sci 45:1618–1630

Maughan PJ, Kolano BA, Maluszynska J, Coles ND, Bonifacio A, Rojas J, Coleman CE, Stevens MR, Fairbanks DJ, Parkinson SE, Jellen EN (2006) Molecular and cytological characterization of ribosomal RNA genes in *Chenopodium quinoa* and *Chenopodium berlandieri*. Genome 49:825–839

Mizui F, Kasai R, Ohtani K, Tanaka O (1988) Saponins from brans of quinoa, *Chenopodium quinoa* Willd. I. Chem Pharm Bull (Tokyo) 36:1415–1418

Mizui F, Kasai R, Ohtani K, Tanaka O (1990) Saponins from brans of quinoa, *Chenopodium quinoa* Willd: II. Chem Pharm Bull (Tokyo) 38:375–377

Mosyakin SL, Clemants SE (1996) New infrageneric taxa and combinations of Chenopodium L. (Chenopodiaceae). Novon 6:398–403

Nuñez L (1986) Evidencias arcaicas de maices y cuyes en Tiliviche: hacia un semisedentarismo en el litoral fertil y quebradas del norte de Chile. Chungará 16–17:25–47

Oliveto LG, Ventura B (2009) Dinámicas poblacionales de los valles rientales del sur de Bolivia y norte de Argentina, siglos XV-XVII: aportes etnohistóriocos y arqueológicos. Poblac Soc 16:119–150

Orsini F, Accorsi M, Gianquinto G, Dinelli G, Antognoni F, Carrasco KBR, Martinez EA, Alnayef M, Marotti I, Bosi S, Biondi S (2011) Beyond the ionic and osmotic response to salinity in *Chenopodium quinoa*: functional elements of successful halophytism. Funct Plant Biol 38:818–831

Pearsall DM (2008) Plant domestication and the shift to agriculture in the Andes. In: Silverman and Isbell (eds) The handbook of South American archaeology. Springer, New York, pp 105–120

Planella MT, Tagle B (2004) Inicios de presencia de cultígenos en la zona central de Chile, periodo Arcaico y Alfarero Temprano. Chungará 36:387–399

Planella MT, Scherson R, McRostie V (2011) New evidence on the use of initial cultigens by the hunter gatherer groups of the Archaic IV period at El Plomo, Alto Maipo, Central Chile. Chungará 43:189–202

Planella MT, Cornejo LE, Tagle B (2005) Las Morrenas 1 Rockshelter: evidence for cultigens among hunter gatherers of the late Archaic period in Central Chile. Chungará 37:59–74

Planella MT, López ML, Bruno MC (2015) Domestication and prehistoric distribution In: Bazile D, Bertero D, Nieto C (eds) State of the Art report on Quinoa around the World 2013. FAO (Santiago de Chile) and CIRAD (Montpellier, Francia), pp 29–41

Prego I, Maldonado S, Otegui M (1998) Seed structure and localization of reserves in *Chenopodium quinoa*. Ann Bot 82:481–488

PROINPA (2003) Variedad quinua jacha grano. Proinpa, Cochabamba, Bolivia

Razzaghi F, Ahmadi SH, Adolf VI, Jensen CR, Jacobsen SE, Andersen MN (2011) Water relations and transpiration of quinoa (*Chenopodium quinoa* Willd.) under salinity and soil drying. J Agron Crop Sci 197:348–360

Razzaghi F, Plauborg F, Jacobsen S-E, Jensen CR, Andersen MN (2012) Effect of nitrogen and water availability of three soil types on yield, radiation use efficiency and evapotranspiration in field-grown quinoa. Agric Water Manag 109:20–29

Richards R, Watt M, Rebetzke G (2007) Physiological traits and cereal germplasm for sustainable agricultural systems. Euphytica 154:409–425

Rojas W (2003) Multivariate analysis of genetic diversity of Bolivian quinoa germplasm. Food Rev Int 19:9–23

Rojas W, Pinto M (2015) Ex situ conservation of quinoa: the Bolivian experience. In: Murphy K, Matanguihan J (eds) Quinoa: improvement and sustainable production. Wiley, Hoboken, NJ, pp 125–158

Rojas W, Pinto M, Alacona C, Gómez Pando L, Lobos PL, Alercia A, Diulgheroff S, Padulosi S, Bazile D (2015) Quinoa genetic resources and ex situ conservation. In: Bazile D, Bertero D, Nieto C (eds) State of the Art report on Quinoa around the World 2013. FAO (Santiago de Chile) and CIRAD (Montpellier, Francia), pp 56–82

Ruiz KB, Aloisi I, Del Duca S, Canelo V, Torrigiani P, Silva H, Biondi S (2016) Salares versus coastal ecotypes of quinoa: Salinity responses in Chilean landraces from contrasting habitats. Plant Physiol Biochem 101:1–13

Ruiz-Carrasco K, Antognoni F, Coulibaly AK, Lizardi S, Covarrubias A, Martínez EA, Molina-Montenegro

MA, Biondi S, Zurita-Silva A (2011) Variation in salinity tolerance of four lowland genotypes of quinoa (*Chenopodium quinoa* Willd.) as assessed by growth, physiological traits, and sodium transporter gene expression. Plant Physiol Biochem 49:1333–1341

Sivori EM (1945) Fotoperiodismo de *Chenopodium quinoa* Willd. Reacción de la cigota y gametófito femenino. Darwiniana 7:541–551

Smith BD (2006) Eastern North America as an independent center of plant domestication. Proc Natl Acad Sci 103:12223–12228

Standley PC (1917) The Chenopodiaceae of the North American flora. Bull Torrey Bot Club 44:411–429

Tapia ME, Mujica A, Canahua A (1980) Origen, distribución geográfica y sistemas de producción de quinua. In: Primera reunión sobre genética y fitomejoramiento de la Quinua. Universidad Nacional Técnica del Altiplano, Instituto Boliviano de Tecnología Agropecuaria, Instituto Interamericano de Ciencias Agrícolas, Centro de Investigación Internacional para el Desarrollo, Puno, Peru. pp A1–A8

Valencia-Chamorro SA (2004) QUINOA. In: Wrigley C (ed) Encyclopedia of grain science. Elsevier, Oxford, pp 1–8

Vavilov NI (1992) Origin and geography of cultivated plants. Cambridge University Press, Cambridge, UK

Walsh BM, Adhikary D, Maughan PJ, Emshwiller E, Jellen EN (2015) Chenopodium polyploidy inferences from Salt Overly Sensitive 1 (SOS1) data. Am J Bot 102:533–543

Ward SM (2000) Response to selection for reduced grain saponin content in quinoa (*Chenopodium quinoa* Willd.). Field Crops Res 68:157–163

Whitehead WT (2007) Exploring the wild and domestic: palethnobotany at Chiripa, a Formative site in Bolivia. Dissertation, University of California

Wilson HD (1981) Genetic variation among South American populations of tetraploid *Chenopodium* sect. *Chenopodium* subsect. *Cellulata*. Syst Bot 6:380–398

Wilson HD (1988a) Quinua biosystematics I: domesticated populations. Econ Bot 42:461–477

Wilson HD (1988b) Quinua biosystematics II: free-living populations. Econ Bot 42:478–494

Wilson HD (1988c) Allozyme variation and morphological relationships of *Chenopodium hircinum* (s.l.). Syst Bot 13:215–228

Wilson HD (1990) Quinua and relatives (*Chenopodium* sect. *Chenopodium* subsect. Celluloid). Econ Bot 44:92–110

Wilson HD, Heiser CB (1979) The origin and evolutionary relationships of "Huazontle" (*Chenopodium nuttalliae* Safford), domesticated Chenopod of Mexico. Am J Bot 66:198–206

Wilson H, Manhart J (1993) Crop/weed gene flow: *Chenopodium quinoa* Willd. and *C. berlandieri* Moq. Theor Appl Genet 86:642–648

Winkel T (2013) Quinoa y quinoeros. IRD editions, Marseille

Wu G (2015) Nutritional properties of quinoa. In: Murphy K, Matanguihan J (eds) Quinoa: improvement and sustainable production. Wiley, Hoboken, NJ, pp 193–205

Zurita-Silva A, Jacobsen SE, Razzaghi F, Alvarez-Flores R, Ruiz KB, Morales A, Silva H (2015) Quinoa drought responses and adaptation In: Bazile D, Bertero D, Nieto C (eds) State of the Art Report on Quinoa Around the World 2013. FAO (Santiago de Chile) and CIRAD (Montpellier, Francia), pp 157–171

3 Botanical Context for Domestication in North America

Eric N. Jellen, David E. Jarvis, Nuri Benet-Pierce, and Peter J. Maughan

Abstract

Pitseed goosefoot, *Chenopodium berlandieri* Moq., is a widespread allotetraploid weed having $2n = 4x = 36$ (AABB subgenomes) found throughout North America from Mexico to Alaska. It is a critical genetic resource for adaptive improvement of its South American cousin and descendant, quinoa (*C. quinoa* Willd.). It has also been important in its own right at various times throughout history and prehistory, having been domesticated at least twice in the Americas north of the Isthmus of Panama. Botanically, *C. berlandieri* belongs to Section *Chenopodium* Subsection *Cellulata* along with its allotetraploid allies *C. quinoa*, weedy South American *C. hircinum* Schrad., and a complex of putative A-genome diploids concentrated in the semiarid southwestern region of North America. This chapter reviews the botanical and ecological context of *C. berlandieri* and its potential as a genetic resource for improving quinoa.

E. N. Jellen (✉) · D. E. Jarvis · P. J. Maughan
Department of Plant and Wildlife Sciences, Brigham Young University, 4105 LSB, Provo, UT 84602, USA
e-mail: jellen@byu.edu

N. Benet-Pierce
Department of Biology, San Diego State University, San Diego, CA 92182, USA

3.1 Introduction

Pitseed goosefoot, *Chenopodium berlandieri* Moq., is an allotetraploid with $2n = 4x = 36$ (AABB sub-genomes). It is the North American free-living member of a New World allotetraploid goosefoot complex that includes cultivated North American *C. berlandieri* subsp. *nuttaliae* (Safford) Wilson & Heiser; South American cultivated *C. quinoa* Willd.; and free-living South American ecotypes *C. hircinum* Schrad. and *C. quinoa* subsp. *milleanum* (Aellen) Aellen or var. *melanospermum* Hunziker (Wilson 1990; Wilson and Manhart 1993). The distinguishing taxonomic characteristic of this species complex is the fruit, consisting of an achene with rounded margins, adhering alveolate pericarp with a honeycombed pattern, and underlying pitted testa. The thick black testa differentiates pitseed goosefoot from its domesticated relatives, which possess a thinner testa that is usually of lighter color. The thick testa in the free-living chenopods inhibits water uptake, thus providing for a reservoir of dormant seed in the soil and accounting for their germination in the wake of human and natural influences that disturb the soil sufficiently to scarify them.

Pitseed goosefoot is of increasing interest as a genetic resource for improving quinoa's adaptation to low-elevation production environments (Jellen et al. 2019). Quinoa was selected under domestication in high-elevation environments

S. M. Schmöckel (ed.), *The Quinoa Genome*, Compendium of Plant Genomes,
https://doi.org/10.1007/978-3-030-65237-1_3

(>3000 masl) in the absence of heat stress and with pressure from a narrow spectrum of pests and pathogens due to the extreme altitude. Coastal quinoa, though selected and produced near sea level in a Mediterranean climate zone, was nonetheless relatively isolated from pathogens and pests of the outside world by the extreme aridity of the Atacama Desert to the north, by the Andean Cordillera and Patagonian Desert to the east, and fronted by the vast expanse of the South Pacific to the west. In contrast, pitseed goosefoot evolved throughout North America, with diverse ecotypes adapted to lowland, highland, subtropical, temperate, and desert environments.

3.1.1 Ecological Context of *C. berlandieri*

The free-living forms of tetraploid *C. berlandieri* have been subdivided into five North American botanical varieties, ignoring uncharacterized natural variation south of the Rio Grande River, in the online Flora of North America (1993+). Of these varieties or ecotypes, var. *zschackei* (Murr) Murr ex Graebner is widespread throughout North America; var. *sinuatum* (Murr) Wahl is found in the semi-arid Southwest; var. *boscianum* (Moq.) Wahl is native to the northern Gulf of Mexico Coast; var. *macrocalycium* (Aellen) Cronquist is from the Atlantic Coast; and var. *berlandieri* is found in interior South Texas. Although Flora of North America (1993+) classified a sixth botanical variety, *bushianum* (Aellen) Cronquist, mounting evidence indicates that this is a separate biological entity, *C. bushianum* Aellen, having a distinct genome composition and hexaploid chromosome number (Bhargava et al. 2005, 2007).

All of the free-living forms of *C. berlandieri* are found in disturbed environments, sometimes in association with the aggressive Eurasian hexaploid weed *C. album* L. (lambsquarters). The surest method of discriminating between the two entities is by examining the pericarps, which are typically adhering and alveolate in the former and adhering or non-adhering but always non-alveolate in the latter. Gentle rubbing of the fruits typically exposes the shiny and, generally, non-pitted testas in a sample of lambsquarters.

3.1.2 Molecular Studies of *C. berlandieri*

There is mounting molecular evidence for the interrelatedness of members of the ATGC and, from targeted and whole-genome DNA sequencing studies, that *C. berlandieri* is most likely the basal or ancestral member of the complex. Five earlier studies by Wilson (1981, 1988a, b, c) and Wilson and Manhart (1993) demonstrated using biochemical marker analyses that the South American ATGC members are closely related to each other and to North American *C. berlandieri* and will spontaneously cross-hybridize to produce varying degrees of fertile progeny.

Storchova et al. (2015) provided strong evidence for common ancestry of ATGC members through Bayesian and Maximum Parsimony phylogenetic analyses of DNA sequences from two *flowering locus T-like* (*FTL*) introns. Moreover, they were able to discriminate between *FTL* versions of the two sub-genomes and reported for the first time that each locus showed affinity to diploid species from different hemispheres. Walsh et al. (2015) sequenced introns 16 and 17 of the *salt overly sensitive 1* (*SOS1*) gene and found similar results. Brown et al. (2015) repeated these findings a third time by comparing sequences from the amylose-synthesis gene *granule bound starch synthase 1* (*GBSS1* or *waxy*). Kolano et al. (2016) compared nuclear ITS sequences and found that all seven *C. berlandieri* genotypes were basally situated relative to the six quinoa genotypes in their study.

Jarvis et al. (2017) compared whole-genome nuclear resequencing data from five *C. berlandieri*, two *C. hircinum*, and 16 *C. quinoa* genotypes in a bootstrapped phylogenetic analysis and showed that *C. berlandieri* was basal to the South American free-living and domesticated genotypes. An additional comparison of whole mtDNA- and cpDNA-based phylogenies in these same genotype sets has reaffirmed these relationships (D Jarvis, personal communication).

3.1.3 Potential Diploid Progenitor Gene Pools of the ATGC in North America

Recent DNA sequencing-based studies of ATGC members and *Chenopodium* diploids (Storchova et al. 2015; Walsh et al. 2015; Brown et al. 2015; Jarvis et al. 2017) have instituted a nomenclature system for the two sub-genomes, with 'A' representing the sub-genome most similar to a large group of New World-native diploids and 'B' representing the sub-genome closest to Eurasian diploids *C. ficifolium* Smith and *C. suecicum* Murr. Kolano et al. (2011) previously reported a minisatellite DNA sequence, 18–24 J, that hybridized abundantly to chromosomes of the B genome in *Chenopodium* diploids, tetraploids, and hexaploids.

Jarvis et al. (2017) demonstrated through comparison of the QQ74 quinoa deep-sequenced reference genome and shallow-sequenced diploids *C. pallidicaule* Aellen (A sub-genome, Andean) and *C. suecicum* (B sub-genome, Central European) that there is abundant synteny and even collinearity of scaffolded sequence contigs between each diploid and distinct sets of orthologous quinoa chromosomes. The three aforementioned single-gene sequencing studies (Storchova et al. 2015; Walsh et al. 2015; Brown et al. 2015) found that A-genome diploids were positioned basally to a clade containing ATGC members, while B-genome diploids were positioned basally to a clade containing ATGC members along with sequenced taxa from the Eurasian *C. album*-complex. Kolano et al. (2016) reported that nuclear ITS sequence in most of the ATGC members is more similar to B-genome diploids *C. ficifolium* and *C. suecicum* than to the A-genome diploids—a predictable result, since only one nucleolar organizer region (NOR) locus has been retained in most ATGC members and it is on a chromosome carrying the 18–24 J B-genome-abundant minisatellite (Kolano et al. 2011).

The question of which New World A-genome is, or is closest to, the original donor of the A genome in the ATGC members is elusive, although studies of cytoplasmic DNA have confirmed that the A-genome diploid was the female parent in the hybridization event that occurred roughly 3.3–6.3 Mya (Kolano et al. 2016; Jarvis et al. 2017). Table 3.1 contains a comprehensive list of known North American- and incomplete list of South American-native diploids presumed to carry variants of the A genome. Kolano et al. (2016) reported that whole-genome DNA of *C. watsonii* A. Nels. was more effective in hybridizing to A-genome chromosomes of *C. berlandieri* and *C. quinoa* than 12 other putative A-genome diploids in genomic in situ hybridization (GISH) experiments. Among their other 12 diploids were South American *C. pallidicaule* and *C. petiolare* Kunth and North American *C. atrovirens* Rydb., *C. desiccatum* A. Nels., *C. fremontii* S. Wats., *C. hians* Standl. (probably not *C. hians* but more likely *C. desiccatum*), *C. incanum* (S. Wats.) A. A. Heller, *C. leptophyllum* (Moq.-Tand.) Nutt. ex S. Wats., *C. neomexicanum* Standl. (collected at a site more typical of *C. arizonicum* Standl. or *C. sonorense* Benet-Pierce & M. G. Simpson), *C. nevadense* Standl., *C. pratericola* Rydb., and *C. standleyanum* Aellen. Morphologically, *C. berlandieri* most closely matches the North American-native diploid members of *Chenopodium* subsect. *Cellulata*, among which are included the seven species of the *C. neomexicanum* complex (Benet-Pierce and Simpson 2017) and *C. watsonii*. Another diploid complex, that of *C. hians*, has tremendous diversity in California and specimens from that state were recently subdivided into twelve taxa by Benet-Pierce and Simpson (2019). Of additional interest is the observation that most of these diploids and several free-living taxa of the ATGC possess the fishy trimethylamine-odor phenotype—a trait that has not been observed in the B-genome diploids *C. ficifolium* and *C. suecicum*.

Table 3.1 Comprehensive list of North American and partial list of South American putative A-genome diploid *Chenopodium* species

Chenopodium L. Species	Habitat
albescens Small	S Texas mesquite woodlands
arizonicum Standl.	Upper Sonoran scrub lands
atrovirens Rydb.	W USA, montane
aureum Benet-Pierce	W USA, montane
brandegeeae Benet-Pierce	S California, montane
cycloides A. Nels.	W Great Plains, sandy soils
desiccatum A. Nels.	W USA, deserts and plains
eastwoodiae Benet-Pierce	Sierra Nevada Mountains
flabellifolium Standl.	San Martin Island, BC, Mexico
foggii Wahl	New England, granitic forest soils
fremontii S. Wats.	W USA, montane woodlands
hians Standl.	W USA, montane woodlands
howellii Benet-Pierce	NE California, Sierra Nevada Mountains
incanum (S. Wats.) A. A. Heller	W USA, deserts
incognitum Wahl	W USA, montane chaparral
lenticulare Aellen	Upper Chihuahuan scrub lands
leptophyllum (Moq.-Tand.) Nutt. ex S. Wats.	W USA, deserts
lineatum Benet-Pierce	Sierra Nevada Mountains
littoreum Benet-Pierce & M. G. Simpson	C California, coastal dunes
luteum Benet-Pierce	Sierra Nevada Mountains
neomexicanum Standl.	Mogollon Plateau, igneous soils
nevadense Standl.	W Great Basin, sodic clay pans
nitens Benet-Pierce and Simpson	W USA, deserts and plains
pallescens Standl.	E Great Plains
pallidicaule Aellen	Andean seed crop (*canahua*)
palmeri Standl.	Upper Sonoran scrub lands
parryi Standl.	Upper Chihuahuan scrub lands
petiolare Kunth[a]	Andean Puna zone
philippianum Aellen	S Andes
pratericola Rydb.	USA, sandy soils
sandersii Benet-Pierce	S California, montane
simpsonii Benet-Pierce	S California, montane
sonorense Benet-Pierce & M. G. Simpson	Sonoran Desert
standleyanum Aellen	C USA, woodlands
subglabrum (S. Wats.) A. Nels.	W Great Plains, sandy soils
twisselmannii Benet-Pierce	Sierra Nevada Mountains
wahlii Benet-Pierce	S California, montane
watsonii A. Nels.	W USA, plains, corrals

[a]Fits taxonomic description of *C. paniculatum* Hook. (Reiche 1911)

3.2 Descriptions of *C. berlandieri* Subspecies and Botanical Varieties

3.2.1 *C. berlandieri* subsp. *berlandieri* var. *boscianum*

Variety *boscianum* is a locally common weed on rock jetties, sand and gravel piles, roadsides, and other disturbed areas where the root zone experiences cyclic tidal drainage on barrier islands and estuaries along the northwestern Gulf of Mexico coast from Laguna Madre to Mobile Bay (Fig. 3.1). This taxon is noteworthy for its production of fishy-smelling trimethylamine throughout the plant (Cromwell 1950), a trait that is dominantly inherited in hybrids with quinoa (personal observations). Variation for timing of flowering and seed set, ranging from spring to very late fall, is a characteristic of var. *boscianum* populations. Occurrence of tropical cyclones is instrumental in seed dispersal of this taxon; Jellen and Maughan (unpublished) encountered populations still bearing substantial quantities of seed in early December 2014—an exceptional year due to the absence of landfalling tropical cyclonic systems in the northern Gulf of Mexico. Many of these populations contained plants that flowered in mid-late summer, presumably in response to shortening daylengths. On the other hand, populations observed on Padre Island, in extreme southeastern Texas, in April exhibited day-neutrality and possessed only faint trimethylamine odor.

Phenotypic variation for var. *boscianum* fruits and typical habitats are shown in Fig. 3.1. Hybrids between the *boscianum* ecotype and cultivated quinoa are >90% fertile in the F_2 generation (Fig. 3.2). At BYU, lines with large and light-colored seeds, heat tolerance, and close resemblance to quinoa panicles have been selfed to the F_9 generation. These plants have consistently produced ample seed in multiple field environments including Bolivia, coastal Peru, northern Guyana, Morocco, the Pacific Northwest, and Great Basin of the USA.

3.2.2 *C. berlandieri* subsp. *berlandieri* var. *berlandieri*

Variety *berlandieri* is found in locally common populations along crop fields in sandy soils of the lower Rio Grande Valley. It also inhabits disturbed roadsides intersected by arroyos, especially on saline or expanding clay substrates, sometimes in the shade of locust trees, and often in association with putative diploid *C. albescens*. Figure 3.3 shows phenotypic variation for fruit morphology and typical habitats of this ecotype.

In the spring of 2018 Jellen, Maughan, and Jarvis (personal communication) collected 30 populations of var. *berlandieri* across South Texas from San Antonio to Brownsville. These plants had a very strong trimethylamine odor, like the northern Gulf Coast populations of var. *boscianum*; few to no bracts in their large inflorescences; and leaves that most often had two prominent basal lobes. In addition, the pericarps were usually black and lacked the yellow-orange area at the stylar base typical of var. *zschackei*.

3.2.3 *C. berlandieri* subsp. *berlandieri* var. *macrocalycium*

Variety *macrocalycium* is an eastern North American seashore (/lakeshore?) plant often found in competition with *C. album*, mostly along the Middle Atlantic and New England Coasts up through the Canadian Maritime provinces. Its phenotype is characterized by its relatively large (to 2 mm diameter) fruits and large calyces that typically spread away from the fruit, exposing it at maturity. It is therefore an intriguing genetic resource for potentially improving quinoa seed size, provided that *macrocalycium* carries complementary alleles contributing to this trait. Figure 3.4 shows variation for fruit type in a small collection of var. *macrocalycium* and also typical habitats where it is found.

In 2008, an experimental population was generated at BYU by crossing Ames 29207 from the beach at Saco, Maine, with the quinoa variety

Fig. 3.1 **a** Fruit morphology of 18 accessions of *C. berlandieri* var. *boscianum* from populations collected along the northern Gulf of Mexico Coast, from west to east. (top row, l to r) BYU 1470, Matagorda, Texas; BYU 1474, Freeport, Texas; BYU 1475, San Luis Pass, Texas; BYU 1301, San Leon, Texas; BYU 1476, Jamaica Beach, Texas; (second row) BYU 1469, Bolivar Peninsula, Texas; BYU 1466, Port Arthur, Texas; BYU 1465, Little Florida Beach, Louisiana; BYU 1455, Cypremort Point, Louisiana; BYU 1458, Golden Meadow, Louisiana; (third row) BYU 1460, Grand Isle, Louisiana; BYU 17131, Metairie, Louisiana; BYU 17134, Yscloskey, Louisiana; BYU 14112, Long Beach, Mississippi; BYU 14113, Gulfport, Mississippi; (bottom row) dispersal unit of BYU 1470; BYU 14114, Biloxi, Mississippi; BYU 14115, Pascagoula, Mississippi; BYU 14111, Mobile Bay, Alabama; dispersal unit of BYU 14111. **b** Photographs of typical *C. berlandieri* var. *boscianum* habitats. (l to r) Rockport, TX; Yscloskey, LA; Coden, AL

Fig. 3.2 Fruit morphology in an F_2 population derived from an interspecific quinoa cv. 'Real-1' X *C. berlandieri* var. *boscianum* (BYU 937 from Galveston Bay, Texas) cross. Top row (l to r): 'Real-1', BYU 937, F_1 hybrid. Rows two, three, and bottom show fruits from random F_2 plants in the population. (Jellen et al. 2019)

'Ingapirca'. The F_1 plant was large, fertile, and vigorous; however, well over half of the F_2 generation plants failed to set seeds under heat-stress conditions in the greenhouse (25C + temperatures) and those that did possessed black seeds and wild-type (highly lax) panicle morphologies. Consequently, var. *macrocalycium* may not be an advantageous taxon for improving quinoa's heat tolerance.

3.2.4 *C. berlandieri* subsp. *berlandieri* var. *sinuatum*

Variety *sinuatum* (Fig. 3.5) is present in locally common populations along roadsides, disturbed (including recently burned) fields, semi-desert scrub and chaparral, and open grasslands throughout southwestern North America (Flora of North America 1993+). Within relatively undisturbed grasslands *sinuatum* can comprise fairly extensive stands, with variation for branching pattern and plant color. In Arizona-New Mexico, it is often found in association with the diploid taxa *C. arizonicum*, *C. neomexicanum*, *C. palmeri*, and *C. sonorense* and in the Davis Mountains of West Texas, with *C. lenticulare*. *Sinuatum* is primarily a summer-fall fruiting ecotype; its flowering is timed to coincide with the mid-summer arrival of the Sonoran-Chihuahuan Monsoon in Arizona-New Mexico-West Texas, although the presence of spring-

Fig. 3.3 **a** Fruit morphology of 18 accessions of *C. berlandieri* var. *berlandieri* from populations collected in South Texas. (top row, 1 to r) BYU 1801, Cotulla, La Salle Co.; BYU 1804, Catarina, Dimmit Co.; BYU 1806, Webb Co.; BYU 1807, Webb Co.; BYU 1809, San Diego, Duval Co.; (second row) BYU 1814, Falfurrias, Brooks Co.; BYU 1817, Brooks Co.; BYU 1818, Linn, Hidalgo Co.; BYU 1819, Lasara, Willacy Co.; BYU 1820, Del Mar Heights, Cameron Co.; (third row) BYU 1822, Rio Hondo, Cameron Co.; BYU 1823, Riviera, Kleberg Co.; BYU 1824, Corpus Christi, Nueces Co.; BYU 1830, Beeville, Bee Co.; BYU 1832, Mustang Hollow, Live Oak Co.; (bottom row) dispersal unit of BYU 1802, Artesia Springs, Cotulla Co.; BYU 1833, George West, Live Oak Co.; BYU 1834, Three Rivers, Live Oak Co.; BYU 1836, Hidden Acres, San Patricio Co.; dispersal unit of BYU 1836. **b** Photographs of typical *C. berlandieri* var. *berlandieri* habitats. (l to r) Horse pasture, Kleberg Co., Texas; edge of cotton field, Hidalgo Co., Texas; roadside under locust tree, Bee Co., Texas

Fig. 3.4 **a** Fruit morphology of six accessions of *C. berlandieri* var. *macrocalycium*. (top row, l to r): calyces of BYU 1488 and BYU 1489; (second row): calyx of BYU 803; BYU 803, Ames 29207 from Saco, Maine; BYU 1488 from Slaughter Beach, Delaware; BYU 1489 from Sea Isle City, New Jersey; (third row): BYU 1490 from Ocean Grove, New Jersey; BYU 14110 from Navarre Beach, Florida; BYU 17129 from Rye Beach, New Hampshire; (bottom row): calyces of BYU 1490, BYU 14110, and BYU 17129. **b** Typical habitats of var. *macrocalycium*: Slaughter Beach, Delaware (left) and Sea Isle City, New Jersey (right, with *C. album*)

Fig. 3.5 **a** Fruit morphology of 14 accessions of *C. berlandieri* var. *sinuatum*. (top row, l to r): BYU 402 from San Diego, California; BYU 1494 from Cottonwood, Arizona; BYU 870 from Mormon Lake, Arizona; BYU 1452 from Big Tujunga Canyon, California; BYU 1493 from Peeples Valley, Arizona; (second row): BYU 17203 from Cochise Co., Arizona; BYU 14101 from San Juan, New Mexico; BYU 14108 from Cochise Co., Arizona; BYU 1511 from White Pine Co., Nevada; BYU 1653 from the Mojave Desert, California; (third row): BYU 17150 from Wild Rose Pass, Texas; BYU 17158 from the Davis Mountains, Texas; BYU 17212 from Sonoita, Arizona; BYU 17242 from Pima Co., Arizona; (bottom row): calyces of BYU 402, BYU 870, BYU 1494, and BYU 17242. **b** Typical habitats of var. *sinuatum* (l to r): Davis Mountains, Texas; Santa Cruz Co., Arizona; Big Tujunga Canyon, California

fruiting genotypes of similar subtropical varieties *berlandieri* and *boscianum* in Southeast Texas suggest that the existence of *sinuatum* populations capable of exploiting the winter-spring rainfall cycle in relatively frost-free lowland areas of southern Arizona and California should be expected. The presence of populations of fruiting var. *sinuatum* growing in Big Tujunga Canyon (along with diploid *C. incognitum* and *C. album*) and in the Santa Monica Mountains under exceptional drought conditions in August 2014—at a time when many of the chaparral perennials were under drought stress-induced dormancy—indicates that this taxon could be a tremendous breeding resource for improving heat and drought tolerance in quinoa.

Wilson and Heiser (1979) reported low fertility in a small number of crosses between *C. quinoa* and *C. berlandieri* var. *sinuatum*. However, a population at BYU derived from a *C. quinoa* cv. 'Ollague' X var. *sinuatum* (BYU 14108) produced 155/178 (89%) fertile F_2's (Fig. 3.6).

3.2.5 *C. berlandieri* subsp. *berlandieri* var. *zschackei*

Variety *zschackei* is the most common and widespread free-living ecotype of *C. berlandieri* in North America. Its distinguishing characteristic is the orange-yellow pigmentation on the pericarp at the stylar attachment point atop the fruit. This variety is found throughout the continent, especially in interior areas and, in the Southwest, at higher elevations. Within the western USA, in particular, it is a fairly common weed of roadsides, construction sites, and pastures (Fig. 3.7). Variety *zschackei* usually flowers from late summer through fall, presumably as a consequence of spring germination rather than a short-day flowering response.

3.2.6 *C. berlandieri* subsp. *jonesianum*

Subspecies *jonesianum* is an extinct cultivated form of pitseed goosefoot discovered in archeological middens throughout the Oak-Savannah and Oak-Hickory Forest regions of North America (Smith and Funk 1985). A comparison of chloroplast DNA from archeological and modern samples supported the hypothesis that this was a local, rather than an introduced, domesticate (Kistler and Shapiro 2011). Specimens had small (<1.5 mm) seed with thin, dark testas. It was a principal food source for the Hopewell and Early Mississippian mound builders and a key component of the Eastern North America (ENA) Crop Complex (Smith 2006). Two additional *C. berlandieri* forms have been described together from the same archeological sites as subsp. *jonesianum* (Smith and Yarnell 2009): common and similarly small-seeded, but thick-testa specimens representing free-living pitseed goosefoot that was likely a weedy form harvested for greens; and a rare, larger (to 2.4 mm diameter), orange-yellow colored and thin-testa type resembling subsp. *nuttaliae*. The appearance of weedy and cultivated pitseed goosefoot types together indicates they likely formed a crop-weed complex—a situation mirroring modern Andean quinoa-*ajara* and Mesoamerican *huauzontle*-pitseed goosefoot complexes (Gremillion 1993; Wilson and Heiser 1979; Wilson 1990; Wilson and Manhart 1993). Both domesticated seed forms no longer appear in middens that postdate 1000 B.C.E., presumably due to the adoption of the Mesoamerican maize-beans-squash complex (Smith and Yarnell 2009).

3.2.7 *C. berlandieri* subsp. *nuttaliae*

Subspecies *nuttaliae* includes an interesting and phenotypically diverse assemblage of highland Mexican leafy garden greens (*quelites*), inflorescence vegetables (*huauzontle*), and seed crops (*chia roja*), along with a companion a free-living type, proposed *C. pueblense* H. S. Reed, with which it likely has continuously hybridized (Wilson and Heiser 1979). While *huauzontle* and chenopod *quelites* continue to be important food crops in central Mexico, *chia roja* is threatened with extinction as a crop due to shifting patterns of rural cultivation and societal upheaval in its

Fig. 3.6 Representative variation for panicle morphology in an F_2 population derived from an interspecific quinoa cv. 'Ollague' X *C. berlandieri* var. *sinuatum* (BYU 14108 from the Chiricahua Mountains, Arizona) cross

Fig. 3.7 **a** Fruit morphology of 15 accessions of *C. berlandieri* var. *zschackei*. (top row, l to r): BYU 457 from Duchesne Co., Utah; BYU 555 from Tooele, Utah; BYU 629 from Niobrara Co., Wyoming; BYU 637 from Muddy Pass, Colorado; BYU 641 from Sanpete Co., Utah; (second row): BYU 661 from Thistle, Utah; BYU 718 from Colfax Co., New Mexico; BYU 862 from Fremont Co., Wyoming; BYU 880 from Ramah, New Mexico; BYU 902 from the Laguna Mts., California; (third row): BYU 1007 from Kyle Cyn., Nevada; BYU 1312 from St. Charles, Missouri; BYU 1448 from Sherman Oaks, California; BYU 14118 from Crystalaire, California; BYU 1672 from Burns, Oregon; (bottom row): calyces of BYU 902, BYU 1312, BYU 14118, and BYU 1672. **b** Representative habitats of var. *zschackei* (l to r): La Sal Mountains, Utah; edge of farm field, Shasta Co., California; roadside in pinyon-juniper woodland, Sacramento Pass, Nevada

center of cultivation in Michoacan State (De la Cruz Torres et al. 2010). Figure 3.8 shows fruit variation in subsp. *nuttaliae*.

Wilson and Heiser (1979) and Wilson (1990) provided and reviewed evidence supporting the hypothesis that subsp. *nuttaliae* was domesticated independently from *C. quinoa*. Molecular evidence supports this hypothesis, given that Kolano et al. (2012, 2016) and Maughan et al. (2006) found that *C. berlandieri* subsp. *nuttaliae* accessions contained one more 5S rRNA locus (three versus two), and some strains have one more NOR locus than *C. quinoa*.

Brown et al. (2015) and Cepeda-Cornejo et al. (2016) identified *waxy* (high amylopectin) landraces of huauzontle. The *waxy* phenotype is attained only when A- and B-genome orthologs of *GBSS1* are mutated; consequently, these would be expected to be rare phenotypes in the absence of strong selective pressure. These authors hypothesized that the amylose-free phenotypes were selected in Mexico because semi-mature *huauzontle* panicles carrying *waxy* seeds would cook more thoroughly in less time and at lower temperatures than panicles containing seed with insoluble amylose. Brown et al. (2015) found single mutations in each sub-genome that were responsible for all of their *waxy* huauzontles: a *gbss1a-tp* A-genome allele carrying an I54T substitution within the plastid-targeting transit peptide portion of the gene; and a *gbss1b-del* B-genome allele carrying a 78-base deletion. In contrast to vegetable *huauzontle*, Andean quinoa seeds were likely selected to be larger and plumper—traits associated with complete starch—containing amylose. Brown et al. (2015) did not find similar *waxy* phenotypes in the six quinoa strains they examined, though they did identify a nonsense *gbss1b-t* mutant allele in coastal cultivar G-205-95.

3.3 Breeding Potential of *C. berlandieri*

Wilson and Heiser (1979) performed intertaxa ATGC hybridization experiments with results indicating that var. *zschackei* hybrids with *C. quinoa* are highly fertile, while *sinuatum* x *quinoa* hybrids are of low fertility but can be backcrossed by applying *C. quinoa* pollen to F_1 stigmas. In a later study, Wilson and Manhart (1993) documented extensive in-field introgression in the Pacific Northwest of the USA between cultivated quinoa and *C. berlandieri* weeds growing in close proximity. Figure 3.2 illustrates phenotypic variation for fruit morphology in a quinoa var. 'Real-1' X var. *boscianum* (BYU 937, Galveston Bay, Texas) segregating F_2 population having 91% self-fertility. Figure 3.6 shows variation for panicle shape in a quinoa var. 'Ollague' X var. *sinuatum* (BYU 14108, Chiricahua Mountains, Arizona) segregating F_2 population having 87% self-fertility.

Given the wide range of environments in which *C. berlandieri* is adapted—ranging from subtropical Mediterranean, subtropical Gulf Coastal, subtropical Sonoran Desert, semi-arid temperate, and humid forest zones—this species consequently represents a significant genetic resource for improving quinoa through intertaxa hybridization and selection. This contrasts with *C. quinoa*, which was selectively adapted by human domestication to high-elevation environments in the Andes; to the narrow, geographically isolated coastal strip of central and southern Chile; to humid intermediate-elevation valleys of northwestern Argentina (Curti et al. 2012); and possibly also in antiquity to the high plains of Argentine Patagonia. The importance of pitseed goosefoot germplasm is magnified for quinoa breeders in countries like the USA that are non-participants in the Convention of Biodiversity and Nagoya Protocol, and therefore have limited access to quinoa germplasm.

Fig. 3.8 **a** Fruit morphology of 16 accessions of *C. berlandieri* subsp. *nuttaliae* from Mexico. Top row (l to r): BYU 567 from Opopeo, Michoacan; BYU 668 from Tecoman, Colima; BYU 1669, line H-18 from ININ-Ocoyoacac; BYU 17176 cv. 'Red Aztec Spinach'; second row: BYU 1670, PI640304; BYU 1668, line H-16 from ININ-Ocoyoacac; BYU 1447, line H-3 from ININ-Ocoyoacac; BYU 1483, PI433230; third row: BYU 1484, *waxy* PI433231; BYU 1485, *waxy* PI568155; BYU 1486, PI568156; BYU 1647, *waxy* line H-4 from ININ-Ocoyoacac; bottom row: BYU 1662, *chia roja* from ININ-Ocoyoacac; BYU 1663, *chia roja* from ININ-Ocoyoacac; BYU 1664, *chia roja* from ININ-Ocoyoacac; BYU 1665, *chia roja* from ININ-Ocoyoacac. *Waxy* phenotyping was reported by Brown et al. (2015) and Cepeda-Cornejo et al. (2016). **b** (l to r) *Chia roja* field near Patzcuaro, Michoacan; green strain of garden *huauzontle* in Toluca, Mexico; pink strain of *huauzontle* in Toluca, Mexico

3.4 Conclusions

Chenopodium berlandieri represents a genetically and ecologically diverse genetic resource for quinoa improvement. It is a species that has been domesticated multiple times as vegetable and seed crops. It also contains a wide variety of interesting genetic mechanisms that allow it to survive and compete in arid, saline, hot, cold, and high-altitude environments. It is also one of a small handful of plants that was domesticated in the Eastern North American Center. As such, pitseed goosefoot should have higher priority in crop genetic resource collection, curation, and maintenance efforts by North American governments.

References

Benet-Pierce N, Simpson MG (2019) The taxonomy of *Chenopodium hians*, *C. incognitum*, and ten new taxa within the narrow-leaved *Chenopodium* group in western North America, with special attention to California. Madroño 66:56–75

Benet-Pierce N, Simpson MG (2017) Taxonomic recovery of the species in the *Chenopodium neomexicanum* (Chenopodiaceae) complex and description of *Chenopodium sonorense* sp. nov. J Torrey Bot Soc 144:339–356

Bhargava A, Shukla S, Ohri D (2007) Genome size variation in some cultivated and wild species of *Chenopodium* (Chenopodiaceae). Caryologia 60:245–250

Bhargava A, Shukla S, Ohri D (2005) Karyotypic studies of some cultivated and wild species of *Chenopodium* (Chenopodiaceae). Genet Resour Crop Evol 53:1309–1320

Brown DC, Cepeda-Cornejo V, Maughan PJ, Jellen EN (2015) Characterization of the *Granule-Bound Starch Synthase I* gene in *Chenopodium*. Plant Genome 8:1–12

Cepeda-Cornejo V, Brown DC, Palomino G, de la Cruz E, Fogarty M, Maughan PJ, Jellen EN (2016) Genetic variation of the granule-bound starch synthase I (*GBSSI*) genes in *waxy* and non-*waxy* accessions of *Chenopodium berlandieri* subsp. *nuttaliae* from Central Mexico. Plant Gen Resour: Charact Util 14:57–66

Cromwell BT (1950) The micro-estimation and origin of trimethylamine in *Chenopodium vulvaria* L. Biochem J 45:578–582

Curti RN, Andrade AJ, Bramardi S, Velasquez B, Bertero HD (2012) Ecogeographic structure of phenotypic diversity in cultivated populations of quinoa from Northwest Argentina. Ann Appl Biol 160:114–125

De la Cruz Torres E, García Andrade JM, Mapes Sanchez C, Lopez Monroy A (2010) Estudio de los recursos genéticos de pseudocereales nativos de México. XXIII Congreso Nacional y III Internacional de Fitogenética, Nayarit, México

Flora of North America Editorial Committee eds (1993 +) Clemants SE, Mosyakin SL, *Chenopodium* Linnaeus, vol 4. In: Flora of North America North of Mexico, 20 + vols, New York and Oxford

Gremillion KJ (1993) Crop and weed in prehistoric Eastern North America: the *Chenopodium* example. Am Antiq 58:496–509

Jarvis DE, Ho YS, Lightfoot DL, Schmockel SM, Li B, Borm T, Ohyanagi H, Mineta K, Michell CT, Saber N, Kharbatia NM, Rupper RR, Sharp AR, Dally N, Boughton BA, Woo YH, Gao G, Schiljen E, Guo X, Negrao S, Al-Babili S, Gehring C, Roessner U, Jung C, Murphy K, Arold S, Gojobori T, van der Linden G, van Loo EN, Jellen EN, Maughan PJ, Tester M (2017) The genome of *Chenopodium quinoa*. Nature 542:307–312

Jellen EN, Jarvis DE, Hunt SP, Mangelsen HH, Maughan PJ (2019) New seed collections of North American pitseed goosefoot (*Chenopodium berlandieri*) and efforts to identify its diploid ancestors through whole-genome sequencing. Ciencia e Investigación Agraria 46:187–196

Kistler L, Shapiro B (2011) Ancient DNA confirms a local origin of domesticated chenopod in eastern North America. J Archaeol Sci 38:3549–3554

Kolano B, Gardunia BW, Michalska M, Bonifacio A, Fairbanks D, Maughan PJ, Coleman CE, Stevens MR, Jellen EN, Maluszynska J (2011) Chromosomal localization of two novel repetitive sequences isolated from the *Chenopodium quinoa* Willd. genome. Genome 54:710–717

Kolano B, McCann J, Orzechowska M, Siwinska D, Temsch E, Weiss-Schneeweiss H (2016) Molecular and cytogenetic evidence for an allotetraploid origin of *Chenopodium quinoa* and *C. berlandieri* (Amaranthaceae). Mol Phylogenet Evol 100:109–123

Kolano B, Tomczak H, Molewska R, Jellen EN, Maluszynska J (2012) Distribution of 5S and 35S rRNA gene sites in 34 *Chenopodium* species (Amaranthaceae). Bot J Linn Soc 170:220–231

Maughan PJ, Kolano BA, Maluszynska J, Coles ND, Bonifacio A, Rojas J, Coleman CE, Stevens MR, Fairbanks DJ, Parkinson SE, Jellen EN (2006) Molecular and cytological characterization of ribosomal RNA genes in *Chenopodium quinoa* and *C. berlandieri*. Genome 49:825–839

Reiche K (1911) Estudios criticos de la Flora de Chile. Anales de la Universidad de Chile 6:148–159. http://www.biodiversitylibrary.org/item/10736#1. Accessed 18 May 2018

Smith BD (2006) Eastern North America as an independent center of plant domestication. Proc Natl Acad Sci (USA) 103:12223–12228

Smith BD, Funk VA (1985) A newly described subfossil cultivar of *Chenopodium* (Chenopodiaceae). Phytologia 57:445–448

Smith BD, Yarnell RA (2009) Initial formation of an indigenous crop complex in North America at 3800 B. P. Proc Natl Acad Sci (USA) 106:6561–6566

Storchova H, Drabesova J, Chab D, Kolar J, Jellen EN (2014) The introns in FLOWERING LOCUS T-LIKE (FTL) genes are useful markers for tracking paternity in tetraploid *Chenopodium quinoa*. Genet Resour Crop Evol 62:913–925

Walsh BM, Adhikary D, Maughan PJ, Emshwiller E, Jellen EN (2015) *Chenopodium* (Amaranthaceae) polyploidy inferences from *Salt Overly Sensitive 1 (SOS1)* data. Am J Bot 102:1–11

Wilson HD (1981) Genetic variation among South American populations of tetraploid *Chenopodium* subsect, *Cellulata*. Syst Bot 6:380–398

Wilson HD (1988a) Quinua biosystematics I: domesticated populations. Econ Bot 42:461–477

Wilson HD (1988b) Quinua biosystematics II: free-living populations. Econ Bot 42:478–494

Wilson HD (1988c) Allozyme variation and morphological relationships of *Chenopodium hircinum* (s.l.). Syst Bot 13:215–228

Wilson HD (1990) *Quinua* and relatives (*Chenopodium* sect. *Chenopodium* subsect. *Cellulata*). Econ Bot 44:92–110

Wilson HD, Heiser CB (1979) The origin and evolutionary relationships of 'Huauzontle' (*Chenopodium nuttaliae* Safford), domesticated chenopod of Mexico. Am J Bot 66:198–206

Wilson H, Manhart J (1993) Crop/weed gene flow: *Chenopodium quinoa* Willd. and *C. berlandieri* Moq. Theor Appl Genet 86:642–648

Quinoa Cytogenetics

4

Bożena Kolano and Maja Orzechowska

Abstract

Chenopodium quinoa has rather a small genome (2.973 pg/2C DNA) and 36 mostly metacentric and submetacentric chromosomes. Like many other plants from the Chenopodiaceae family, it is a polysomatic species. Quinoa is an allotetraploid and its diploid parental taxa, which were suggested by GISH and molecular phylogenetic analyses, belong to the two different evolutionary lineages that are assigned as genome A and B. Two families of rRNA gene loci, 35S and 5S rDNA, were located on the different chromosome pairs for most of the analyzed *Chenopodium* species. In the quinoa karyotype, two pairs of 5S rDNA loci and only one pair of 35S rDNA loci were detected. The nrITS sequences were of a B-genome ancestry and the 35S rRNA gene locus was observed in the chromosome of the B subgenome. The chromosomal localization of other repetitive sequence, clone *12–13P* containing minisatellite repeats, was mainly restricted to pericentromeric regions of all quinoa chromosomes; however, the obtained hybridization signals showed variable intensity suggesting a difference in copy number of the repeat among the chromosomes of the karyotype. Recent analysis showed that sequences similar to the *12–13P* clone are also present in other diploid and polyploid species from *Chenopodium s.s.*. Hybridization signals of retrotransposons were enriched in pericentromeric regions of the quinoa chromosomes and were ordinarily excluded from the distal parts of the chromosomes. Two other repeats (*pTaq10* and *18–24 J*) showed also dispersed distribution in chromosomes of *C. quinoa* although they did not show homology to any known mobile elements. The amplification of the *pTaq10* repeat appears to be characteristic for chenopods with the A genome. However, hybridization signals of *pTaq10* were found on each *C. quinoa* chromosome. A second dispersed repetitive sequence, *18–24 J*, was present in most analyzed *Chenopodium s.s.* species. In quinoa genome hybridization signals of *18–24 J* repeat was mainly restricted to the chromosomes of B subgenome.

The genomes of eukaryotes are composed of multiple chromosomes. The chromosome is composed of a linear DNA molecule and associated proteins that pack the long DNA thread into a more denser and compact structure (Heslop-Harrison and Schwarzacher 2011). The number and structure of a chromosome complement (karyotype) is unique for each plant species

B. Kolano (✉) · M. Orzechowska
Institute of Biology, Biotechnology and Environmental Protection, Faculty of Natural Sciences, University of Silesia, Katowice, Katowice 40-007, Poland
e-mail: bozena.kolano@us.edu.pl

S. M. Schmöckel (ed.), *The Quinoa Genome*, Compendium of Plant Genomes,
https://doi.org/10.1007/978-3-030-65237-1_4

(Heslop-Harrison and Schwarzacher 2011). In plants, the nuclear DNA contains in addition to single- or low-copy coding sequences also different classes of repetitive DNAs, which can be found in hundreds or even thousands units in the genome (Heslop-Harrison and Schwarzacher 2011). Repetitive DNAs are largely responsible for the "C-value paradox" (the complexity of an organism is not associated with the size of its genome; Hawkins et al. 2008). To date, the significance of repetitive DNA, with respect to structure and function, is still poorly understood, although biological roles have been suggested for its specific families, for example, gene regulation, recombination, the maintenance of chromosome structure or a centromere function (Li et al. 2017).

Chromosomal diversification within genera has been a focal point of plant evolutionary studies (Schubert and Vu 2016). There are two prime reasons for this interest. First, a chromosomal change imparts a partial or complete barrier to an interspecific gene exchange. Second, chromosomal characteristics may provide clues to the relationships of species. Several mechanisms account for the high variability of number, length and shape of chromosomes, as well as in DNA content among eukaryotes (Schubert and Vu 2016). Polyploidy and dysploidy are considered to be the most important mechanisms responsible for high variation in chromosome number in plant (De Storme and Mason 2014). Polyploidy refers to the presence of more than two genomes and is a driving force in plant evolution that can induce various genetic and epigenetic alternation leading to reorganization of the polyploid nuclei (Leitch and Leitch 2008). It is now believed that a polyploidization event occurred in all angiosperms at least once in their history and some evolutionary lineages show evidence of several such events (Weiss-Schneeweiss et al. 2013; Jiao et al. 2011; Soltis and Soltis 2009). Polyploidy may induce a variety of chromosomal, genic and epigenetic changes that can take place quickly over a several generations and result in diploid-like chromosome pairing, segregation and gene function (Weiss-Schneeweiss et al. 2013; Soltis and Soltis 2009; Leitch and Leitch 2008). The shape and size of chromosomes can be modified by structural rearrangements (e.g., reciprocal translocations, pericentric inversions, etc.) or by sequence amplification/loss (Schubert and Vu 2016). Chromosome rearrangements can also change the chromosome numbers, a phenomenon called ascending or descending dysploidy (increase or decrease in chromosome numbers, respectively; Schubert and Vu 2016; De Storme and Mason 2014).

Studying the karyotype of selected species involves analyzing chromosome numbers, centromere position and the number and position of secondary constrictions and other chromosomal markers, e.g., banding patterns and the position of various DNA sequences that are uncovered by fluorescence *in situ* hybridization (FISH). Using FISH with single-copy (BAC-clones) and various repetitive DNA sequences (e.g., rDNA or satellite sequences), detailed physical chromosomal maps can be constructed that enable evolutionary patterns and processes to be determined (Kolano et al. 2013b; Hřibová et al. 2010; Lysak et al. 2006).

4.1 Genome Constitution of Polyploid *C. Quinoa* and Related Species

Quinoa belongs to *Chenopodium sensu stricto* (*Chenopodium s.s.*; Fuentes-Bazan et al. 2012). All species of this genus have the basic chromosome number $x = 9$. The diploid *Chenopodium s.s.* species ($2n = 2x = 18$) can be divided into three evolutionary lineages: (1) American diploids with the A genome (e.g., *C. hians*, *C. pallidicaule*); (2) the Old World species with the B genome (e.g., *C. ficifolium* and *C. suecicum*) and (3) *C. vulvaria*, which was recovered as a sister taxon to the rest of the *Chenopodium s.s.* diploids (Kolano et al. 2015; Walsh et al. 2015). In *Chenopodium s.s.*, as in most plant groups, polyploidization has a significant impact on speciation, and several polyploid species representing three different polyploidy levels (tetraploid e.g., *C. berlandieri*; $2n = 4x = 36$; hexaploid e.g., *C. album*, $2n = 6x = 52$ and decaploid e.g., *C. frutescens* $2n = 10x = 90$) have been described (Kolano et al. 2012b; Jellen et al. 2011; Lomonosova 2005). Polyploidy is generally classified into two types—allopolyploidy, which origin involves interspecific hybridization, and autopolyploidy consisting of more than two the same genomes. Whether autopolyploids or allopolyploids are more widespread among plants is a question that still remains unresolved (Weiss-Schneeweiss et al. 2013). Quinoa, which displays a disomic inheritance, is an allotetraploid, with $2n = 4x = 36$ chromosomes (Fig. 4.1a; Ward 2000). The allotetraploid nature of *C. quinoa* and the closely related *C. berlandieri* was supported by molecular phylogenetic studies that were based on different plastid and nuclear markers (nrITS, 5S rDNA NTS; *Flowering Locus T-Like* and *Salt Overly Sensitive 1*) and genomic *in situ* hybridization (GISH), which enable the parental subgenomes to be differentiated in the karyotype of a polyploid (Kolano et al. 2016; Storchova et al. 2015; Walsh et al. 2015). A combination of molecular phylogenetic analyses and GISH showed that the two parental taxa of *C. quinoa* and *C. berlandieri* belong to two various genome pools. One of the parental species contributing the maternal genome was similar to extant American diploids (genome A, maternal genome), whereas, the second ancestral species was similar to contemporary species from Eurasia (B genome, paternal genome; Kolano et al. 2016; Storchova et al. 2015; Walsh et al. 2015). The paternal species that contributed the B subgenome for *C. quinoa* resembled the present *C. ficifolium* or a closely related species (Fig. 4.1b), whereas the origin of the A-subgenome progenitor seems to be more vague (Kolano et al. 2016; Storchova et al. 2015; Walsh et al. 2015). The ancestral species of the A subgenome of *C. quinoa* and *C. berlandieri* possible resembled extant *C. standleyanum*, *C. incanum* or *C. fremontii* based on molecular phylogenetic studies (Kolano et al. 2016; Storchova et al. 2015; Walsh et al. 2015). In contrast, GISH suggested species similar to *C. watsonii* or *C. nevadense* for *C. quinoa* and *C. watsonii* for *C. berlandieri* as the A-genome progenitors (Fig. 4.1b; Kolano et al. 2016). B-genome diploid (*C. ficifolium* or related species) was also inferred to be one of the ancestral species of Eurasian hexaploids (*C. album, C. giganteum, C. pedunculare*; Fig. 4.1c; Krak et al. 2016; Walsh et al. 2015; Kolano et al. 2019).

4.2 Karyotype and Chromosome Banding

C. quinoa has 36 small and poorly differentiated chromosomes. Metacentric and submetacentric chromosomes whose lengths range between approximately 1–3 µm dominate in its karyotype (Palomino et al. 2008; Bhargava et al. 2006; Kolano et al. 2001). Palomino et al. (2008), basing on chromosome morphology, distinguished nine groups of chromosomes supporting the tetraploid origin of quinoa genome. The numerous small chromosomes of *C. quinoa* and other chenopods make cytogenetic studies that are based on procedures such as simple chromosome staining followed by karyotype analysis very challenging. Also, the scarcity of chromosome markers has made it difficult to match chromosomes in homologous pairs. Other techniques, such as chromosome banding, are useful tools for chromosome identification and they often permit conclusions about chromosomal evolution to be drawn (Guerra 2000). Various banding techniques allow to reveal a series of consistent landmarks along the length of metaphase chromosomes that allow identification of homologous chromosome pairs. Three different banding techniques are most often used in plant cytogenetics: C-banding, double-fluorescent banding (CMA_3/DAPI) and silver staining (Guerra 2000).

C-banding has been used to stain constitutive heterochromatin in metaphase chromosomes (Tanaka and Taniguchi 1975). In plant species that have relatively large genomes and appreciable amounts of constitutive heterochromatin (e.g., *Triticum* or *Avena*), this technique produces rich banding patterns that often allows

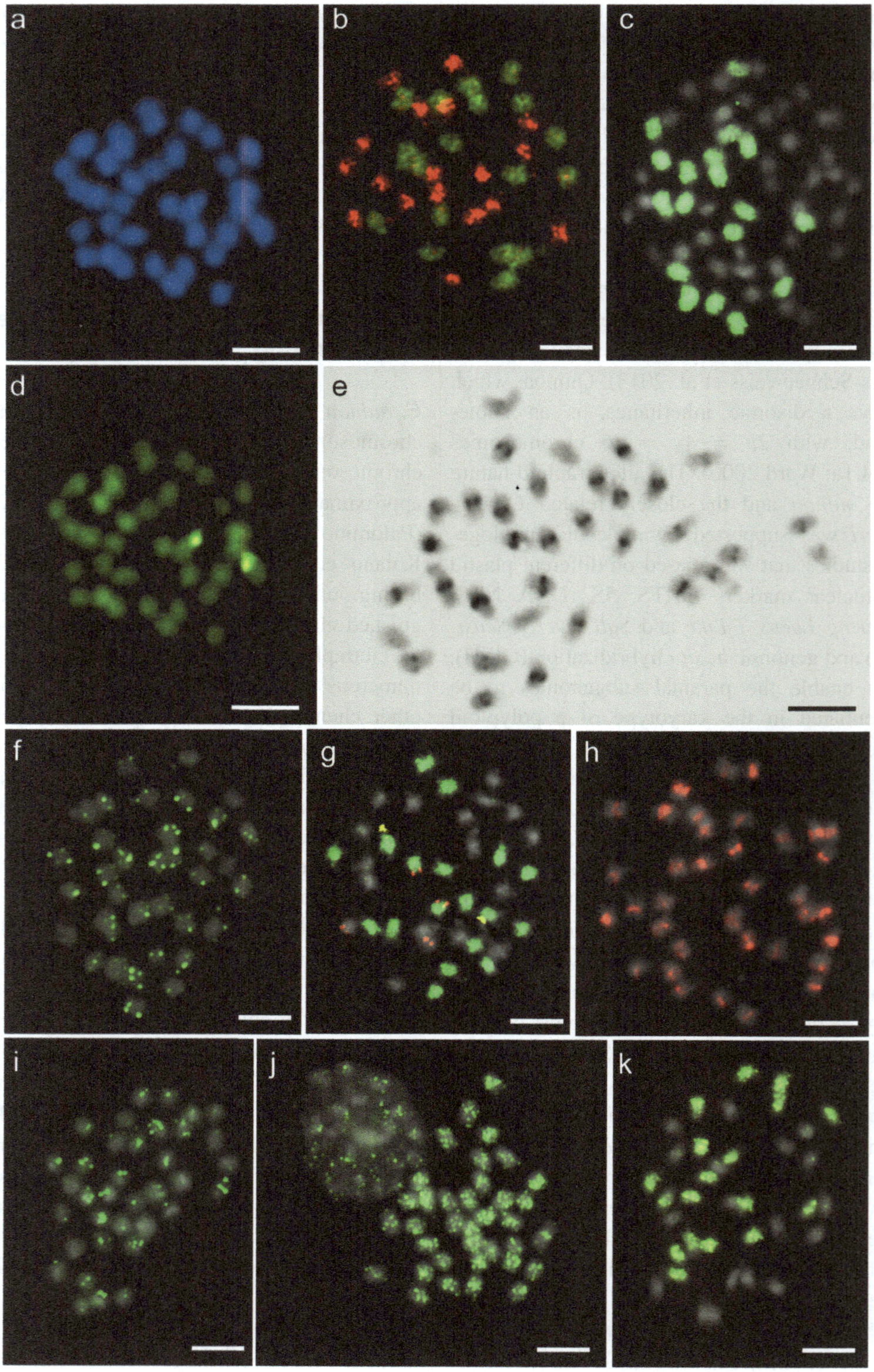
a
b
c
d
e
f
g
h
i
j
k

◀ **Fig. 4.1** Somatic metaphase chromosomes of (**a**, **b**, **d**–**k**) *Chenopodium quinoa* and (**c**) *C. album*; **a** DAPI staining (from Kolano et al. 2001); **b** double GISH with gDNA isolated from the B-genome diploid *C. ficifolium* (green) and the A-genome diploid *C. watsonii* (red); **c** GISH with gDNA isolated from the B-genome diploid *C. ficifolium*; **d** differential fluorescent staining with chromomycin A3, CMA_3^+ bands indicate large GC-rich regions of chromatin (from Kolano et al. 2001); **e** C-bands indicate constitutive heterochromatin; **f** FISH with telomeric repeats; **g** GISH/FISH with gDNA isolated from *C. ficifolium* (green), 35S rDNA (yellow) and 5S rDNA (red); **h** FISH with the *12-13P* repeat; **i** FISH with the *rt* clone of Ty1-*copia* retrotransposons; **j** FISH with dispersed repetitive sequence *pTaq10*; **k** FISH with dispersed repetitive sequences *18–24 J*. Scale bar = 5 μm

identification of single chromosomes and can display morphological or karyotypic variations, such as large chromosomal rearrangements and aneuploidies (Jellen 2016; Gill and Kimber 1974). However, in the *C. quinoa* karyotype, as in many other plants with a small genome size, C-bands and constitutive heterochromatin are present only in the pericentromeric regions of chromosomes (Fig. 4.1e). Other types of heterochromatin bands could be demonstrated using a GC-specific fluorochrome—chromomycin A_3 (CMA_3). Double-fluorescent staining with CMA_3 and DAPI (4′,6-Diamidino-2-phenylindole dihydrochloride) allowed to distinguish GC-rich and AT-rich chromatin region (Schweizer 1876). Two CMA_3^+ bands were revealed in one pair of homologous chromosomes of quinoa (Fig. 4.1d). These CMA_3^+ bands co-localized with secondary constrictions and 35S rRNA gene loci (Kolano et al. 2001). A similar organization of GC-rich chromatin (colocalization with 35S rDNA locus) was also observed in other analyzed chenopods (*C. album, C. ficifolium, C. berlandieri*) and appears to be very often observed in many other angiosperms (e.g., *Amaranthus*; Kolano et al. 2001, 2013b; Guerra 2000; Maragheh et al. 2019). Silver staining, a marker of the transcriptional activity of 35S rRNA genes, revealed the presence of two distally located silver-positive bands in one pair of quinoa chromosomes. One or two nucleoli were observed in the interphase nuclei. However, most frequently, one nucleoli was present (Kolano et al. 2001). Thus, classical banding methods have not been very effective as chromosome markers and have allowed to identify only one pair of homologous chromosomes in *C. quinoa*. The elaboration of new chromosome markers using fluorescent *in situ* hybridization (FISH) with various DNA sequences was necessary for better understanding of chenopod karyotype structure and evolution.

4.3 Genome Size

Genome size (typically measured in picograms or as the total number of nucleotide base pairs in megabases) is one of the fundamental biological characters of all living organisms. The DNA 1C-value for a species is the amount of nuclear DNA in the unreplicated haploid genome of a gamete (Greilhuber et al. 2005). Genome size varies approximately 2,400-fold in angiosperms, ranging from 0.065 pg/1C of DNA in *Genlisea margaretae* to 152.23 pg/1C in *Paris japonica* (Pellicer et al. 2010; Greilhuber et al. 2006). *Chenopodium s.s.* consists of plants with small or very small genomes, which range between 0.47 pg/1C DNA (diploid *C. vulvaris*) and 3.22 pg/1C DNA (decaploid *C. frutescens*; Mandak et al. 2016; Kolano et al. 2015; Bhargava et al. 2007). The analysis of genome size in a phylogenetic background enabled to hypothesize about the direction and trends of genome size evolution during the diversification of the diploid species of *Chenopodium s.s.* The distribution of genome sizes in this genus reflects the phylogenetic grouping into three different evolutionary lineages. The Eurasian diploids (genome B) have 70% larger genomes than the diploids from the Americas (genome A). The smallest genome size among *Chenopodium s.s.* was estimated for *C. vulvaria* (Kolano et al. 2015).

There is very little data on the evolution of the genome sizes of polyploid chenopods. The mean *C. quinoa* genome size is 2C = 2.973 pg (Kolano et al. 2012a). Polyploidization can generate quick and often directional changes to the subgenomes of allopolyploids that often result in genome downsizing (Leitch and Bennett 2004). This phenomenon, reported for many polyploid angiosperms, is usually due to the elimination of repetitive sequences or duplicated genes (Leitch and Bennett 2004). However, the genome size of the allotetraploid *C. quinoa* and other analyzed polyploid chenopods (*C. berlandier* and *C. album*) fit almost perfectly into the estimated additive values of the hypothetical parental species (Kolano et al. 2016; Mandák et al. 2012).

Although genome size is known to be quite constant feature of a species there are more and more studies showing significant intraspecific variation in this characteristic (e.g., *Tanacetum vulgare* or *Arabidopsis thaliana;* Schmuths et al. 2004; Keskitalo et al. 1998). An intraspecific genome size polymorphism also exists in *C. quinoa* (Kolano et al. 2012a). A statistically significant variation in genome size was found among the 20 accessions of *C. quinoa* collected in Peru, Bolivia, Ecuador, Argentina and Chile. The greatest intraspecific difference found between those accessions with the largest genome size and the smallest genome size was nearly 6%. The largest genome had the accession from Chile (1.539 pg/1C) and the smallest was found in the Peruvian accession (1.452 pg/1C). The intraspecific variation in genome size found in *C. quinoa* was rather small when compared to that observed in *Tanacetum vulgare* (27%; Keskitalo et al. 1998); however, it is consistent with other studies that investigated genome size polymorphisms in species that have small genomes, such as *A. thaliana* (Schmuths et al. 2004). The intraspecific polymorphism in genome size was reported to be correlated with various ecological conditions prevailing in specific areas (e.g., altitude or latitude and longitude; Knight et al. 2005). The cultivated populations of quinoa groups into two main clusters—the Andean highland and the coastal lowland from Chile (Christensen et al. 2007). No correlation was observed between the genome size and the geographical origin of the accessions (although the quinoa accessions analyzed in this study belonged to both the highland and lowland groups), possible due to the relatively large variation of genome size revealed among the highland quinoa populations collected on the Altiplano near Lake Titicaca in Bolivia and Peru (Kolano et al. 2012a).

The main reasons for the intraspecific and interspecies genome size polymorphism are polyploidy, the presence of B chromosomes and a different amplification of the repetitive DNA sequences (Hawkins et al. 2008). All of the analyzed *C. quinoa* accessions were tetraploid with $2n = 4x = 36$ chromosomes and there were no B chromosomes in the analyzed accessions. Thus, the main reason for intraspecific genome size polymorphism could be an amplification or reduction in the copy number of the repetitive DNA sequences in the genomes of the analyzed quinoa accessions (Greilhuber 2005).

4.4 Repetitive Sequence Organization and Evolution

Angiosperms vary in the proportion of DNA repetitive sequences in their genomes. Species with small genomes have a relatively small amount of repetitive DNAs, whereas in larger genomes, such as *Zea mays*, at least 85% of the genomes are repetitive sequences (Ibarra-Laclette et al. 2013 Schnable et al. 2009). Repetitive DNA elements can be divided into two major groups based on their genomic organization—dispersed and tandem repeats. Dispersed repetitive DNA elements are distributed throughout the whole chromosomes, intermingled with other sequences although some regions of chromosomes can show reduced or increased number of copies. The second group includes sequences that are organized in tandem repeating units in which the monomers are arranged adjacent to each other to form tandem arrays (Kubis et al. 1998).

4.4.1 Tandem Repetitive Sequences

Tandem repetitive sequences include rRNA genes, telomeric repeats and a very heterogenic group of satellite DNAs (Kubis et al. 1998). *C.*

quina, as many angiosperm species, has arabidopsis-type telomeric sequence. After probing with telomeric repeats, small but discrete signals were observed in all chromosome ends however the hybridization signals had various intensities, revealing the copy number variation among the chromosomes of the same complement (Fig. 4.1f).

rRNA genes are the most extensivly studied repetitive sequences in *Chenopodium s.s.* The 35S and 5S rDNA units consist of conserved genic regions encoding for 35S rRNA (18S–5.8S–25S rDNA) and 5S ribosomal RNA (5S rDNA) and fast evolving transcribed and non-transcribed spacer regions that are arranged as tandem arrays at one or more loci (Volkov et al. 2004). The coding sequences of rRNA genes as evolutionarily highly conserved regions are most often used as chromosome markers in cytogenetic analyses of non-model organisms (Weiss-Schneeweiss et al. 2013;). On the other hand, the non-coding sequences of rRNA genes evolve very quickly and they are often used in phylogenetic analyses (Álvarez and Wendel 2003). Combining a phylogenetic analysis with cytogenetic studies allows to improve the understanding of the evolution of 35S and 5S rDNA loci organization. Our studies indicated that the 35S rDNA and 5S rDNA sequences are organized in a low number of loci, mostly in the subterminal position of the chromosome arm in *Chenopodium* species (Kolano et al. 2015; Kolano et al. 2012b). An analysis of the chromosomal organization of rDNA loci in a phylogenetic context showed that the ancestral states for the diploid *Chenopodium s.s.* taxa were one 35S rDNA locus and one 5S rDNA locus per haploid chromosome set, both located in subterminal positions on different chromosomes. This pattern was also observed in several extant diploids from *Chenopodium s.s.* (e.g., *C. watsonii* or *C. nevadense*; Kolano et al. 2016; Kolano et al. 2015). The number of 5S rDNA loci increased from ancestral state (one locus) to two in the common ancestor of the B-genome diploids since most of them have two 5S rDNA loci (Kolano et al. 2015). Although all of the A-genome diploids had one locus of each rRNA gene, rearrangements of rDNA loci were found in some species. The pattern of 5S rDNA loci observed in two South American species, *C. pallidicaule* and *C. petiolare,* is likely the results of translocation from subterminal to interstitial position in chromosomes. The repositioning of the 35S rDNA locus to the chromosome with a subterminally located 5S rDNA appears to be most likely explanation of the chromosomal organization of rDNA loci revealed in *C. standleyanum* (Kolano et al. 2015, 2016).

The rDNA loci pattern in allopolyploid species are showed to evolve quickly and often the number of rDNA loci in an allopolyploid is not the sum of the loci of its ancestral species (Weiss-Schneeweiss et al. 2013). Polyploids often show genome rearrangements that involve the loss/acquisition of the 35S rDNA repeats or the silencing of rDNA sites as well as interlocus recombinations and complete or near-complete repeat replacements (Sochorová et al. 2017; Weiss-Schneeweiss et al. 2013; Kovarik et al. 2004). The direction of the conversion and homogenization of 35S rDNA can be incline toward the one of the ancestral taxa and may vary in the allopolyploids of the same parentage but independent origin (Sochorová et al. 2017; Weiss-Schneeweiss et al. 2013; Kovarik et al. 2004).

Neither *C. quinoa* nor closely related *C. berlandieri* showed an additive number of rRNA gene loci; but they both experienced a reduction in the number of 35S rDNA loci (Fig. 4.1g). *C. quinoa* has one locus of 35S rDNA in a subterminal position of chromosome (Maughan et al. 2006). Similarly, most analyzed *C. berlandieri* accessions also exhibited only one subterminally located locus of 35S rDNA (this species showed limited polymorphism in 35S rDNA loci number and some accession had two loci of 35S rDNA; Maughan et al. 2006). In both *C. quinoa* and *C. berlandieri*, only the 35S rDNA loci placed in the chromosomes of the B subgenome (paternal parent) were observed. The nrITS sequences were also of a B-genome ancestry (Fig. 4.1g; Kolano et al. 2016). The 35S rDNA loci from the maternal A subgenome was lost in these two allotetraploid (Kolano et al. 2016). A different pattern of 35S rDNA loci evolution was exchibited by the Eurasian allohexaploids *C. album* and *C. giganteum*. The number of 35S rDNA loci in their genomes seems to be equal to the sum of the loci of their ancestral taxa. However, sequence analysis indicated that nrITS was in most analyzed accessions the subject of homogenization and that only one maternal type of nrITS existed in this hexaploid taxa (Krak et al. 2016; Kolano et al. 2019).

Unlike 35S rDNA, the interlocus homogenization of 5S rDNA has not been yet described for any polyploid systems that were studied (Weiss-Schneeweiss et al. 2013; Weiss-Schneeweiss et al. 2012). However, the number and position of 5S rDNA loci may be changed in the karyotype of a polyploid relative to its putative ancestral taxa (Weiss-Schneeweiss et al. 2012; Clarkson et al. 2005). The number of 5S rDNA loci in analyzed American allotetraploids (*C. quinoa* and *C. berlandieri*) and Eurasian allohexaploids (*C. album* and *C. giganteum*) seems to be additive as compared with their ancestor species and the 5S rDNA NTS sequences have not undergone homogenization (Kolano et al. 2016, 2019). In the *C. quinoa* karyotype, the chromosomes of B subgenome bear the subterminal locus of 5S rDNA like in the B-genome diploids. The second 5S rDNA locus in the A subgenome was observed in an interstitial position in place of subterminal position observed in its hypothetical ancestor species (Fig. 4.1g). The observed incongruence can suggest two scenarios: (i) different species (related to *C. watsonii* or *C. nevadense)* which possessed an interstitial 5S rDNA locus could have been the ancestral taxon or (ii) the localization of 5S rDNA was changed in chromosomes of A subgenome of *C. quinoa* after polyploidization. In *C. berlandieri* karyotype the localization of the 5S rDNA loci was unchanged compared to its putative diploid parents (Kolano et al. 2016).

A clone *12–13P* containing minisatellite repeats, was localized in pericentromeric regions of all quinoa chromosomes using FISH. The obtained hybridization signals showed variable intensity, thus suggesting a differences in copy number of the repeats among the chromosomes of the karyotype (Fig. 4.1h; Kolano et al. 2011). The minisatelite repeats from *12–13P* showed a high sequence similarity to the *Beta corolliflora* minisatellite pBC1447 (Gao et al. 2000). Minisatellites are short, usually 10 to 100 bp, tandemly repetitive sequences, which are predominantly localized in the subterminal or pericentromeric regions of chromosomes (Shcherban 2015; Gao et al. 2000). Recent analysis showed that sequences similar to the *12–13P* clone were present in the genomes of most diploid and polyploid species that belong to all *Chenopodium s.s.* lineages, while they were absent in the remaining lineages of *Chenopodium sensu lato* (*Chenopodiastrum, Oxybasis, Blitum, Dysphania, Lipandra*). This suggests that the *12–13P* sequence may be specific to *Chenopodium s. s.* (Orzechowska et al. 2018). In all analyzed species hybridization signals of *12–13P* were present in pericentromeric regions of all chromosomes (Orzechowska et al. 2018).

4.4.2 Disperse Repetitive Sequences

Disperse repetitive sequences primarily include transposable elements (DNA transposons and retrotransposons; Bennetzen and Wang 2014). Retrotransposons that multiple their copy number when transpose to new genome locations are particularly abundant in plant genomes and it has been clear for some time that these transposable elements have the major impact on plant genome size variation (Lisch 2013). To date, the chromosomal organization of the conserved domain of the reverse transcriptase (*rt*) of the LTR retrotransposons (both Ty1-*copia* and Ty3-*gypsy*) has been analyzed in the *C. quinoa* genome (Kolano et al. 2013a). The number of copies of the particular LTR retroelement families can differ remarkably from a few copies to thousands of copies in plant genomes (Du et al. 2010). *C. quinoa* has a relatively small genome and among isolated retrotransposon sequences only one family of Ty1-*copy* retrotransposons was highly amplified. All of the other analyzed *rt* clones were present in the quinoa genome in a low number of copies (Kolano et al. 2013a).

Previous surveys of LTR retrotransposons divided the Ty1-*copia* elements isolated from plant genomes into four major evolutionary lineages (*Tork, Sire, Oryco* and *Retrofit*). Among Ty3-*gypsy* retrotransposons six major evolutionary lineages were distinguished (*Del, Reina, Galadriel, CRM, Athila* and *Tat*; Llorens et al. 2009). To date, four Ty1-*copia* lineages were found in the quinoa genome but among isolated clones dominate elements from *Tork* and *Retrofit* lineages. Most of the elements from the Ty3-*gypsy* superfamily that have been identified belong to *Del* lineage (Kolano et al. 2013a). The chromosomal organization of retrotransposons varies and depends on the retrotransposon families and the host species. The retrotransposons in the *C. quinoa* chromosomes showed an uneven distribution (Fig. 4.1i). Relatively strong hybridization signals were observed on approximately half of the chromosomes, whereas the rest of the chromosomes had only a very weak or no signal. The hybridization signals of retrotransposons were mainly observed in the pericentromeric regions of the chromosomes and were usually not detectable in the distal parts of the chromosomes. In the interphase nuclei, the hybridization signals were mainly observed in the heterochromatic regions (Kolano et al. 2013a).

Following the insertion of a retrotransposons into a genome, nucleotide substitutions accumulate in the elements and they undergo fragmentation over time. These sequences have a dispersed organization in a genome, similar to the transposable elements; however, their nucleotide sequences show a very low or no similarity to the transposable elements (Menzel et al. 2008; Kubis et al. 1998). Two sequences with disperse organization and without homology to known retrotransposons were described in the *C. quinoa* genome. One of these, *pTaq10*, was spread throughout the 36 chromosomes (Fig. 4.1j). Discrete hybridization signals (although weak) were shown in the pericentromeric, interstitial or terminal localizations on the chromosome arms (Kolano et al. 2008a). Hybridization signals of the *pTaq10* clone were present in the interphase nuclei, particularly in the heterochromatic regions. Although such localization of *pTaq10* may suggested that these repeats were present in genomes of both hypothetical ancestors of *C. quinoa*, however our recent studies suggested that *pTaq10* is present only in species with the A genome and that it is not present in the B-genome diploids (Orzechowska et al. unpublished). FISH analyses indicated that the hybridization signals for 5S rDNA and *pTaq10* partly overlapped, thus suggesting partial co-localization of these two repetitive sequences in some chromosome regions. Fiber-FISH, allowing mapping with higher resolution, showed that in *C. quinoa* chromosomes most of the 5S rDNA arrays were free of *pTaq10* repeats, except for a part of the interstitial 5S rDNA locus where *pTaq10* and 5S rDNA repeats co-localized (Kolano et al. 2008a). This interstitial 5S rDNA locus is placed very close or even partly overlaps with the pericentromeric heterochromatin, which implies that at least part of the locus is transcriptionally inactive (Kolano et al. 2008a).

Another dispersed repetitive sequence with no homology to any known retroelements is clone *18–24 J,* which is highly amplified in only half of the quinoa chromosome complements (18 chromosomes; B subgenome; Fig. 4.1k). Similar results were obtained in the related species *C. berlandieri* and the Eurasian hexaploid *C. album* (Kolano et al. 2011). Such a localization could suggest that *18–24 J* is specific to one parental genome of these species. However, a recent study on the *18–24 J* repeat, which comprised a putative ancestral species of *C. quinoa*, revealed the presence of the *18–24 J* repeat in the B-genome diploids as well as in most of the A-genome diploids. Moreover, this repeat was also present in genomes of species which are more distantly related to quinoa (*C. vulvaria* and *Lipandra polysperma*; Orzechowska et al. 2018). The pattern of the *18–24 J* chromosomal localization in *C. quinoa* and the related allotetraploid *C. berlandieri* might be explained in two ways. First, the level of the *18–24 J* repeat amplification in the A-genome ancestral species may have been very low, as has been observed in the case

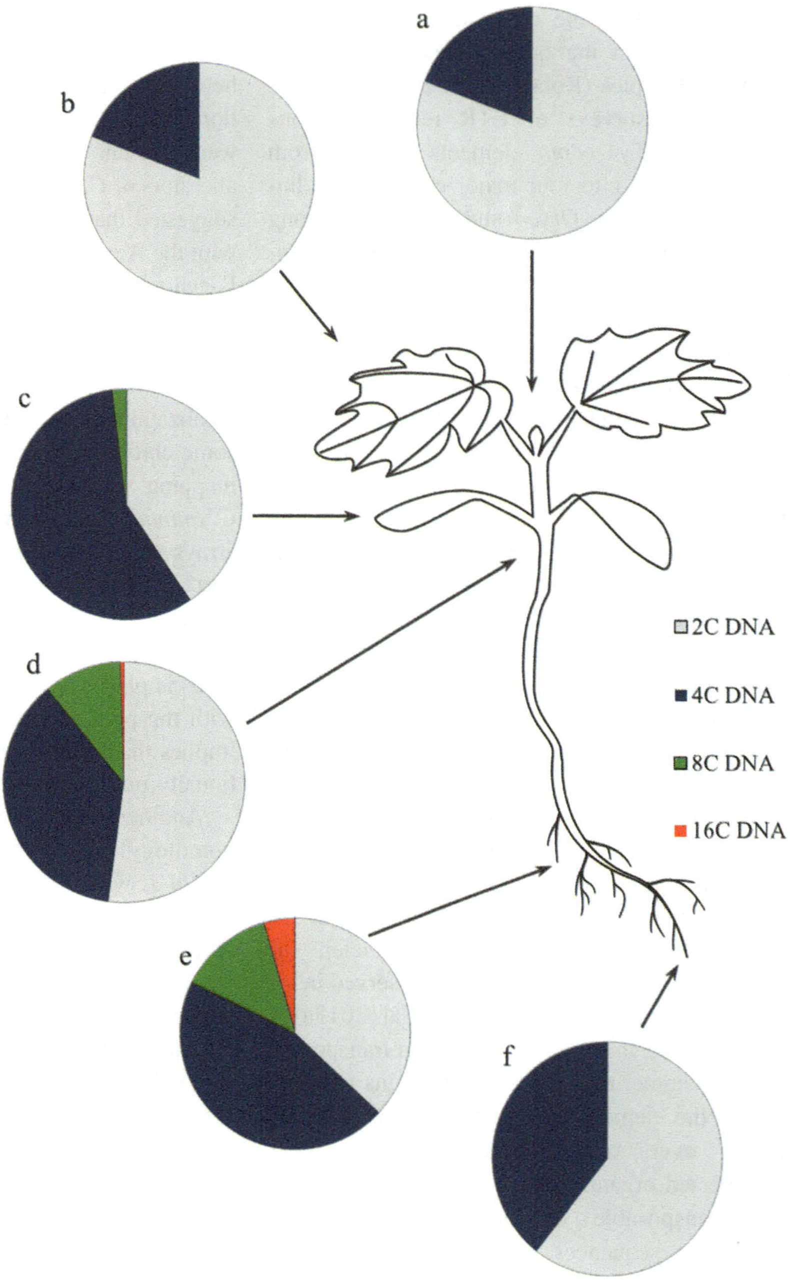

Fig. 4.2 Comparison of the endopolyploidy patterns in quinoa seedlings; **a** shoot apex; **b** young leaf; **c** cotyledons; **d** hypocotyl; **e** differentiated part of a root; **f** root apex (based on data published in Kolano et al. 2009)

of some putative ancestral species of the A subgenome (e.g., *C. watsonii;* Kolano et al. 2016; Storchova et al. 2015; Walsh et al. 2015; Orzechowska et al. 2018). On the other hand, *18–24 J* repeats may have been eliminated from the A subgenome during the evolution of the allopolyploid genomes, because another putative ancestral species, *C. nevadense*, has shown a high amplification of *18–24 J* (Kolano et al. 2016; Storchova et al. 2015; Walsh et al. 2015; Orzechowska et al. 2018). A preferential elimination of various repetitive sequences (retroelements, rDNA) was reported for a number of allopolyploids, e.g., tetraploid *Nicotiana tabacum* or *Melampodium* species (Weiss-Schneeweiss et al. 2012; Kovarik et al. 2004).

4.5 *Chenopodium quinoa* Is a Polysomatic Plant

Polysomaty concern the plant organ/tissue which consists of cells which differ in ploidy levels (endopolyploid cells; Traas et al. 1998). Endopolyploidy is a term that described the outcome of the multiplication of nuclear DNA without cell division and plays an important role in determining cell fate and in interactions with symbiotic and pathogenic organisms (De Veylder et al. 2011). In single plant, different organs can vary in patterns of endopolyploidization. Some species revealed endopolyploid cells in most organs, whereas in other taxa, endopolyploidyzation is very limited and present only in few organs. Moreover, endopolyploidy patterns are usually correlated with plant development (Maluszynska et al. 2012).

C. quinoa, similar to many other species from Chenopodiaceae (e.g., *C. album, Spinacia oleracea, Beta vulgaris*), is a polysomatic plant (Kolano et al. 2008b, 2009; Barow and Meister 2003). Endopolyploid nuclei appear very early in quinoa plant development, already in matured embryo radicle (Kolano et al. 2009). During successive seedling development, the pattern of endopolyploidization differed between organs and changed over time (Fig. 4.2). The highest level of endopolyploidization was present in the hypocotyls and primary roots of quinoa seedlings whereas in the leaves and the shoot apex the endopolyploid cells were absent (Fig. 4.2; Kolano et al. 2009). The lack of endopolyploidization at the shoot apex, reported for many angiosperms, could be one of the mechanisms that protect the genetic stability of the germ line (Maluszynska et al. 2012; Kudo and Kimura 2001). The high degree of endopolyploidyzation in hypocotyls and primary roots possible is connected with large amount of vascular tissue in these organs (Kolano et al. 2009). The increase of endopolyploidy observed in root and hypocotyl during germination and seedling development seems to be correlated with their quick elongation and with vascular tissue development (Alarcón and Salguero 2017; Kolano et al. 2009).

References

Alarcón MV, Salguero J (2017) Transition zone cells reach G2 phase before initiating elongation in maize root apex. Biol Open 6:909–913

Álvarez I, Wendel JF (2003) Ribosomal ITS sequences and plant phylogenetic inference. Mol Phylogenet Evol 29:417–434

Barow M, Meister A (2003) Endopolyploidy in seed plants is differently correlated to systematics, organ, life strategy and genome size. Plant Cell Environ 26:571–584

Bennetzen JL, Wang H (2014) The contributions of transposable elements to the structure, function, and evolution of plant genomes. Annu Rev Plant Biol 65:505–530

Bhargava A, Shukla S, Ohri D (2006) Karyotypic studies on some cultivated and wild species of *Chenopodium* (Chenopodiaceae). Genet Resour Crop Evol 53:1309–1320

Bhargava A, Shukla S, Ohri D (2007) Genome size variation in some cultivated and wild species of *Chenopodium* (Chenopodiaceae). Caryologia 60:245–250

Christensen SA, Pratt DB, Pratt C, Nelson PT, Stevens MR, Jellen EN et al (2007) Assessment of genetic diversity in the USDA and CIP-FAO international nursery collections of quinoa (*Chenopodium quinoa* Willd.) using microsatellite markers. Plant Genet Resour 5:82–95

Clarkson JJ, Lim KY, Kovarik A, Chase MW, Knapp S et al (2005) Long-term genome diploidization in allopolyploid *Nicotiana* section *Repandae* (Solanaceae). New Phytol 168:241–252

De Storme N, Mason A (2014) Plant speciation through chromosome instability and ploidy change: Cellular mechanisms, molecular factors and evolutionary relevance. Curr Plant Biol 1:10–33

De Veylder L, Larkin JC, Schnittger A (2011) Molecular control and function of endoreplication in development and physiology. Trends Plant Sci 16:624–634

Du J, Tian Z, Hans CS, Laten HM, Cannon SB et al (2010) Evolutionary conservation, diversity and specificity of LTR-retrotransposons in flowering plants: insights from genome-wide analysis and multi-specific comparison. Plant J 63:584–598

Fuentes-Bazan S, Uotila P, Borsch T (2012) A novel phylogeny-based generic classification for *Chenopodium sensu lato*, and a tribal rearrangement of Chenopodioideae (Chenopodiaceae). Willdenowia 42 (1):5–24

Gao D, Schmidt T, Jung C (2000) Molecular characterization and chromosomal distribution of species-specific repetitive DNA sequences from *Beta corolliflora*, a wild relative of sugar beet. Genome 43:1073–1080

Gill BS, Kimber G (1974) Giemsa C-banding and the evolution of wheat. Proc Natl Acad Sci USA 71:4086–4090

Greilhuber J (2005) Intraspecific variation in genome size in angiosperms: identifying its Existence. Ann Bot 95:91–98

Greilhuber J, Borsch T, Müller K, Worberg A, Porembski S, Barthlott W (2006) Smallest angiosperm genomes found in Lentibulariaceae, with chromosomes of bacterial size. Plant Biol 8:770–777

Greilhuber J, Doležel J, Lysák MA, Bennett MD (2005) The origin, evolution and proposed stabilization of the terms 'genome size' and 'C-Value' to describe duclear DNA contents. Ann Bot 95:255–260

Guerra M (2000) Patterns of heterochromatin distribution in plant chromosomes. Genet Mol Biol. 23:1029–1041

Hawkins JS, Grover CE, Wendel JF (2008) Repeated big bangs and the expanding universe: directionality in plant genome size evolution. Plant Sci 174:557–562

Heslop-Harrison JS, Schwarzacher T (2011) Organisation of the plant genome in chromosomes. Plant J 66:18 33

Hřibová E, Neumann P, Matsumoto T, Roux N, Macas J et al (2010) Repetitive part of the banana (*Musa acuminata*) genome investigated by low-depth 454 sequencing. BMC Plant Biol 10:204–204

Ibarra-Laclette E, Lyons E, Hernandez-Guzman G, Perez-Torres CA, Carretero-Paulet L et al (2013) Architecture and evolution of a minute plant genome. Nature 498:94–98

Jellen EN (2016) C-banding of plant chromosomes. In: Kianian SF, Kianian PMA (eds) Plant Cytogenetics: methods and protocols. Springer, New York, pp 1–5

Jellen EN, Kolano B, Sederberg MC, Bonifacio A, Maughan PJ (2011) *Chenopodium*. In: Kole C (ed) Wild crop relatives: genomic and breeding resources. Springer, Berlin Heidelberg, pp 34–62

Jiao YN, Wickett NJ, Ayyampalayam S, Chanderbali AS, Landherr L et al (2011) Ancestral polyploidy in seed plants and angiosperms. Nature 473:97–U113

Keskitalo M, Lindén A, Valkonen JPT (1998) Genetic and morphological diversity of Finnish tansy (*Tanacetum vulgare* L., Asteraceae). Theor Appl Genet 96:1141–1150

Knight CA, Molinari NA, Petrov DA (2005) The large genome constraint hypothesis: evolution, ecology and phenotype. Ann Bot 95:177–190

Kolano B, Bednara E, Weiss-Schneeweiss H (2013a) Isolation and characterization of reverse transcriptase fragments of LTR retrotransposons from the genome of *Chenopodium quinoa* (Amaranthaceae). Plant Cell Rep 32:1575–1588

Kolano B, Gardunia BW, Michalska M, Bonifacio A, Fairbanks D., Maughan PJ, Coleman CE, Stevens MR, Jellen EN, Maluszynska J (2011) Chromosomal localization of two novel repetitive sequences isolated from the *Chenopodium quinoa* Willd. genome. Genome 54(9):710–717

Kolano B, McCann J, Orzechowska M, Siwinska D, Temsch E et al (2016) Molecular and cytogenetic evidence for an allotetraploid origin of *Chenopodium quinoa* and *C. berlandieri* (Amaranthaceae). Mol Phylogenet Evol 100:109–123

Kolano B, McCann J, Oskedra M, Chrapek M, Rojek M, Nobis A, Weiss-Schneeweiss H (2019) Parental origin and genome evolution of several Eurasian hexaploid species of Chenopodium (Chenopodiaceae). Phytotaxa 392:163–185

Kolano B, Pando LG, Maluszynska J (2001) Molecular cytogenetic studies in *Chenopodium quinoa* and *Amaranthus caudatus*. Acta Soc Bot Pol 70:85–90

Kolano B, Plucienniczak A, Kwasniewski M, Maluszynska J (2008a) Chromosomal localization of a novel repetitive sequence in the *Chenopodium quinoa* genome. J Appl Genet 49:313–320

Kolano B, Saracka K, Broda-Cnota A, Maluszynska J (2013b) Localization of ribosomal DNA and CMA3/DAPI heterochromatin in cultivated and wild *Amaranthus* species. Sci Hortic 164:249–255

Kolano B, Siwinska D, Gomez Pando L, Szymanowska-Pulka J, Maluszynska J (2012a) Genome size variation in *Chenopodium quinoa* (Chenopodiaceae). Plant Syst Evol 298:251–255

Kolano B, Siwinska D, Maluszynska J (2008b) Comparative cytogenetic analysis of diploid and hexaploid *Chenopodium album*. Acta Soc Bot Pol 77:293–298

Kolano B, Siwinska D, Maluszynska J (2009) Endopolyploidy patterns during development of *Chenopodium quinoa*. Acta Biol Cracov Bot 51:85–92

Kolano B, Siwinska D, McCann J, Weiss-Schneeweiss H (2015) The evolution of genome size and rDNA in diploid species of *Chenopodium s.l.* (Amaranthaceae). Bot J Linn Soc 179:218–235

Kolano B, Tomczak H, Molewska R, Jellen EN, Maluszynska J (2012b) Distribution of 5S and 35S rRNA gene sites in 34 *Chenopodium* species (Amaranthaceae). Bot J Linn Soc 170:220–231

Kovarik A, Matyasek R, Lim KY, Skalicka K, Koukalova B et al (2004) Concerted evolution of 18–5.8–26S rDNA repeats in *Nicotiana allotetraploids*. Biol J Linn Soc 82:615–625

Krak K, Vit P, Belyayev A, Douda J, Hreusova L et al (2016) Allopolyploid origin of *Chenopodium album s. str.* (Chenopodiaceae): a molecular and cytogenetic insight. Plos One 11

Kubis S, Schmidt T, Heslop-Harrison JS (1998) Repetitive DNA elements as a major component of plant genomes. Ann Bot 82:45–55

Kudo N, Kimura Y (2001) Patterns of endopolyploidy during seedling development in cabbage (*Brassica oleracea* L.). Ann Bot 87:275–281

Leitch AR, Leitch IJ (2008) Genomic plasticity and the diversity of polyploid plants. Science 320:481–483

Leitch IJ, Bennett MD (2004) Genome downsizing in polyploid plants. Biol J Linn Soc 82:651–663

Li SF, Su T, Cheng GQ, Wang BX, Li X et al (2017) Chromosome evolution in connection with repetitive sequences and epigenetics in plants. Genes 8:290

Lisch D (2013) How important are transposons for plant evolution? Nat Rev Genet 14:49–61

Llorens C, Muñoz-Pomer A, Bernad L, Botella H, Moya A (2009) Network dynamics of eukaryotic

LTR retroelements beyond phylogenetic trees. Biol Direct 4:41

Lomonosova MN (2005) Chromosome numbers of Chenopodiaceae species from Russia and Kazakhstan. Botanicheskii Zhurnal 90:1132–1134

Lysak M, Berr A, Pecinka A, Schmidt R, McBreen et al (2006) Mechanisms of chromosome number reduction in *Arabidopsis thaliana* and related *Brassicaceae species*. Proc Natl Acad Sci USA 103:5224–5229

Maluszynska J, Kolano B, Sas Nowosielska H (2012) Endopolyploidy in plants. In: Leitch IJ, Greilhuber J, Dolezel J, Wendel J (eds) Plant Genome Diversity. Springer, Wien, pp 99–119

Mandak B, Krak K, Vit P, Pavlikova Z, Lomonosova MN et al (2016) How genome size variation is linked with evolution within *Chenopodium sensu lato*. Perspect Plant Ecol 23:18–32

Mandák B, Trávníček P, Paštová L, Kořínková D (2012) Is hybridization involved in the evolution of the *Chenopodium album* aggregate? An analysis based on chromosome counts and genome size estimation. Flora 207:530–540

Maragheh FP, Janus D, Senderowicz M, Haliloglu K, Kolano B (2019) Karyotype analysis of eight cultivated *Allium* species. J Appl Genet 60:1–11

Maughan P, Kolano BA, Maluszynska J, Coles ND, Bonifacio A et al (2006) Molecular and cytological characterization of ribosomal RNA genes in *Chenopodium quinoa* and *Chenopodium berlandieri*. Genome 49:825–839

Menzel G, Dechyeva D, Wenke T, Holtgräwe D, Weisshaar B et al (2008) Diversity of a complex centromeric satellite and molecular characterization of dispersed sequence families in sugar beet (*Beta vulgaris*). Ann Bot 102:521–530

Orzechowska M, Majka M, Weiss-Schneeweiss H, Kovařík A, Borowska-Zuchowska N, Kolano B (2018) Organization and evolution of two repetitive sequences, 18-24 J and 12-13P, in the genome of Chenopodium (Amaranthaceae). Genome 61:643–652

Palomino G, Hernández LT, de la Cruz Torres E (2008) Nuclear genome size and chromosome analysis in *Chenopodium quinoa* and *C. berlandieri* subsp. *nuttalliae*. Euphytica 164:221

Pellicer J, Fay MF, Leitch IJ (2010) The largest eukaryotic genome of them all? Bot J Linn Soc 164:10–15

Schnable PS, Ware D, Fulton RS, Stein JC, Wei FS et al (2009) The B73 maize genome: complexity, diversity, and dynamics. Science 326:1112–1115

Schmuths H, Meister A, Horres R, Bachmann K (2004) Genome size variation among accessions of *Arabidopsis thaliana*. Ann Bot 93:317–321

Schubert I, Vu GTH (2016) Genome stability and evolution: attempting a holistic view. Trends Plant Sci 21:749–757

Schweizer D (1876) Reverse fluorescent chromosome banding with chromomycin and DAPI. Chromosoma 58:307–324

Shcherban AB (2015) Repetitive DNA sequences in plant genomes. Russ J Genet 5:159–167

Sochorová J, Coriton O, Kuderová A, Lunerová J, Chèvre AM et al (2017) Gene conversion events and variable degree of homogenization of rDNA loci in cultivars of *Brassica napus*. Ann Bot 119:13–26

Soltis PS, Soltis DE (2009) The role of hybridization in plant speciation. Annu Rev Plant Biol 60:561–588

Storchova H, Drabesova J, Chab D, Kolar J, Jellen EN (2015) The introns in FLOWERING LOCUS T-LIKE (FTL) genes are useful markers for tracking paternity in tetraploid *Chenopodium quinoa* Willd. Genet Resour Crop Evol 62:913–925

Tanaka R, Taniguchi K (1975) A banding method for plant chromosomes. Jpn J Genet 50:163–167

Traas J, Hülskamp M, Gendreau E, Höfte H (1998) Endoreduplication and development: rule without dividing? Curr Opin Plant Biol 1:498–503

Volkov R, Medina F, Zentgraf U, Hemleben V (2004) Organization and molecular evolution of rDNA nucleolar dominance and nucleolus structure. In: Esser K, Luttge U, Beyschlag W, Murata J (eds) Progress in botany. Springer, Berlin Heidelberg New York, pp 106–146

Walsh BM, Adhikary D, Maughan PJ, Emshwiller E, Jellen EN (2015) *Chenopodium* polyploidy inferences from Salt Overly Sensitive 1 (SOS1) Data. Am J Bot 102:533–543

Ward SM (2000) Allotetraploid segregation for single-gene morphological characters in quinoa (*Chenopodium quinoa* Willd.). Euphytica 116:11–16

Weiss-Schneeweiss H, Blöch C, Turner B, Villaseñor JL, Stuessy TF et al (2012) The promiscuous and the chaste: frequent allopolyploid speciation and its genomic consequences in american daisies (*Melampodium* sect. *Melampodium*; Asteraceae). Evolution 66:211–228

Weiss-Schneeweiss H, Emadzade K, Jang TS, Schneeweiss GM (2013) Evolutionary consequences, constraints and potential of polyploidy in plants. Cytogenet Genome Res 140:137–150

A Chromosome-Scale Quinoa Reference Genome Assembly

5

Bo Li and Damien J. Lightfoot

Abstract

Chenopodium quinoa has been considered as a future crop for its nutritional value and tolerance to abiotic stresses. A high quality genome assembly would greatly accelerate the molecular breeding and the fundamental research of quinoa. Long-read single-molecule real-time (SMRT) technology was applied to sequence a costal Chilean quinoa accession. With the assistance of optical, chromosome-contact and genetic maps, the first chromosome-scale reference genome of quinoa was published in early 2017. This informative resource will significantly help to elucidate the complex structure of the quinoa genome and therefore to identify the genetic loci underlying important agronomic traits.

B. Li (✉)
School of Life Sciences, Lanzhou University, Lanzhou, China
e-mail: li_bo@lzu.edu.cn

B. Li
Ministry of Education Key Laboratory of Cell Activities and Stress Adaptations, Lanzhou, China

D. J. Lightfoot
Biological and Environmental Science and Engineering, King Abdullah University of Science and Technology, Thuwal, Kingdom of Saudi Arabia

5.1 Introduction

Quinoa (*Chenopodium quinoa* Willd.), an herbaceous species belonging to the Amaranthaceae, is a pseudocereal crop that originated from the Andean region of South America (Maughan et al. 2004; Zurita-Silva et al. 2014). Cultivation of quinoa dates back more than 7,000 years, with the grain of the plant being a staple food for Pre-Columbian cultures of the Incas (Bazile et al. 2015) (Chap. 1). Quinoa has received increasing global attention as a highly nutritious food source in recent years (Maradini Filho et al. 2015) (Chaps. 2 and 3). Specifically, quinoa grain is high in vitamins (A, B1, B2, B9, C, and E), minerals (Ca, Fe, Mg, P, and Zn), fibre, lipids, linolenate, natural antioxidants (polyphenols), and carbohydrates (Maradini Filho et al. 2015; Verena et al. 2016). It is also gluten-free, with a low glycemic index and an excellent balance of essential amino acids (particularly methionine and lysine) (Maradini Filho et al. 2015; Verena et al. 2016).

Quinoa and its wild relatives differ, agro-morphologically, in their growth habit, color, panicle shape, grain morphology (color, shape, and size), growth cycle, and grain yield (Bazile et al. 2015) (Chaps. 2 and 3), with more than 16,000 distinct wild varieties identified to date (FAO 2013). It has been shown that many quinoa varieties are able to grow and maintain yield under high abiotic stresses (Jacobsen et al. 2003).

S. M. Schmöckel (ed.), *The Quinoa Genome*, Compendium of Plant Genomes,
https://doi.org/10.1007/978-3-030-65237-1_5

The largely uncharacterized phenotypic and genetic variation, and high-stress tolerances of quinoa suggest that there is a valuable and large genetic pool available for researchers to improve quinoa for the purposes of food security and human nutrition (Zurita-Silva et al. 2014).

Despite the potential of the species to help alleviate food security and malnutrition concerns, quinoa is considered to be a neglected and under-developed crop. Conventional breeding of quinoa commenced more than 50 years ago, but advances have been limited (Rojas et al. 2009). Quinoa is an allotetraploid species ($2n = 4x = 36$), with flow cytometry-based methods estimating a genome size of between 1.42 and 1.50 Gb (Kolano et al. 2012; Palomino et al. 2008). The allotetraploid genome consists of two distinct sub-genomes (termed "A" and "B") (Palomino et al. 2008; Yasui et al. 2016) (Chap. 6). Genetic tools, including microsatellite markers, Single Nucleotide Polymorphism (SNP) markers, Expressed Sequence Tag (EST) libraries, Bacterial Artificial Chromosome (BAC) libraries, Recombinant Inbred Lines (RILs) and linkage maps have been previously developed (Coles et al. 2005; Jarvis et al. 2008; Mason et al. 2005; Maughan et al. 2004, 2012). Research utilizing these tools has identified loci associated with important agronomic traits, such as grain yield (Gómez et al. 2011), harvest index (Bertero et al. 2004), grain saponin content (Chap. 8) and tolerance to stresses such as salt, drought, heat and diseases (e.g., downy mildew) (Jacobsen et al. 2003; Zurita-Silva et al. 2014) (Chap. 9).

Genetic markers are vital for modern breeding techniques, which will enable the improvement of quinoa genotypes (Maughan et al. 2012). The first set of DNA markers in quinoa was reported in 1993 with the use of Randomly Amplified Polymorphic DNA (RAPD) markers, revealing 26 polymorphic markers among the 16 quinoa accessions studied (Fairbanks et al. 1993). Using Amplified Fragments Length Polymorphism (AFLP) technology, Maughan et al. (2004) published the first quinoa linkage map, identifying 35 linkage groups, spanning 1,020 cM. The first set of Simple Sequence Repeat (SSR) markers was developed in 2005, which included 208 SSR markers from 31 quinoa accessions (Mason et al. 2005). Further development of SSR markers led to the production of a new linkage map that comprised 38 linkage groups, and covered over 900 cM (Jarvis et al. 2008). Maughan et al. (2012) used SNP markers to generate an improved linkage map that contained 20 linkage groups, and covered 1404 cM. The development of markers of improved quality and density, as facilitated by new sequencing technologies, will allow for improved research and breeding outcomes. In this regard, the development of a high-quality reference genome assembly would further facilitate these advances.

With modern DNA sequencing technologies, assembling complete genomes has become more feasible and affordable. As such, the development of higher density molecular marker maps in quinoa has become possible, and has allowed for the improvement of agronomic traits, such as those described by Zurita-Silva et al. (2014). Besides, the introgression of new alleles from wild quinoa species, and the domestication of new quinoa species can both be accelerated (Chap. 7). The increased availability of newer sequencing technologies allowed for the production of a preliminary quinoa genome assembly (Yasui et al. 2016). This initial assembly is a valuable resource, and has been utilized by the research community. However, the assembly is highly fragmented (scaffold N50 of 87 kb) and is not complete, containing only 72% of the estimated 1.5 Gb genome. As such, a more contiguous and complete genome assembly is required to assist future genetic, genomic and evolutionary investigations of quinoa.

5.2 Quinoa Accession PI 614886

We selected the Chilean accession PI 614886 (Fig. 5.1), also known as NSL 106399 or QQ74, as the accession for sequencing to produce a reference-quality assembly. PI 614886 was collected from the Chilean costal region, and is publicly available from the Germplasm Resources Information Network (GRIN; http://www.arsgrin.gov/index.html) of the United States

Fig. 5.1 A photo of a quinoa plant, accession PI 614886, grown in a glasshouse at King Abdullah University of Science and technology (KAUST), Kingdom of Saudi Arabia. Photo credit to Dr. David Jarvis, KAUST

Department of Agriculture (USDA) Agricultural Research Service (ARS).

PI 614886 was selected for sequencing because it has been shown to maintain a relatively high and consistent yield under saline and drought conditions. These higher tolerances were demonstrated at the International Center for Biosaline Agriculture (ICBA) in Dubai, United Arab Emirates. In this study, field trials were performed over two consecutive cropping seasons with 20 accessions assessed for their tolerances to local growth conditions (e.g., salinity and drought) (Rao and Shahid 2012). When plants were irrigated with highly saline water (EC_w 2.8 dS cm^{-1}), PI 614886 was shown to be the most tolerant and highest yielding accession (Rao and Shahid 2012). Using microsatellite markers, PI 614886 was confirmed to be a coastal (lowland) variety (Christensen et al. 2007).

5.3 Primary Sequencing by PacBio SMRT® Sequencing Technology

To produce a high-quality genome assembly of quinoa, we utilized a strategy based on long-read Pacific Biosciences (PacBio) Single Molecule, Real-Time (SMRT®) technology. The resulting primary assembly consisted of 4,232 scaffolds (scaffold N50 of 1.66 Mb) (Jarvis et al. 2017). The total assembly size was 1.32 Gb, which is in line with previous estimates of genome size by Palomino et al. (2008) and Kolano et al. (2012).

5.3.1 PacBio SMRT Sequencing Technology

Plant genomes are challenging to assemble into contiguous and complete genome assemblies. This is due to their large genome size, high proportion of repetitive regions and their polyploid nature. Assembly strategies based on short-read Illumina technology have become routine, and are financially and computationally affordable for most research groups. However, the genome assemblies that are produced by short-read sequencing are often highly fragmented and incomplete. This is likely due to the inability of Illumina short reads to adequately cover, and distinguish between, repetitive regions in the genome. To overcome these challenges, PacBio developed SMRT® technology. This technology produces long sequence reads thereby, generating individual reads that can span repetitive regions in the genome. Each SMRT® sequencing cell contains up to one million sequencing "factories" that are termed Zero-Mode Waveguides (ZMWs). Each ZMW has a well with an attached DNA polymerase. DNA sequencing libraries are added to the SMRT® cell such that one DNA molecule binds to the polymerase in each ZMW. The surface-bound DNA polymerase then synthesizes a DNA chain, incorporating labeled nucleotides. As each fluorescently labeled nucleotide is added to the DNA chain, illumination and signal reading occur. SMRT® sequencing could produce reads longer than 20 kb, generating 1 Gb data in each SMRT® cell (RSII system).

To perform PacBio SMRT® sequencing of a target genome, A SMRTbell library needs to be prepared. Following isolation of double-stranded DNA, the DNA is capped with hairpin loops at both ends, which allows the SMRTbell template to be structurally linear and topologically circular. This enables the efficient loading of DNA into ZMWs. The hairpin adaptors provide common primer-binding sites for DNA polymerase, which reduces sequencing bias between inserts of different sizes. The circular format of DNA enables multiple sequencing "loops" of the same DNA fragment, which allows for a high intramolecular consensus accuracy.

5.3.2 Quinoa Genomic DNA Preparation and PacBio Sequencing

A single quinoa plant was grown in soil in a growth chamber with a photoperiod of 10/14 h (light/dark) and a photon irradiance of 120 μmol $m^{-2}s^{-1}$ until maturity. The plant was placed in the dark for two days prior to harvesting of leaf and flower tissues. DNA preparation was performed according to the "Preparing *Arabidopsis* Genomic DNA for Size-Selected ~20 kb SMRTbell Libraries" protocol (http://www.pacb.com/wp-content/uploads/2015/09/Shared-Protocol-Preparing-Arabidopsis-DNA-for-20-kb-SMRTbell-Libraries.pdf) and purified twice using Beckman Coulter Genomics AMPure XP magnetic beads. Approximately 30 g of plant tissue was used, which produced a DNA yield of 175 μg. Thermo Fisher Scientific Qubit Fluorometry was used to quantify the DNA and a sample of the DNA was inspected with pulsed field gel electrophoresis. The DNA appeared as a high molecular weight band of greater than 50 kb in size, indicating that it is appropriate for use in PacBio sequencing.

Using the PacBio RS II system, quinoa genomic DNA was sequenced on 100 SMRT® cells with P6-C4 chemistry by DNALink (Seoul, Republic of Korea). From the 100 SMRT® cells, a total of 15,029,200 reads were collected, representing 78,829,079,710 bp of single-molecule sequencing data (Table 5.1). Filtering was performed to exclude low quality reads and contaminants. This was achieved by assessing the quality score for each read (Options-filter = 'MinReadScore = 0.80, MinSRL = 500, MinRL = 100'). An error correction was then performed (CUTOFF option setting with GENOME_SIZE 750 Mb*30). A total of 6,037,280 PacBio post-filtered reads remained, representing a total of 75,132,015,080 bp of sequencing data with an average read length of 12.44 kb (Table 5.1). The number of reads was substantially decreased after filtering, while the total size of data output was only slightly reduced (Table 5.1), indicating that the low quality reads were mostly short, and only account for a very small proportion of all data.

De novo genome assembly was performed using the SMRT®-assembly pipeline (https://github.com/PacificBiosciences/smrtmake) with the genome size set at 750 Mb. Within this pipeline, the Celera Assembler generated a draft assembly using the error-corrected reads (Denisov et al. 2008). This draft assembly was then polished using the quiver algorithm, and the resulting PacBio assembly was 1.32 Gb in size, consisting of 4,232 contigs (contig N50 = 1.66 Mb) (Jarvis et al. 2017).

The PacBio reads that were used for genome assembly were long, with 38.4% of the reads being at least 10 kb, with an average of 17,004 bp (Table 5.2). This represented 53.7 Gb of data, which equates to 71.6% of the original sequencing data from the SMRT® sequencing, thus providing a substantial amount of coverage of the genome size.

5.4 Scaffolding of the Assembly with BioNano Genomics (Optical Maps)

BioNano genome mapping was performed using the BioNano Irys system at Brigham Young University (Provo, UT, USA). With BioNano mapping, DNA is labeled and then linearized on an IrysChip for direct imaging and reading. As the DNA sample is relatively intact, BioNano Irys system allows for obtaining long-range genomic spacing information. The BioNano

Table 5.1 PacBio RS II read quality report

Pre-filter	Number of sub-read read bases	78,829,079,710 bp
	Number of sub-read reads	15,029,200
	Sub-read length	5,245 bp
	Sub-read quality	0.352
Post-filter	Number sub-read bases	75,132,015,080 bp
	Number sub-read reads	6,037,280
	Sub-read length	12,444 bp
	Sub-read quality	0.845

Table 5.2 Sub-read size

Region (kb)	Number of bases/cell	Number of bases	Number of reads	Mean sub-read length
>10 ~	537,002,681	53,700,268,075 (71.6%)	3,158,132 (38.1%)	17,004
Total	750,282,020	75,028,202,002	8,280,426	9,061

sequencing reads provide a pattern of fluorescence that can be utilized to scaffold contigs generated by PacBio sequencing.

To produce the material for BioNano sequencing, quinoa plants were grown in soil in a greenhouse for 3 weeks, and then placed in dark for 2 days prior to harvesting. High molecular weight DNA was then isolated from young leaf tissue using the BioNano DNA isolation kit, with DNA labeling performed using the BioNano Prep DNA labeling kit. Specifically, the DNA was digested with a single-stranded nicking endonuclease Nt.BspQI, and was labeled with a fluorescent-dUTP nucleotide analogue using *Taq* polymerase. Nicks were ligated with *Taq* DNA ligase, and the backbone of the labeled DNA was stained using the intercalating dye YOYO-1. The labeling of the DNA with fluorescent nucleotides in specific sequence motifs (as dictated by Nt. BspQI) produced identifiable and unique genome-wide patterns. Labeled DNA was imaged automatically using the BioNano Irys system and de novo assembled into consensus physical maps using the BioNano IrysView v2.5.1 analysis software (https://bionanogenomics.com/support/download-form/?file=http://bnxinstall.com/IrysView/irysview.htm&title=IrysView). This allowed for scaffolding of the primary contigs produced by PacBio sequencing. Single molecules longer than 150 kb (with more than 8 labels) were used in the final de novo assembly. The *p*-values were set for the initial assembly (10^{-8}), extension of the assembly (10^{-9}), and chimera detection (10^{-15}).

PacBio-BioNano hybrid scaffolds were identified using IrysView's hybrid scaffold alignment subprogram with a *p*-value of (10^{-8}) for initial and final alignments, and of (10^{-13}) for chimera detection and merging. As a result, the assembly size of the PacBio-BioNano integrated genome was increased from 1.32 to 1.39 Gb.

5.5 Scaffolding of the Assembly with Dovetail Genomics (Chromatin-Contact Maps)

To scaffold the PacBio sequences, high-throughput short-read sequencing is often utilized to revel the 3-dimentional structure of chromosomes. In the nucleus, chromatin-DNA segments that are physically close to each other can be captured and used as mapping anchors (Dixon et al. 2012; Lieberman-Aiden and Dekker 2009). After fixation of the chromosomes, DNA segments in close proximity are likely ligated together and sequenced in pairs, with the number of intra-chromosome ligations negatively correlate with their genetic distances (Putnam et al. 2015). The "Hi-C" is a method developed to probe the 3-dimentional architecture of the massive genomes by constructing a spatial proximity map (Lieberman-Aiden and Dekker 2009). It and its related methods can reveal chromatin contacts at multi-megabase level, constructing genome-wide interaction maps in a relatively high resolution.

5.5.1 Chicago Library Preparation and Sequencing

Utilizing Hi-C data, the HiRise pipeline was developed to de novo scaffold DNA fragments using short-read data (such as Illumina Hiseq) generated from proximity ligated DNA with in vitro reconstituted chromatin (Putnam et al. 2015). Proximity ligation of DNA in its native chromatin-bound state can be utilized to generate DNA read pairs, which are then used to estimate the relative distances between regions of genomic sequence. These distance estimates are then utilized to scaffold assemblies (Putnam et al. 2015). As demonstrated in Fig. 5.2, Chicago libraries are prepared from purified DNA that has been reconstituted with chromatin in vitro. In this process, formaldehyde is used to fix chromatin, followed by restriction enzyme digestion to generate "sticky ends", which are then filled in with biotinylated (blue circles in Fig. 5.2) and thiolated (green squares in Fig. 5.2) nucleotides. The filled in blunt ends are then ligated and the cross links are removed. The libraries are then treated with exonuclease to remove biotinylated nucleotides. As a result, the thiolated nucleotides are at the ends of library fragments, preventing further digestion of molecule ends. Scaffolding can therefore use the signals of thiolated nucleotides as indicators of the physical positions between DNA library fragments within chromosomes.

The same high molecular weight genomic DNA that was prepared for PacBio sequencing was used to prepare a Chicago library. Specifically, 2 μg of genomic DNA (mean fragment size longer than 100-kb) was reconstituted into chromatin in vitro and fixed with formaldehyde. Fixed chromatin was digested with the restriction enzyme *Dpn*II. The 5' overhangs were filled in with biotinylated nucleotides, and free blunt ends were ligated. After ligation, DNA was purified from protein, and was treated to remove biotin. DNA was sheared to ~350 bp mean fragment size. Libraries were generated using NEBNext Ultra enzymes and Illumina-compatible adapters. Biotin-containing fragments were then isolated using streptavidin beads before PCR enrichment of the library. The prepared library was sequenced on an Illumina HiSeq 2500 in "rapid run" mode to produce 122 million 2 × 100-kb read pairs, providing a 51.6-fold physical coverage (1–50 kb pairs).

5.5.2 Scaffolding the PacBio-BioNano Assemblies with HiRise

HiRise is a software pipeline designed specifically for processing Chicago sequencing data to scaffold genome assemblies (Putnam et al. 2015). It was used to scaffold the quinoa assembly by incorporating Chicago sequence data with the PacBio-BioNano assembly (in FASTQ format). Specifically, an alignment was performed between Chicago sequence data and the input assembly (PacBio-BioNano assembly) by using a modified SNAP read mapper (http://snap.cs.berkeley.edu). Chicago read pairs were separated within input scaffolds and were analyzed by HiRise. A likelihood model was provided to misjoins and to score prospective joins.

Chicago-HiRise sequencing and assembly improved the length of the assembly scaffolds. As a result, the longest scaffold increased from 11.56 to 23.82 Mb when compared with the PacBio-BioNano assembly. Correspondingly, the total number of scaffolds decreased from 4,014 to 3,486. More importantly, the N50 scaffold length increased from 2.45 Mb to 3.84 Mb, indicating that the Chicago-HiRise assembly had a substantially improved contiguity compared to the PacBio-BioNano assembly (Fig. 5.3).

When very short scaffolds (< 1 kb) were excluded, the Chicago-HiRise assembly scaffold N50 number decreased from 216 to 70 scaffolds, while the scaffold N50 length increased from 1.66 Mb to 5.36 Mb. Likewise, the scaffold N90 number decreased from 1,065 to 349 scaffolds while the N90 length increased from 0.14 Mb to 0.25 Mb.

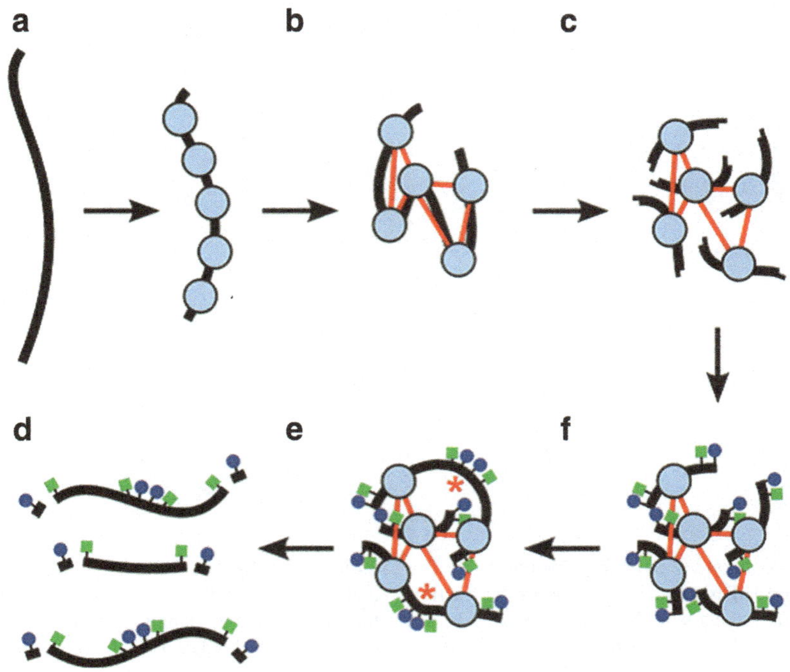

Fig. 5.2 The process of preparing a Chicago library (Putnam et al. 2015). **a** Chromatin (nucleosomes in blue) is reconstituted in vitro upon naked DNA (black strand). **b** Chromatin is fixed with formaldehyde (thin red lines are crosslinks). **c** Fixed chromatin is cut with a restriction enzyme, generating free sticky ends (performed on streptavidin-coated beads). **d** Sticky ends are filled in with biotinylated (blue circles) and thiolated (green squares) nucleotides. **e** Free blunt ends are ligated (ligations indicated by red asterisks). **f** Crosslinks are reversed with proteins removed to yield library fragments. Library fragments are then digested with an exonuclease to remove the terminal biotinylated nucleotides. The thiolated nucleotides protect the interior of the library fragments from being digested

5.6 Validating the Reference Genome Assembly

Short-read Illumina sequencing data was used to validate the Chicago-HiRise assembly of the quinoa genome by (1) mapping the sequencing reads back to the final genome assembly and (2) by producing a de novo short-read assembly and comparing this assembly to the Chicago-HiRise assembly. DNA was prepared from leaf tissue of a single soil-grown plant using the Qiagen DNeasy Plant Mini Kit. To prepare the 500-bp paired-end (PE) libraries, the NEBNext Ultra DNA Library Prep Kit for Illumina was used. Short-read Illumina sequencing was performed using an Illumina HiSeq 2000 machine at the Bioscience Core Laboratory at KAUST, Saudi Arabia. Sequencing reads were processed with Trimmomatic (v0.33) (Bolger et al. 2014) to remove adapter sequences, leading and trailing bases with a quality score below 20, and reads with an average per base quality of 20 over a 4-bp sliding window. Sequencing reads shorter than 75 nucleotides after trimming were removed. The remaining high-quality reads were assembled with Velvet (v1.2.10) (Zerbino and Birney 2008) using a k-mer of 75. The assembly comprised 838,071,669 bp in 1,040,940 contigs with an N50 of 2,175 bp (in 95,338 contigs), with the largest contig being 45,571 bp.

The Chicago-HiRise assembly was validated using the Illumina short-reads (66X coverage) and Bacterial artificial chromosome End Sequences (BESs) data (Poulsen and Johnsen 2004) by determining the proportion of those two data sets

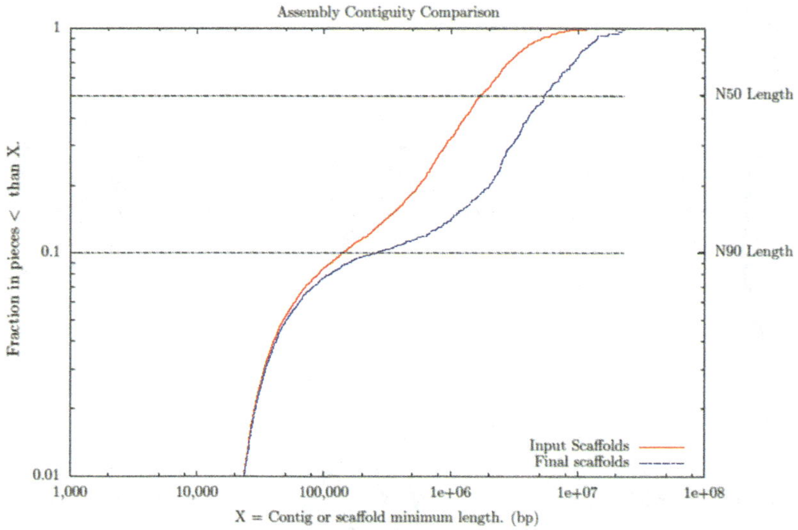

Fig. 5.3 A comparison of assembly contiguity before and after Chicago-HiRise. Each curve shows the fraction of the total length of the assembly present in scaffolds of a given length or smaller. The fraction of the assembly is indicated on the y-axis and the scaffold length in base-pairs is given on the x-axis. The two dashed lines mark the N50 and N90 length of each assembly. This plot excludes scaffolds less than 1 kb in size

that map to the final assembly. Of the 851,664,032 Illumina short-reads generated, 99.4% was successfully mapped back to the final assembly with 97.7% properly paired. To further validate the genome, a total of 2,106 BESs were aligned to the reference quinoa genome assembly using BLASTN. After filtering for hits with E-values $< 1 \times 10^{-100}$ and inserts that were too large (> 370 kb) or too small (< 10 kb) in size, 109 scaffolds with a total of 286 Mb (representing 22% of the genome) were validated by the BES data.

The short-read assembly was compared to the reference assembly using BLASTN. When considering only the top two BLASTN matches for each of the short-read contigs, a total of 59,305,355 bp (47%) of the quinoa final assembly was covered by the short-read assembly. Given the highly repetitive nature of the quinoa genome and the limitations of short-read sequencing, it is very likely that repetitive regions in the Illumina assembly would have been collapsed during the assembly process. Thus, it is not unexpected that such a BLAST search of the short-read assembly back onto the reference assembly would yield a much lower overall coverage. To extend the validation further, Multiple BLASTN hits from the short-read assembly onto the reference assembly were allowed. BLASTN hits with an E-value of less than 5×10^{-4} were filtered out. This increased the total covered bases to 1,203,491,061 bp, which represents 86.6% of the total quinoa genome.

5.7 Pseudomolecules–Chromosomes

To integrate scaffolds from the Chicago-HiRise assembly into pseudomolecules, three genetic maps were utilized. A previously published linkage map (Maughan et al. 2004) together with two new linkage maps (Jarvis et al. 2017) generated from genetic analysis of two mapping populations were included.

For identification of genes underling saponin biosynthesis (see Chap. 8) in seeds, two mapping populations, Kurmi × 0654 and Atlas × Carina

Red, were developed (Jarvis et al. 2017). These populations were used to produce new linkage maps with SNP markers. Parents and a number of individual F_3 progeny (45 for Kurmi × 0654 and 94 for Atlas × Carina Red) were sequenced using Illumina technology. The reads from all lines were trimmed first using Trimmomatic and mapped onto the Chicago-HiRise genome assembly using the programs BWA (Li and Durbin 2010) or TopHat (Trapnell et al. 2009). SNPs calling was performed using SAMtools mpileup v1.1 (Li et al. 2009). Markers were grouped using the JoinMap v4.1 maximum likelihood algorithm. The order of markers on resulting linkage groups of Kurmi × 0654 map was set to be the starting order, and genetic distances were determined by regression mapping in JoinMap. Genetic maps for each linkage group were produced (Fig. 5.4). To combine the published map with these two newly generated maps, the Kurmi × 0654 map was again used as a reference for anchoring marker position and scaling. The final integrated linkage map contained 6,403 unique markers, spanning a length of 2,034 cM (Fig. 5.4). The map consists of 18 linkage groups, corresponding to the haploid chromosome number of quinoa.

Using the integrated linkage map, the order and orientation of scaffolds with each linkage group were determined. Concatenation of these sorted scaffolds allowed assembly of pseudomolecules. The positions of the scaffold-based assembly in coordinates of pseudomolecules assembly were recorded in an "A Golden Path" (AGP) file. Scaffold gaps in the AGP file were filled with 100 "N"s. The coordinates in the AGP file were used to re-coordinate annotation files. By anchoring 565 scaffolds into the integrated linkage map with correct order and orientation, pseudomolecules were assembled, representing 1.18 Gb (85%) of the total genome assembly. At this stage, the pseudomolecules were referred as "chromosomes", representing a high-quality quinoa reference genome. In contrast to the previously published quinoa draft genome with 24,000 scaffold and 25% missing data (Yasui et al. 2016), the current pseudomolecule-assembly of quinoa is substantially improved.

5.8 Characterization of the Quinoa Genome

Having produced a high-quality quinoa reference genome, gene identification and genome annotation are required in order to produce a resource that can be utilized by the research community. For this, characterization ("masking") of repetitive regions was first performed. Following that, prediction and validation of gene models was performed using Illumina RNA-seq and PacBio Iso-seq.

5.8.1 Identification of Repetitive Sequences

A large proportion of the quinoa genome consists of repetitive regions of sequence, accounting for 64% of the genome (Jarvis et al. 2017). Repeat families were independently de novo identified and classified using the software package RepeatModeler. RepeatModeler utilizes the programs RECON and RepeatScout for the de novo identification of repeats within the genome. After the classification process, the output data file from RepeatModeler was used as a custom repeat library by the program RepeatMasker (http://www.repeatmasker.org) to discover and identify repeats within the genome. Among these repeats, the Long Terminal Repeat (LTR) transposable element class is the predominant type, accounting for 36.8% of the genome (Table 5.3) (Jarvis et al. 2017).

Other major classes of repetitive sequences that were identified include Long Interspersed Nuclear Element (LINE, 8.41% of the genome), Terminal Inverted Repeat (TIR, 6%), simple repeat (1.81%), and unclassified sequences (10.50%) (Jarvis et al. 2017).

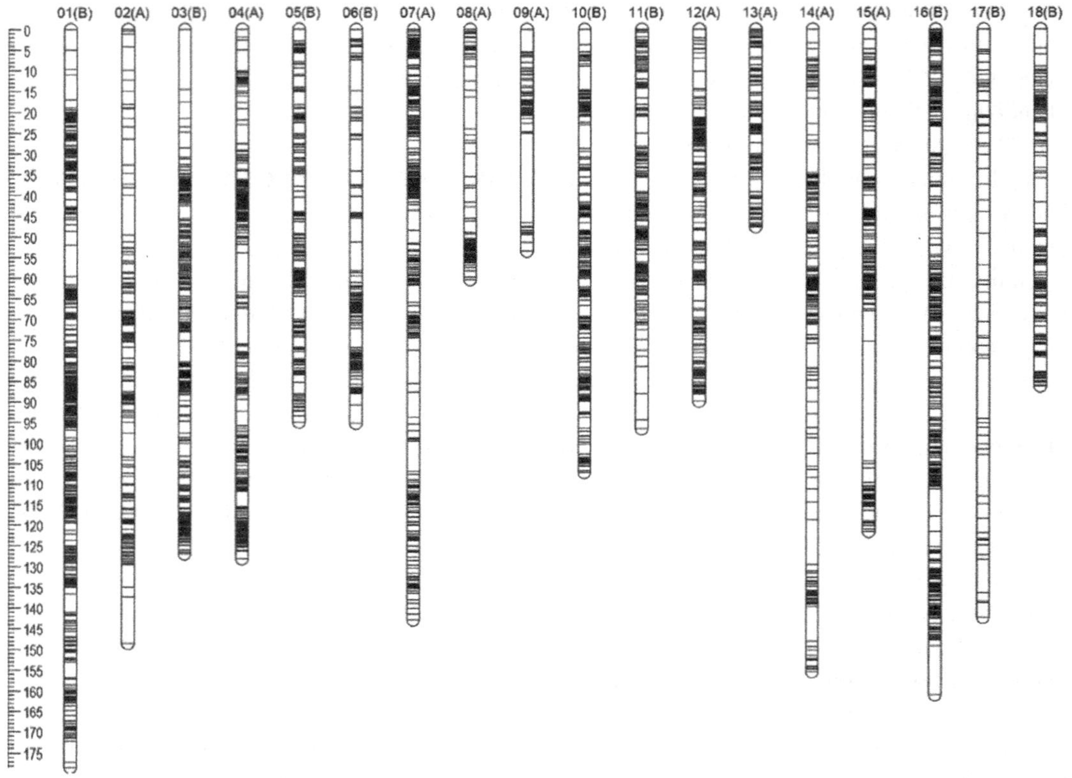

Fig. 5.4 The final integrated linkage map of quinoa. A linkage map of quinoa was generated by integrating maps from three independent quinoa populations. Mapped markers are represented as black bands. Putative assignment of sub-genomes A or B is indicated as letters next to the linkage group numbers. Scale bar, cM. Figure from Jarvis et al. (2017)

Table 5.3 A summary of LTR transposons and their families identified

LTR transposon and their families	Base pairs	% of the genome
All	502,546,481	36.84
Cassandra	278,710	0.02
Caulimovirus	317,559	0.02
Copia	113,928,967	8.22
ERV1	323,293	0.02
Gypsy	394,369,811	28.46
Ngaro	109,720	0.01

5.8.2 RNA Sample Preparation and Sequencing of the Transcriptome

To identify as much of the transcriptional spectrum of quinoa as possible, RNA was extracted from different tissues (roots, leaves, petioles, apical meristems, lateral meristems, stems, flowers, seeds, and seedlings), and treatments (heat, drought, or low-phosphate stresses) (Jarvis et al. 2017). All plants were grown in growth chambers for 3 weeks at 20 °C with 12 h daily light. Drought was applied by withholding water from plants for 7 days. Heat was applied by transferring plants to a new growth chamber with conditions of 12 h light at 37 °C and 12 h dark at 32 °C for 7 days. To apply low-phosphate stress to quinoa plants, a hydroponic system as described in Conn et al. (2013) was utilized. Seeds were sown on germination medium containing 0.7% agar and grown for 3 weeks in tanks containing nutrient solution (Conn et al. 2013). Then plants were either transferred to tanks containing fresh nutrient solution (control) or tanks containing nutrient solution lacking KH_2PO_4 (low phosphate, supplemented with a compensatory amount of KCl) for 5 days. When harvest, plant materials were snap-frozen in liquid nitrogen, and were ground for RNA extraction. Total RNA was isolated using a Zymo Directzol RNA MiniPrep Kit. RNA quality was assessed using an Agilent 2100 BioAnalyzer. To perform Illumina RNA-seq, sequencing libraries were prepared using the NEBNext Ultra Directional RNA Library Prep Kit for Illumina. PE sequencing (100-bp) was performed using an Illumina HiSeq 2000 machine at KAUST Bioscience Core Laboratories. To perform PacBio Iso-seq, RNA samples were purified with Beckman Coulter Genomics AMPure XP magnetic beads. RNA quality was assessed with standard agarose gel electrophoresis and Thermo Fisher Scientific Qubit Fluorometry. RNA was fractionated into 3 libraries consisting of differently sized RNA (1 to 2 kb, 2 to 3 kb and 3 to 6 kb). A total of 9 SMRT® cells for each RNA sample was run on the PacBio RS II system with the P6-C4 chemistry by DNALink (Seoul, Republic of Korea). A total of 836,322 reads covering 2,292,247,217 bp and 699,876 reads covering 1,696,155,060 bp was produced for the shoot and root RNA libraries, respectively. Sequencing reads were processed with the RS_IsoSeq protocol of SMRT® Analysis (v2.2), and polished consensus sequences were produced with the ToFU pipeline.

RNA-seq reads and Iso-seq full-length transcripts were mapped onto the reference genome using Bowtie 2 (Langmead and Salzberg 2012) and GMAP (Wu and Watanabe 2005), respectively. The BAM2hints from the MAKER package (Cantarel et al. 2008) was used to give hints with locations of potential intron-exon boundaries using the alignments.

5.8.3 Ab Initio Gene Model Prediction

The AUGUSTUS was used as the main ab initio prediction software for the prediction of gene models for the quinoa genome assembly (Stanke et al. 2008). To ensure a comprehensive prediction, coding sequences from *Amaranthus hypochondriacus, Beta vulgaris, Spinacia oleracea,* and *Arabidopsis thaliana* (102,149 genes in total) were obtained and concatenated for a master list of genes. To train the AUGUSTUS model, 50% of the genes on the master list were used, with the other half being used for validation. Prediction-optimization was performed twice. The intron-exon boundary hints provided from RNA-seq and Iso-seq were used to predict genes in the repeat-masked reference genome. To enhance the quality of the prediction, peptide sequences from *B. vulgaris* (EST form) were fed to MAKER package. Putative function of genes was annotated by a BLAST search of the peptide sequences against the UniProt database. PFAM domains and InterProScan ID to the gene models were assigned using the MAKER package. Gene model AUR62017258 (includes exons, introns, UTRs) was presented as an example of the output of the annotation process (Fig. 5.5). As shown, Iso-seq reads (grey lines) span across the full length of the gene, demonstrating a potential to identify splicing variants of genes.

GC rich regions of a plant genome are likely to be regions that have a more active gene expression and regulation (Šmarda et al. 2014; Sumner et al. 1993). Gene density and GC content at genome level were therefore plotted together using DensityMap (Guizard et al. 2016) (Fig. 5.6), providing an overview of gene and gene expression across all chromosomes. Relatively more genes can be found toward the end of the chromosomes, with areas of low gene density in the middle of the chromosomes (centromeres regions). The GC content was found to be similar across all chromosomes, with no notable pattern can be observed.

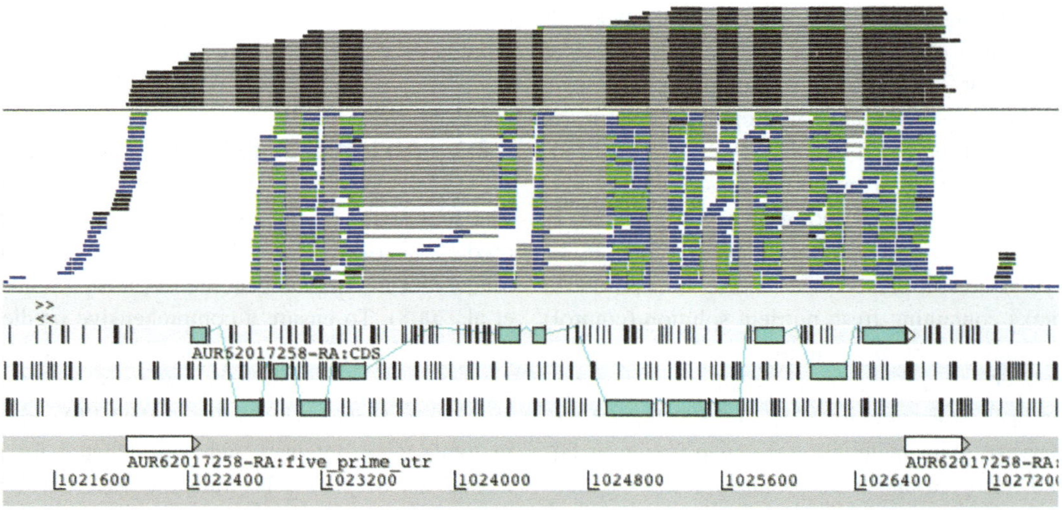

Fig. 5.5 A demonstration of gene model prediction (AUR62017258). Iso-seq full-length transcripts and RNA-seq reads that mapped to AUR62017258 gene model were shown on the top and middle panel, respectively. Full-length Iso-seq reads were able to span the 5' UTR, all exons, and the 3' UTR in a single read. Light grey lines show the intros as indicated by the split of reads. Blue and green lines show the sense the antisense reads. Figure from Jarvis et al. (2017)

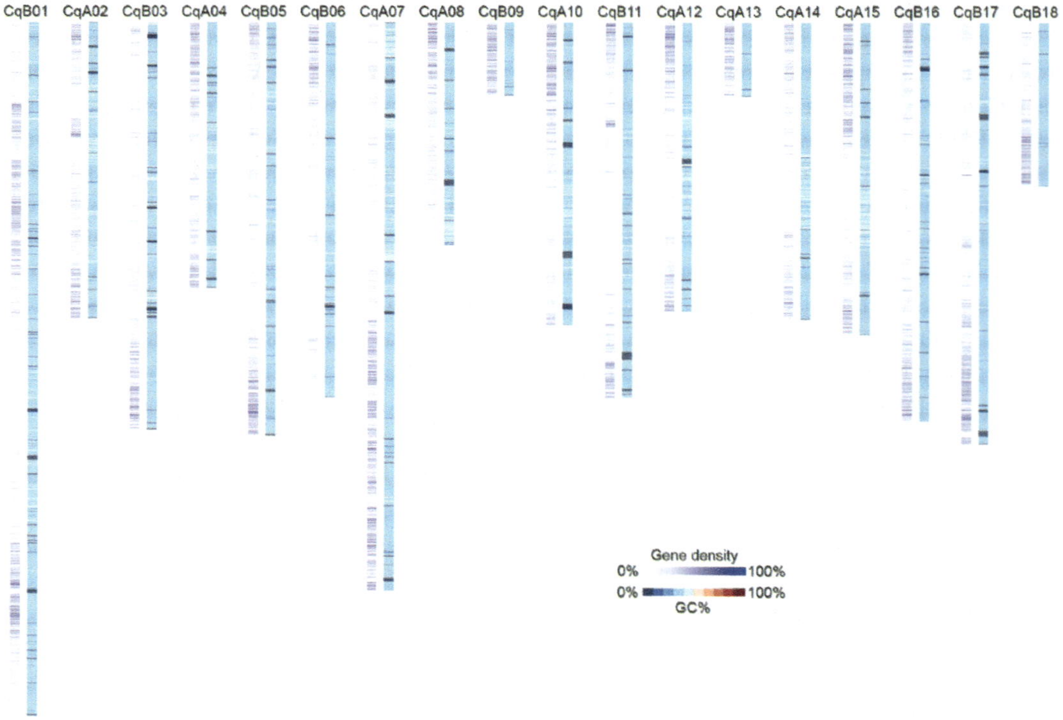

Fig. 5.6 Gene density and GC content in 100-kb windows across the quinoa genome. For each chromosome, the left bar presents gene density while the right bar presents GC content. Figure from Jarvis et al. (2017)

The annotation identified a total of 57.06 Mb of coding sequence, which covered 44,776 gene models. The total number of genes is similar to the gene number in upland cotton (*Gossypium hirsutum*), another allotetraploid plant species (Li et al. 2015). When comparing to quinoa's ancestral diploids, *Chenopodium pallidicaule* and *Chenopodium suecicum*, the number of genes of quinoa is approximately equal to the sum of the gene number in the diploids—17,961 for *C. pllidicaule* and 21,861 for *C. suecicum*. This is in keeping with the hypothesis that tetraploid quinoa developed from a hybridization of the two diploids (Jarvis et al. 2017). The average length of a gene in the quinoa annotation is 1.27 kb, with the longest gene being 15,933 bp. Among these annotated genes, 6,864 single-exon genes were identified. To evaluate the quality of the gene prediction, Annotation Edit Distance (AED) scores were generated with the MAKER package. The result showed that 33,365 out of 44,776 genes had an AED of less than or equal to 0.3, which indicates that the gene models are well supported, as AED scores range from 0 to 1, with "0" indicating a very high level of evidence for a given annotation (Yandell and Ence 2012).

5.8.4 Annotation Validation with BUSCO

Genome annotation completeness was assessed using "Benchmarking Universal Single-Copy Orthologs" (BUSCO, available at http://busco.ezlab.org.), a quantitative assessment method based on evolutionarily informed expectations of gene content (Simão et al. 2015). With the BLAST E-value cutoff set to 10^{-5}, out of the 956 genes in the Plantae BUSCO dataset, 97.3% genes were identified in the quinoa gene annotation (Jarvis et al. 2017). The fact that only 26 BUSCO genes are missing indicates a good quality of both the genome assembly and the gene annotation.

5.8.5 MicroRNA Gene Prediction

MicroRNAs (miRNA) are short ($\sim$21 nucleotides), non-translated RNAs that play important regulatory roles in gene expression of plants (Jones-Rhoades et al. 2006). As it is found to be involved in the processes such as development, reproduction,and stress-response, miRNA genes of quinoa genome were annotated. The transcriptomic data generated by PacBio Iso-seq and RNAseq was used to prepare the libraries. An Illumina HiSeq 2000 machine was used to perform the sequencing. To exclude the adapter sequence and reads with low quality, trim galore v0.4.0 (http://www.bioinformatics.babraham.ac.uk/projects/trim_galore/) was used to trim the reads. The output data included 102,858,577 reads, with the length of each sequence being between 19 and 26 nucleotides. Using ShortStack with default settings (Axtell 2013), 523,752 loci were detected to have clustered mappings of small RNAs. Among them, 483,702 were predicted to be from RNAi mediated processing of small RNAs. The algorithm implemented in ShortStack stringently detected 67 candidate miRNA with canonical secondary hairpin structures. Only 17 out of 165 known microRNAs are detected by the ShortStack algorithm with a high specificity (high confidence miRNA candidates). The stem-loop precursors of identified miRNA were matched to known miRNA families using Rfam scan (http://rfam.xfam.org, Rfam 12.1, April 2016), and multiple putative miRNA loci were found to be likely from a highly conserved, highly expressed mir166 (Rfam ID RF00075) (Jarvis et al. 2017).

5.9 Analysis of Quinoa Sub-genome Structure

To investigate the sub-genome structure of the tetraploid *C. quinoa*, the genome assembly scaffolds of *C. pallidicaule* and *C. suesicum* were mapped onto the tetraploid *C. quinoa* genome

Table 5.4 A comparison between sub-genomes A and B

	Sub-genome A	Sub-genome B
Number of scaffolds	156	410
Size	202.6 Mb	646.3 Mb
Genetic size	946 cM	1,087 cM
Physical size	524 Mb	660 Mb
Number of 18-24 J	172	6322

assembly to putatively identify chromosome relationships between the tetraploid and the ancestral diploids. The 18 quinoa chromosomes could be assigned to two sub-genomes A and B, corresponding to the ancestral *C. pallidicaule* and *C. suesicum* genomes, respectively. The sub-genomes were assigned in scaffolds, which is approximately 72% of the chromosome-scale genome assembly in size. To be specific, sub-genome A contained 156 scaffolds, consisting of 202.6 Mb while sub-genome B contained 410 scaffolds, consisting of 646.3 Mb.

To confirm the sub-genome analysis, the presence of the 18–24 J mini-satellite marker in the two sub-genomes was investigated. It was previously reported that this marker was substantially more abundant in *C. suesicum* (B sub-genome) than in *C. pallidicaule* (A sub-genome) (Kolano et al. 2011). An analysis of the sub-genome assignment confirmed that the 18–24 J mini-satellite marker was much more frequent in the B sub-genome (6322), relative to the A sub-genome (172) thereby adding support to the assignments (Table 5.4). Furthermore, the smaller genetic and physical size of the A sub-genome (Table 5.4) corresponds to the smaller genome size of *C. pallidicaule* (452 Mb) relative to *C. suesicum* (815 Mb) estimated by k-mer analysis.

Comparative analysis between quinoa genome and genomes of *A. hypochondriacus,* and *B. vulgaris* is discussed in detail in Chap. 6.

5.10 ChenopodiumDB

A free-access database, ChenopodiumDB (http://www.cbrc.kaust.edu.sa/chenopodiumdb/), was launched to make the genome sequences, protein sequences, and gene annotation of quinoa available to the public. The database also contains sequence data for other species in genus *Chenopodium*, serving as an integrated repository for population genetics associated with this genus.

References

Axtell MJ (2013) ShortStack: comprehensive annotation and quantification of small RNA genes. RNA 19:740–751

Bazile D, Bertero D, Nieto C (2015) State of the art report on quinoa around the world in 2013. FAO and CIRAD, Roma (Italy)

Bertero HD, Ajdela V, Correa G, Jacobsen SE, Mujica A (2004) Genotype and genotype-by-environment interaction effects for grain yield and grain size of quinoa (*Chenopodium quinoa* Willd.) as revealed by pattern analysis of international multi-environment trials. Field Crops Res 89:299–318

Bolger AM, Lohse M, Usadel B (2014) Trimmomatic: a flexible trimmer for Illumina sequence data. Bioinformatics 30:2114–2120

Cantarel BL, Korf I, Robb SM, Parra G, Ross E, Moore B, Holt C, Sánchez AA, Yandell M (2008) MAKER: an easy-to-use annotation pipeline designed for emerging model organism genomes. Genome Res 18:188–196

Christensen SA, Pratt DB, Pratt C, Nelson PT, Stevens MR, Jellen EN, Coleman CE, Fairbanks DJ, Bonifacio A, Maughan PJ (2007) Assessment of genetic diversity in the USDA and CIP-FAO international nursery collections of quinoa (*Chenopodium quinoa* Willd.) using microsatellite markers. Plant Genet Resour 5:82–95

Coles ND, Coleman CE, Christensen SA, Jellen EN, Stevens MR, Bonifacio A, Rojas-Beltran JA, Fairbanks DJ, Maughan PJ (2005) Development and use of an expressed sequenced tag library in quinoa (Chenopodium quinoa Willd.) for the discovery of single nucleotide polymorphisms. Plant Sci 168:439–447

Conn SJ, Hocking B, Dayod M, Xu B, Athman A, Henderson S, Aukett L, Conn V, Shearer MK,

Fuentes S (2013) Protocol: optimizing hydroponic growth systems for nutritional and physiological analysis of *Arabidopsis thaliana* and other plants. Plant Methods 9:4

Denisov G, Walenz B, Halpern AL, Miller J, Axelrod N, Levy S, Sutton G (2008) Consensus generation and variant detection by Celera Assembler. Bioinformatics 24:1035–1040

Dixon JR, Selvaraj S, Yue F, Kim A, Li Y, Shen Y, Hu M, Liu JS, Ren B (2012) Topological domains in mammalian genomes identified by analysis of chromatin interactions. Nature 485:376–380

Fairbanks DJ, Waldrigues A, Ruas CF, Ruas PM, Maughan PJ, Robison LR, Andersen WR, Riede CR, Pauley CS, Caetano LG (1993) Efficient characterization of biological diversity using field DNA extraction and random amplified polymorphic DNA markers. Int J Biol Macromol 46:27–36

FAO (2013) Genebank standards for plant genetic resources for food and agriculture. Genebank Stand Plant Genet Resour Food Agric 11:1–16

Gómez MB, Castro PA, Mignone C, Bertero HD (2011) Can yield potential be increased by manipulation of reproductive partitioning in quinoa (*Chenopodium quinoa*)? Evidence from gibberellic acid synthesis inhibition using Paclobutrazol. Funct Plant Biol 38:420–430

Guizard S, Piégu B, Bigot Y (2016) DensityMap: a genome viewer for illustrating the densities of features. BMC Bioinf 17:204

Jacobsen SE, Mujica A, Jensen CR (2003) The resistance of quinoa (*Chenopodium quinoa* Willd.) to adverse abiotic factors. Food Rev Int 19:99–109

Jarvis DE, Ho YS, Lightfoot DJ, Schmöckel SM, Li B, Borm TJA, Ohyanagi H, Mineta K, Michell CT, Saber N, Kharbatia NM, Rupper RR, Sharp AR, Dally N, Boughton BA, Woo YH, Gao G, Schijlen EGWM, Guo X, Momin AA, Negrão S, Al-Babili S, Gehring C, Roessner U, Jung C, Murphy K, Arold ST, Gojobori T, Linden CGvd, van Loo EN, Jellen EN, Maughan PJ, Tester M, CGVD (2017) The genome of *Chenopodium quinoa*. Nature 542:307–312

Jarvis DE, Kopp OR, Jellen EN, Mallory MA, Pattee J, Bonifacio A, Coleman CE, Stevens MR, Fairbanks DJ, Maughan PJ (2008) Simple sequence repeat marker development and genetic mapping in quinoa (*Chenopodium quinoa* Willd.). J Genet 87:39–51

Jones-Rhoades MW, Bartel DP, Bartel B (2006) MicroRNAs and their regulatory roles in plants. Annu Rev Plant Biol 57:19–53

Kolano B, Gardunia BW, Michalska M, Bonifacio A, Fairbanks D, Maughan PJ, Coleman CE, Stevens MR, Jellen EN, Maluszynska J (2011) Chromosomal localization of two novel repetitive sequences isolated from the Chenopodium quinoa Willd. genome. Genome 54:710–717

Kolano B, Siwinska D, Gomez PL, Szymanowskapulka J, Maluszynska J (2012) Genome size variation in *Chenopodium quinoa* (Chenopodiaceae). Plant Syst Evol 298:251–255

Langmead B, Salzberg SL (2012) Fast gapped-read alignment with Bowtie 2. Nat Methods 9:357–359

Li F, Fan G, Lu C, Xiao G, Zou C, Kohel RJ, Ma Z, Shang H, Ma X, Wu J (2015) Genome sequence of cultivated Upland cotton (*Gossypium hirsutum* TM-1) provides insights into genome evolution. Nat Biotechnol 33:524–530

Li H, Durbin R (2010) Fast and accurate long-read alignment with Burrows-Wheeler transform. Bioinformatics 26:589–595

Li H, Handsaker B, Wysoker A, Fennell T, Ruan J, Homer N, Marth G, Abecasis G, Durbin R (2009) The sequence Alignment/Map format and SAMtools. Bioinformatics 25:2078–2079

Lieberman-Aiden E, Dekker J (2009) Comprehensive mapping of long-range interactions reveals folding principles of the human genome. Science 326:289–293

Maradini Filho AM, Pirozi MR, Da SBJ, Pinheiro Sant'Ana HM, Paes Chaves JB, Js DRC (2015) Quinoa: nutritional, functional and antinutritional aspects. Crit Rev Food Sci Nutr 57:1618–1630

Mason SL, Stevens MR, Jellen EN, Bonifacio A, Fairbanks DJ, Coleman CE, Mccarty RR, Rasmussen AG, Maughan PJ (2005) Development and use of microsatellite markers for germplasm characterization in quinoa (Willd.). Crop Sci 45:1618–1630

Maughan PJ, Bonifacio A, Jellen EN, Stevens MR, Coleman CE, Ricks M, Mason SL, Jarvis DE, Gardunia BW, Fairbanks DJ (2004) A genetic linkage map of quinoa (*Chenopodium quinoa*) based on AFLP, RAPD, and SSR markers. Theor Appl Genet 109:1188–1195

Maughan PJ, Smith SM, Rojasbeltrán JA, Elzinga D, Raney JA, Jellen EN, Bonifacio A, Udall JA, Fairbanks DJ (2012) Single Nucleotide Polymorphism identification, characterization, and linkage mapping in Quinoa. Plant Genome 5:114–125

Palomino G, Hernández LT, Eulogio DLCT (2008) Nuclear genome size and chromosome analysis in *Chenopodium quinoa* and *C. berlandieri* subsp. nuttalliae. Euphytica 164:221–230

Poulsen TS, Johnsen HE (2004) BAC End Sequencing. In: Zhao S, Stodolsky M (eds) Bacterial artificial chromosomes: Volume 1 Library construction, physical mapping, and sequencing. Humana Press, Totowa, NJ, pp 157–161

Putnam NH, O'Connell BL, Stites JC, Rice BJ, Blanchette M, Calef R, Troll CJ, Fields A, Hartley PD, Sugnet CW (2015) Chromosome-scale shotgun assembly using an *in vitro* method for long-range linkage. Genome Res 26:342–350

Rao NK, Shahid M (2012) Quinoa- A promising new crop for the Arabian Peninsula. Am-Eurasian J Agric Environ Sci 12:1350–1355

Rojas W, Valdivia R, Padulosi S, Pinto M, Soto JL, Alcocer E, Guzman L, Estrada R, Apapza V, Bravo R (2009) From neglect to limelight: issues, methods and approaches in enhancing sustainable conservation and use of Andean grains in Peru and Bolivia. J Agric Rural Dev Tropics Subtropics 92:1–32

Simão FA, Waterhouse RM, Ioannidis P, Kriventseva EV, Zdobnov EM (2015) BUSCO: assessing genome assembly and annotation completeness with single-copy orthologs. Bioinformatics 31:3210–3212

Šmarda P, Bureš P, Horová L, Leitch IJ, Mucina L, Pacini E, Tichý L, Grulich V, Rotreklová O (2014) Ecological and evolutionary significance of genomic GC content diversity in monocots. Proc Natl Acad Sci 111:E4096–E4102

Stanke M, Diekhans M, Baertsch R, Haussler D (2008) Using native and syntenically mapped cDNA alignments to improve *de novo* gene finding. Bioinformatics 24:637–644

Sumner AT, de la Torre J, Stuppia L (1993) The distribution of genes on chromosomes: A cytological approach. J Mol Evol 37:117–122

Trapnell C, Pachter L, Salzberg SL (2009) TopHat: discovering splice junctions with RNA-Seq. Bioinformatics 25:1105–1111

Verena N, Juan D, Ruth UC (2016) Assessment of the nutritional composition of quinoa (*Chenopodium quinoa* Willd.). Food Chem 193:47–54

Wu TD, Watanabe CK (2005) GMAP: a genomic mapping and alignment program for mRNA and EST sequences. Bioinformatics 21:1859–1875

Yandell M, Ence D (2012) A beginner's guide to eukaryotic genome annotation. Nat Rev Genet 13:329–342

Yasui Y, Hirakawa H, Oikawa T, Toyoshima M, Matsuzaki C, Ueno M, Mizuno N, Nagatoshi Y, Imamura T, Miyago M (2016) Draft genome sequence of an inbred line of *Chenopodium quinoa*, an allotetraploid crop with great environmental adaptability and outstanding nutritional properties. DNA Res 23:535–546

Zerbino DR, Birney E (2008) Velvet: Algorithms for de novo short read assembly using de Bruijn graphs. Genome Res 18:821–829

Zurita-Silva A, Fuentes F, Zamora P, Jacobsen SE, Schwember AR (2014) Breeding quinoa (*Chenopodium quinoa* Willd.): potential and perspectives. Mol Breed 34:13–30

Structural and Functional Genomics of *Chenopodium quinoa*

6

Elodie Rey and David E. Jarvis

Abstract

Quinoa (*Chenopodium quinoa* Willd.), also known as 'the mother grain' of the Incas, is a pseudo-cereal crop originating from the Andes, mainly cultivated for its seeds and consumed in a similar way to rice and other staple grains. Although it is primarily a subsistence crop in Andean regions, quinoa is gaining international importance due to the exceptional nutritive value of its grains and its ability to maintain yields in harsh environmental conditions. As a consequence, breeding programs are rapidly expending, and a better knowledge of the structure and function of quinoa genome is becoming increasingly needed in order to support and fasten breeding efforts and make quinoa more productive and better adapted to its novel culture environments. The recent release of several novel sequence resources such as the genome reference sequence of coastal quinoa accession QQ74 and the re-sequencing of several wild and cultivated quinoas will certainly contribute to this aim. In this chapter, we review the current molecular resources available for the structural characterization of quinoa allotetraploid genome and discuss future prospects for the functional characterization of genes underlying traits of agronomic importance.

6.1 Introduction

Quinoa (*Chenopodium quinoa* Willd.), also known as 'the mother grain' of the Incas, is a pseudo-cereal crop originating from the Andes, mainly cultivated for its seeds and consumed in a similar way to rice and other staple grains. Quinoa seeds are highly nutritious, and contain high protein content including all essential amino acids, high content of several minerals and health-promoting compounds such as flavonoids (Ruiz et al. 2014). Quinoa seeds have a good balance between oil (4–9%), protein (averaging 16%), and carbohydrates (64%) (Bhargava et al. 2006; Vega-Galvez et al. 2010) and also lacks gluten, which makes it a good food source for consumers with celiac disease (Jacobsen 2003). Though it is used as a cereal, it is a dicotyledonous annual species originating from the region surrounding the Bolivian and Peruvian Altiplano, where it was presumably domesticated by ancient Andean civilizations approximately 7,000 years ago

E. Rey (✉)
Biological and Environmental Sciences and Engineering Division (BESE), King Abdullah University of Science and Technology (KAUST), Thuwal 23955-6900, Saudi Arabia
e-mail: elodie.rey@kaust.edu.sa

D. E. Jarvis
Department of Plant and Wildlife Sciences, College of Life Sciences, Brigham Young University, Provo, UT 84602, USA
e-mail: david_jarvis@byu.edu

S. M. Schmöckel (ed.), *The Quinoa Genome*, Compendium of Plant Genomes,
https://doi.org/10.1007/978-3-030-65237-1_6

(Dillehay et al. 2007). Quinoa belongs to the Amaranthaceae family (formerly Chenopodiaceae) together with the other economically important crops spinach (*Spinacia oleracea* L.) and sugar beet (*Beta vulgaris* L.). Quinoa species diversity is distributed around Lake Titicaca, with major dispersion cores located in Bolivia and Peru which define five main ecotypes (highlands, inter-Andean valley, Salares, Yungas, and coastal lowlands) (Ruiz et al. 2014).

Quinoa's rich genetic diversity results from a mode of production that is fragmented and localized across the diverse eco-environments of the Andean region: from Ecuador to southern Chile and from sea level to more than 3,500 m altitude in the Andean Altiplano. Cultivation of quinoa for several thousand years in the harsh and diverse environments of the Andes, reinforced by the tradition of seed exchange by local Andean farmers, has shaped its capacity to maintain its modest productivity on marginal lands (low nutritive soils, under water shortage, and high salinity conditions). The species exhibits good tolerance to several abiotic stresses including frost, high soil salinity, and drought (Jacobsen 2003). Hence, although quinoa crop productivity remains low compared to that of other major crops such as wheat or maize, it has been identified as a good candidate to enhance food security, especially in the context of the predicted increases in global soil salinization and aridity (Ruiz et al. 2014). As such, quinoa is a good model crop to study the mechanisms leveraged by plants to cope with high salinity and drought (Orsini et al. 2011; Ruiz-Carrasco et al. 2011; Pulvento et al. 2012; Adolf et al. 2013; Shabala et al. 2013).

Most quinoa is produced in Peru and Bolivia, which respectively accounted for 53% and 44% of the global production of 148,000 tones in 2016 (www.fao.org), with other Andean countries like Ecuador, Chile, Argentina, and Colombia accounting for approximately 3%. Although it is primarily a subsistence crop produced by small-scale local farms for indigenous Andean populations, a new international market for quinoa is rapidly expanding due to the exceptional nutritive value of its grains and its ability to maintain yields in harsh environmental conditions (Maughan et al. 2007). As a result, both research and production have been developing steadily beyond Andean production countries and now occurs in more than 70 countries, including France, England, Sweden, Denmark, Holland, Italy, Kenya, India, and the USA (www.fao.org). However, important breeding improvements are still needed to transform this subsistence crop into a major crop that can reduce dependence on other staples like wheat and rice. Many traits needs to be improved in order to increase quinoa production as it expands into new environments, such as yield related traits (seed size, grain yield), time to flowering, plant maturity, growth habit, quality traits (seed color, seed saponins content), and resistance to pathogens (Maughan et al. 2007; Zurita-Silva et al. 2014). In order to facilitate the development of quinoa as a crop, a better understanding of the genetic basis underlying these traits in quinoa is needed. In addition, as quinoa production expands into new areas, it will be increasingly important to understand genotype-by-environment effects on yield and nutritional properties (Jacobsen 2003).

Building on the work of breeding programs initiated in the mid-1970s in Andean countries (McElhinny et al. 2007) and in the 1980s in the USA and Europe (Zurita-Silva et al. 2014), some of the first genetic and genomic resources were produced in the early 2000s—mostly molecular markers generated to support the characterization of quinoa species diversity and molecular mapping of quinoa seed quality traits. Later, the Food and Agriculture Organization (FAO) of the United Nations declared 2013 as The International Year of Quinoa (IYQ2013) (FAO 2012) helping quinoa to gain further international scientific interest. Several whole genome sequencing projects were initiated for coastal lowland (Jarvis et al. 2017) and highland (Yasui et al. 2016; Zou et al. 2017) quinoa accessions, as well as for several wild and cultivated quinoas and related species. These novel resources are meant to provide a better characterization of the allotetraploid quinoa genome, facilitate the identification of the molecular basis of important agronomical traits and further enhance breeding programs needed for quinoa genetic improvement (Jacobsen 2003; Danial et al. 2007).

In this chapter, we review the molecular resources available for the structural characterization of quinoa genome and discuss future prospects for the functional characterization of genes underlying traits of agronomic importance.

6.2 Structural Genomics

6.2.1 Organization of Quinoa Genome

Cytological studies established that quinoa is an allotetraploid species with a basic set of 9 chromosomes ($2n = 4x = 36$; AABB) resulting from the hybridization between ancestral A- and B-genome diploid species (Palomino et al. 1990; Bhargava et al. 2006; Kolano et al. 2012a). The allotetraploid origin of quinoa was confirmed based on genetic, molecular, and cytogenetic analyses using a combination of three different types of DNA markers (chloroplastic (cp) DNA, nuclear ribosomal Internal Transcribed Spacer (nrITS), and 5S ribosomal DNA (rDNA) non-transcribed sequences (NTS)) as well as Genomic in situ hybridization (GISH) cytogenetic analysis (Wilson 1990; Ward 2000; Maughan et al. 2006; Kolano et al. 2012b; Štorchová et al. 2015). Several studies have shown that quinoa chromosomes are mostly inherited through disomic and minimal tetrasomic segregations (Simmonds 1971; Ward 2000; Maughan et al. 2004). The quinoa genome size was estimated at 1.453 Gb (mean genome size of quinoa accessions is $2C = 2.973 \pm 0.043$ pg) (Kolano et al. 2012a), which is considered as a small genome compared to most plant species (the angiosperm mean genome size is $1C = 5.7$ Gb; (Dodsworth et al. 2015). In the set of *Chenopodium* accessions studied by Palomino et al. (2008), very little correlation was found between chromosome length (in μm) and DNA content (in pg), suggesting different degrees of chromosome condensation between species (Palomino et al. 2008).

The small size of quinoa chromosomes as well as their high degree of condensation and low level of sub-genome-specific banding patterns makes cytogenetic and conventional karyotype analysis difficult in this species. Consequently, beyond the species' chromosome number, the general distribution of heterochromatin vs. euchromatin and distribution of a small number of highly repetitive sequences, including the rRNA genes, we know relatively little about quinoa cytogenetics. C-banding of quinoa somatic chromosomes reveals mostly isobrachial chromosomes, organized with heterochromatin concentrated in the pericentric region while euchromatin extends distally to the telomeres (Gill et al. 1991). Fluorescent in situ hybridization (FISH) markers designed from quinoa retrotransposons have been used to evidence important genomic divergence among the progenitors of the *C. album* and *C. berlandieri*/*C. quinoa* hexaploid complexes (Karcz et al. 2005). However, these markers were unsuccessful at discriminating the two sub-genomes of quinoa at the cytological level. These difficulties to obtain a clear karyotype of quinoa is a limitation to the study of chromosome behavior such as pairing and recombination between homoeologous chromosomes, which is an important measure for the determination of species boundaries and taxonomic classification. As a result, the taxonomic identification and classification of chenopods are primarily based on growth habit and agro-ecological environment which is largely complicated by the high phenotypic plasticity of *Chenopodium* species (Aellen and Just 1943; Wilson 1981b).

6.2.2 Molecular Tools

Understanding the origins, evolution, and phylogenetic relationships within quinoa accessions is of major interest in order to better exploit the phenotypic and genetic diversity for genetic improvement of quinoa. With this aim, most research reported to date has dealt with the development of molecular markers to investigate the genetic relationships among the different taxa and has been reviewed in detail (Maughan et al. 2007; Zurita-Silva et al. 2014).

6.2.2.1 Isozyme and Protein Markers

The first molecular characterization of quinoa genetic diversity consisted of the analysis of the allelic variation at 21 isozyme loci within 99 populations of domesticated quinoa and wild species (*C. hircinum* and wild quinoa ajara) (Wilson 1981, 1988a, b, c). For the first time, this study revealed the presence of two distinctive groups: a coastal type defining populations from southwestern Chile, and an Andean type defining populations extending from northwestern Argentina to southern Colombia. Moreover, the Andean ecotypes further split into northern and southern phylogenetic groups. These data also support the hypothesis of the center of diversity of quinoa species originating in the Altiplano, in an area surrounding Lake Titicaca in Bolivia and Peru where the largest number of quinoa landraces was found by Gandarillas (1968). Another study used a protein-based approach for the identification of quinoa cultivars through the characterization of seed storage proteins which can be used in breeding programs for improved protein quantity and quality (Fairbanks et al. 1990).

6.2.2.2 DNA-Based Molecular Markers

Subsequently, the development of DNA-based markers such as Random Amplified Polymorphic DNA (RAPD) and Amplified Fragment Length Polymorphism (AFLP), simple sequence repeat (SSR)-based and single nucleotide polymorphism (SNP) markers, have facilitated the assessment of genetic variation among quinoa accessions and other *Amaranthaceae* species. Fairbanks et al. (1990) used 30 RAPD markers to identify DNA polymorphisms among 16 quinoa accessions, while Bonifacio (1995) used these markers to confirm hybridization between quinoa and *C. nuttalliae* Safford, as well as intergeneric crosses obtained by crossing quinoa with three Altriplex species. Another study by Ruas et al. (1999) used 33 RAPD markers (399 alleles) in order to investigate the genetic relationship among 19 *Chenopodium* taxa, composed of 10 quinoa (two wild and eight cultivated) accessions and 6 related *Chenopodium* species. The results enabled the discrimination of the different germplasm into five groups according to the species classification: a *C. quinoa* group composed of the eight cultivated quinoa varieties and the two wild accessions ('ajara' ecotypes) suggesting little differentiation between sympatric domesticated cultivars and wild quinoa accessions, a second group composed of the North American *C. berlandieri* and *C. album*, and a fourth and fifth group, respectively, representing the clusters of *C. pallidicaule* and *C. ambrosioides* species. Later on, 230 AFLP, 6 RAPD and 19 microsatellite markers were used to create the first linkage map of quinoa described further in Sect. 6.2.3 below (Maughan et al. 2004). Additional RAPD studies include the characterization of 87 cultivated and weedy quinoa individuals (representative of eight quinoa field populations from the Altiplano and interandean valleys of Bolivia) using 10 markers (38 alleles) (del Castillo et al. 2007), and the characterization of 55 accessions belonging to 14 species of chenopods (including *C. berlandieri* subsp. *nuttalliae*, diploid, tetraploid and hexaploidy *C. album* and *C. giganteum* species) (Rana et al. 2010).

The next molecular marker technology that presents even greater transferability of markers between *Chenopodium* species were simple sequence repeat (SSR)-based markers. These markers have been first developed in quinoa by Mason et al. (2005) who identified 208 polymorphic SSR markers from the sequences of three SSR-enriched genomic libraries (CA, ATT, ATG) from a panel of 31 cultivated quinoa accessions. Mason et al. (2005) observed a number of alleles ranging from 2 to 13, with an average of 4 alleles detected per locus, as well as heterozygosity ranging from 0.2 and 0.9 with a mean value of 0.57, and a high level (67%) of transferability of these SSR markers between the other cultivated *Chenopodium* species *C. pallidicaule*, *C. giganteum*, and *C. berlandieri* subsp. *nuttalliae*. Further studies made use of SSR markers for the characterization of additional quinoa accessions (Christensen et al. 2007; Jarvis et al. 2008; Fuentes et al. 2009, 2012; Tartara et al. 2012). In their work, Christensen et al. (2007) characterized 151 quinoa accessions originating from Peru, Ecuador, Bolivia, Argentina, and Chile using 36 SSR markers (420 alleles) displaying levels of heterozygosity ranging from 0.45 to 0.94. The authors observed limited diversity in the germplasms from Argentina and Ecuador while a high genetic diversity was observed in the material originating from the central Andean highlands. These results identified the Altiplano as a center of diversity and a probable point of introduction for Ecuadorian accessions, whereas the Chilean highland and lowland zones would be at the origin to the introduction into Argentina. Fuentes et al. (2009) characterized 59 Chilean quinoa accessions and reported a higher level of genetic diversity in the Chilean coastal lowland germplasm (with heterozygosity values ranging from 0.07 to 0.90) than previously postulated. This observation was explained by gene flow between quinoa and the weedy populations of *C. album* and *C. hircinum* grown in sympatry in the lowland fields of Chile. In 2019, Salazar et al. used 15 SSR markers to characterize the genetic diversity within 84 accessions cultivated in the Ecuadorian Andes. Their study revealed extensive allelic richness (196 alleles) and a high level of genetic heterozygosity (HE = 0.71). A population structure analysis derived from these SSR markers revealed three distinct clusters among this Ecuadorian germplasm. These clusters displayed little correlation with geographic origin, suggesting that the structure of Ecuadorian diversity wasn't shaped through geographic patterning, but rather originated from independent genetic lineages representing ancestral landrace populations which have been disseminated throughout Ecuador.

In the absence of genome sequence resources, the first single nucleotide polymorphism (SNP) markers were developed in quinoa from expressed sequence tag (EST) libraries described in Sect. 6.2.3 below (Coles et al. 2005). In this work, 51 polymorphic markers were obtained between six quinoa accessions (the Bolivian quinoa variety 'Real' from which the ESTs were derived, two Altiplano ecotypes, 'Chucapaca' and '0654', and three coastal ecotypes, 'KU2', 'G-205-95DK', and 'NL-6'), comprising 38 SNPs and 13 insertions-deletions (Indels), with an average of one SNP per 462 base-pair (bp) and one Indel per 1,812 bp. An additional 81 interspecific SNPs were identified that showed polymorphism between quinoa accessions and *C. berlandieri* and were found at a higher (2.5X) frequency than the SNPs identified among just the quinoa accessions. All 132 SNPs of this study were validated by reverse-strand sequencing, and proof of concept of their utility for germplasm characterization was demonstrated by phenetical analysis of the five quinoa accessions which confirmed the results obtained in previous analysis performed with isozymes, RFLPs, RAPDs and microsatellites markers. Several years later, a large-scale set of 14,178 newly identified SNP markers was generated through a genomic reduction approach and allowed the development of 511 polymorphic SNP assays for quinoa based on KASPar genotyping chemistry (Maughan et al. 2012). These SNP assays were used to screen a diversity panel of 113 quinoa accessions from Bolivia, Chile, Peru, Argentina, Brazil, Europe, North and Central America, as well as eight accessions representing several related Chenopodium taxa (i.e., *C. berlandieri, C. hircinum* Schrad., *C. ficifolium* Sm *and C. watsonii* A. Nelson). The genotyping results revealed a minor allele frequency (MAF) of the SNPs ranging from 0.02 to 0.50 across the quinoa accessions, with an average MAF of 0.28. The structure analysis of the quinoa diversity panel revealed two major subgroups corresponding to the Andean and coastal quinoa ecotypes. These markers showed high degree of transferability to related species, with 34% of the SNPs amplifying all eight taxa and 74% amplifying in six of the eight taxa. However, due to the design of this SNP essay which was initially developed to differentiate between eight quinoa accessions, the authors highlighted the limits of this genotyping platform for phylogenetic studies at the genus level (Zurita-Silva et al. 2014; Vidueiros et al. 2015).

The development of molecular markers for quinoa has enabled genetic analysis of quinoa diversity at the ecotype, population, germplasm, and genus levels with an increasing scale of the samples analyzed and resolution of the analysis. Taken together, these analyses consistently confirmed the structure of quinoa species as two distinct germplasm pools: an Andean highland quinoa pool that includes its associated weedy complex, highly diverse and considered the primary center of diversity and origin of quinoa species, and a second pool of quinoa originating from the central and southern Chilean coastal lowlands and constituting a second center of major quinoa diversity (Jellen et al. 2011).

6.2.2.3 EST and BAC Libraries

In the absence of whole genome sequencing, ESTs and bacterial artificial chromosome (BAC) libraries represented the first source of genomic sequences available for quinoa.

A first set of 424 ESTs was generated from cDNA libraries derived from two tissues (immature seeds and flowers) (Coles et al. 2005). Comparison of these sequences to public protein databases identified 349 EST sequences with significant homology to protein-coding genes from related model plant species. While most of the annotated ESTs had ambiguous or unclassified functions, several were present in relatively high abundance and had putative functions related to plant defense.

These publicly available EST libraries have been used notably for the development of SNP molecular markers (Coles et al. 2005) as reported in the previous Sect. (6.2.2.2). Subsequently, this sequence resource has been exploited for the cloning and analysis of genes. In their study, Balzotti et al. (2008) identified and characterized the gene encoding the 11S globulin seed storage protein in quinoa, which represents a major step toward the understanding of quinoa seeds' exceptional composition of protein and essential amino acids. Another major achievement making use of the EST sequence resource was the cloning of the Salt Overly Sensitive 1 (SOS1) gene, an potential factor contributing to quinoa salt stress tolerance (Maughan et al. 2009). Finally, these EST databases represent a rich source of genomic sequence information that can be exploited for the development of genetic markers for various applications.

Another important sequence resource has been the development of bacterial artificial chromosome (BAC) libraries. In their work, Stevens et al. (2006) produced a BAC library of quinoa containing 74,880 clones that were made publicly available at the Arizona Genomic Institute at the University of Arizona, Tucson, AZ, USA. The library was constructed from a first pool of 26,880 clones issued from the digestion using BamHI with an average clone insert size of 113 kb, and a second pool of 48,000 clones issued from the digestion using EcoRI with an average clone insert size of 130 kb. On the basis of a calculated genome size of 967 Mb, this library was estimated to cover 9.0 equivalents of the haploid nuclear genome. The authors demonstrated the relevance of this BAC library for gene cloning by identifying two positive clones representing different genetic loci encoding the quinoa 11S seed storage protein. BAC libraries will continue to represent an important resource for gene cloning projects, as they provide reliable sequence information as well as support for the construction of physical maps of the quinoa genome.

6.2.3 Mapping the Quinoa Genome

The first genetic linkage map of quinoa was developed from a population of 80 F2 individuals derived from the cross between KU-2, a Chilean lowland cultivar, and 0654, a Peruvian Altiplano cultivars (Maughan et al. 2004). The map was built from 255 DNA markers (AFLPs, RAPDs and SSRs) and resulted in 35 linkage groups spanning 1,020 centimorgans (cM) with an average density of 4.0 cM per marker. The higher number of linkage groups than haploid chromosome number of quinoa ($n = 18$) indicates that this map was still not saturated and several regions of the genome remain undetected, while the predicted total length was estimated to reach 1700 cM.

The second quinoa genetic map reported by Jarvis et al. (2008) was developed from a recombinant inbred lines (RIL) population consisting of 82 F5 individuals issued from the cross between KU-2 and 0654 cultivars described above. The map was constructed using 200 SSRs and 70 AFLPs and resulted in 38 genetic linkage groups spanning 913 cM.

The third quinoa genetic map was developed from two advanced (F8) RIL mapping populations sharing a common paternal parent (0654 highland quinoa cultivar) composed of 128 individuals (Maughan et al. 2012). The linkage analysis of 451 polymorphic SNP markers yielded 29 genetic linkage groups spanning 1,404 cM with an average of 3.1 cM per marker. The greater completeness of this map that is closer to the total length of the predicted quinoa map (1,700 cM) owed to the genotyping of a larger mapping population with a higher marker density than the two previous studies.

This later genetic map represents an important milestone in the process of getting the first genome sequence assembled into pseudomolecules (Jarvis et al. 2017). Indeed, to combine scaffolds into pseudomolecules, the genetic map of quinoa developed by Maughan et al. (2012) was integrated with two new linkage maps. These two maps were generated from two populations segregating for the presence of saponins in the seeds: 45 individual F3 progeny from the cross between Kurmi (sweet) and 0654 (bitter) and 94 sweet F2 lines from the cross between Atlas (sweet) and Carina Red (bitter). The resulting genetic map was built from 6,403 unique markers and resulted in 18 linkage groups spanning a total length of 2,034 cM, which corresponds to the haploid chromosome number of quinoa.

6.2.4 Whole Genome Sequencing (WGS) of Quinoa Species

Although quinoa has a moderate genome size compared to most plant genomes, its allotetraploid nature, relatively high number of linkage groups (n = 18) and high level of heterozygosity represent a high degree of complexity for the production of a genome sequence reference. Despite the difficulty, two draft genome assemblies of the highland quinoa cv. REAL as well as a genome reference assembly of the lowland quinoa cv. QQ74, shallow sequencing of various nuclear genomes, transcriptome sequencing and cytoplasmic DNA sequence assemblies have been produced. In this part, we will review these different resources and how they enabled a better characterization of quinoa genome.

6.2.4.1 Draft Genome Sequence Assemblies

The first effort to sequence the genome of quinoa was produced by Yasui et al. (2016) who generated the draft genome sequence (Cqu_r1.0) of the Japanese inbred quinoa accession Kyoto-d (Kd). The sequencing was performed using a combination of 290 Gb of Illumina HiSeq 2500 high-throughput next-generation sequencing and 45 Gb of long-read Pacific Biosciences (PacBio) RS II sequencing, corresponding to 196X and 31X coverage of the quinoa genome, respectively. The resulting de novo genome assembly contained 24,847 scaffolds, with a total length of 1.1 Gb, an N50 length of 86,941 bp, and 24% missing data. A k-mer frequency distribution curve allowed the size of quinoa genome to be estimated at about 1.5 Gb, in good agreement with that of 1,448 Mb from cytometry analysis (Palomino et al. 2008). The transcriptome sequencing and de novo assembly of 12 different tissues of quinoa were used to support the ab initio prediction of 226,647 protein-coding sequences (CDSs) (Cqu_r1.0_cds) consisting of 190.5 Mb, while similarity searches predicted 150,029 genes among which 62,512 were functionally annotated thanks to their homology to other protein-coding genes in UniProtKB database. Similarity searches against the NCBI non-redundant (NR) protein database allowed the inference of 668.2 Mb of transposable element (TE) space, including 132.7 Mb (12.2% of Cqu_r1.0) of known repeat sequences with predominance of Class I LTR elements (7.2% of Cqu_r1.0) and novel repeat sequences accounting for a total length of 535.5 Mb (49.2% of Cqu_r1.0). However, the fragmented nature of the draft sequence as well as the absence of clear identification and sequencing of the diploid progenitors of the A and B sub-genomes of quinoa prevented the discrimination between quinoa sub-genomes and inference of homoeology.

A second draft genome was published in 2017 by Zou et al. who sequenced an inbred line of the highland quinoa cv. Real, which is one of the most widely cultivated and consumed quinoas in the world. Similar to the previous draft, the sequencing was performed using a combination of Illumina short-read sequencing with various library insert sizes, and long-read PacBio sequencing, representing 146 × and 33 × coverage of the quinoa genome, respectively. The final assembly (Cq_real_v1.0) resulted in a total sequence length of 1,337 Mb with a scaffold N50 of 1.16 Mb with the largest scaffold being 5.4 Mb. The assembly covers 90.2% of the expected nuclear genome size estimated through k-mer analysis (1,482 Mb), with 90% of the assembly falling into 1,087 scaffolds that are at least 423 Kb in length, of which 485 were anchored to the genetic map from Maughan et al. (2012). The transcriptome of five different quinoa tissues was sequenced through Illumina mRNA-sequencing and further de novo assembled into 234,311 transcripts (>200 bp) of which 98.9% aligned to the genome assembly with more than 95% identity. Gene annotation was performed using a combination of ab initio prediction, homology searches and transcriptome assembly and resulted in the identification of 54,438 protein-coding genes models with an average length of 3,548 bp and 4.8 exons per gene, out of which 95.6% were assigned functions. The authors also identified non-coding RNAs, including 192 micro RNAs, 1,310 ribosomal RNAs, 2,934 transfer RNAs, and 5,922 small nuclear RNAs. The annotation of the repeat elements was performed using a combination of in silico prediction and homology searches and revealed 64.5% of the quinoa genome being repetitive, 85.6% of which were transposable elements (TEs). The most abundant transposon families were the long terminal repeat (LTR) transposons Gypsy and Copia, as well as the DNA transposon CMC, making up 33.58, 11.69, and 3.64% of the total genome content, respectively. Moreover, the authors identified a total of 392,764 SSRs, with motif sizes ranging from mononucleotide to hexanucleotide and representing 38 distinctive families. This genomic sequence represents an important resource for the development of molecular markers for phylogenetic analysis within the highland quinoa subgroup, as well as for inference of highland specific haplotypes and further comparisons with coastal quinoas.

6.2.4.2 Pseudomolecule Level Genome Sequence Assembly

The Genome Reference Sequence of QQ74
In 2017, Jarvis et al. published the first quinoa genome reference sequence of the coastal Chilean quinoa QQ74 (accession PI 614886). The sequence assembly generated is a chromosome-scale reference genome sequence, which was produced using a combination of single-molecule real-time (SMRT) sequencing technology from PacBio RS II system, BioNano Genomics Irys optical maps, and chromosome-conformation capture from Dovetail Genomics. About 75 Gb of PacBio sequences representing 52 × coverage of the quinoa genome size was assembled and integrated into hybrid scaffolds after mapping the BioNano optical maps against the assembly. PacBio + BioNano hybrid scaffolds were further merged into super-scaffolds based on the mapping of Dovetail Chicago in vitro Hi-C sequencing reads (representing 51.6X physical coverage of the quinoa genome) against the hybrid scaffolds and using HiRise software. The resulting 1.39 Gb assembly contained 3,486 scaffolds, with a scaffold N50 of 3.84 Mb and 90% of the assembled genome contained in 439 scaffolds. Of this assembly, 1.18 Gb (85%) was anchored to the integrated genetic map composed of 6,403 unique markers spanning 2,034 cM across 18 linkage groups (see Sect. 6.2.3), allowing for the concatenation of scaffolds onto 18 pseudomolecules based on their order and orientation as determined from the integrated linkage map.

Annotation and Composition of Quinoa Genome

Protein coding genes and micro (mi)-RNA genes of quinoa were annotated using a combination of ab initio prediction and transcript evidence approaches. Transcript evidence was provided by the PacBio isoform sequencing (Iso-Seq) and short-reads Illumina Hi-seq 2000 sequencing of six different tissue samples (apical meristems, lateral meristems, whole seedlings, flowers and immature seeds, leaves petioles, stems) for protein-coding genes, while for mi-RNA genes, RNA samples were isolated from roots and shoots of hydroponically grown plants in control conditions or exposed to low phosphate and were sequenced on Illumina Hi-seq 2000. The annotation yielded 44,776 gene models covering a total coding region of 57,064,233 bp with an average gene length of 1,274 bp. Additionally, 523,752 small-RNA loci were detected across the genome, of which 483,702 were likely to be from RNAi mediated processing of small RNAs. The de novo identification and annotation of the repeat elements revealed a fraction of 64% of the genome to be repetitive, including a large proportion (36%) of long terminal repeat (LTR) and Gypsy (28%) transposable elements.

Sub-genome Evolution

In the same paper, Jarvis and colleagues (2017) also reported the whole genome shotgun Illumina sequencing and assembly of two diploids, *C. pallidicaule* and *C. suecicum* from the A and B genetic pools, respectively. These novel resources represent an improvement to other quinoa sequencing projects, as they enabled the identification of the sub-genomes in the quinoa reference sequence by analyzing the coverage of each diploid's reads on the scaffolds of the quinoa genome sequence. Nine quinoa chromosomes were assigned to each sub-genome, with the B sub-genome accounting for a larger proportion of the genetic (1,087 cM) and physical (660 Mb) sizes than the A sub-genome (946 cM, 524 Mb). These sub-genome sizes are in agreement with the sizes of *C. suecicum* (815 Mb) and *C. pallidicaule* (452 Mb) diploid genomes estimated from k-mer analyses. Functional annotation of protein-coding genes and repetitive elements was also performed in a similar way as for the tetraploid quinoa, and allowed for further analysis of sub-genome evolution. The calculation of the rate of synonymous substitutions per synonymous site (Ks) between orthologous gene pairs identified individually in quinoa, *C. suecicum*, and *C. pallidicuale* identified a peak of orthologous gene pairs in quinoa genome for Ks $\sim$ 0.1. This observation reflects the duplication of genes due to the hybridization of ancestral diploid species as no similar peak was detected for the diploids. Using the mutation rates calculated for *Arabidopsis thaliana* and core eukaryotes, the tetraploidization event was dated between 3.3 and 6.3 million years ago (MYA). Using a set of 5,807 homoeologous gene pairs retained in single copy in each diploid A- and B- genome and in the genome of the close relative *B. vulgaris* (1:1:1), Jarvis and colleagues showed the high level of synteny between the sub-genomes of quinoa, and identified several major chromosomal rearrangements that have occurred between the A and B sub-genomes relative to the chromosomal structure of the common ancestor of *B. vulgaris* and quinoa. While allopolyploidization is usually known to induce genome size changes to the sub-genomes of the newly formed allopolyploids as a result of the mobilization (expansion/elimination) of repetitive or single copy DNA sequences (Renny-Byfield et al. 2011), only limited and similar numbers of lost genes (1,031 and 849) were observed from the A and B sub-genomes of quinoa relative to *C. pallidicaule* and *C. suecicum* diploids, respectively. These results were in agreement with previous cytogenetic studies of allotetraploid quinoa which exhibited additive genome size values compared to the presumed ancestral species (Kolano et al. 2012a).

6.2.4.3 Reduced-Coverage Sequencing

Together with the reference genome of quinoa published in 2017, reduced-coverage ($\sim$12X) genome sequencing of 22 additional allotetraploid samples from the goosefoot complex were produced by Illumina sequencing, including 15 additional quinoa samples representing the two major recognized groups of quinoa (highland and coastal), five accessions of the North American *C. berlandieri,* and one accession each of *C. hircinum* from the Pacific and Atlantic Andean watersheds. These sequences were mapped to the reference genome and SNP calling was performed, allowing the identification of 7,809,381 SNPs between the different accessions and the reference quinoa genome, including 2,668,694 SNPs that are specific to quinoa. Out of these, 3,132 SNPs were retained for reconstructing the phylogenetic tree in which *C. berlandieri* was found to be the basal member of the species complex, while the Chilean *C. hircinum* and the Argentinian *C. hircinum* samples were found basal to coastal and highland quinoa ecotypes, respectively. These new findings suggest that quinoa could result from two independent domestication events, one in highland and the second in coastal environments. This hypothesis is in contrast to the long-held belief that quinoa was domesticated from *C. hircinum* in a single event in the Altiplano, from which coastal quinoa was later derived. In order to confirm the mono- vs poly-phyletic origin of tetraploid quinoa, further phylogenetic analysis must be realized using a deeper sampling of quinoa and *C. hircinum* diversity in order to clarify whether their clustering results from shared ancestry or from the cultivation of these two species in sympatry, especially in Chile, allowing for interspecific hybridization and gene flow between them (Wilson 1981; Wilson and Manhart 1993; Mujica and Jacobsen 2006). Finally, the set of SNPs developed in this study will be useful in further studies assessing genetic diversity and identifying genomic regions associated with desirable traits.

In 2017, Zhang et al. performed the re-sequencing and assembly of the genome of the quinoa accession Riobamba using 23 × coverage Illumina sequencing, and also performed shallow sequencing (7–8× genome coverage) of 11 quinoa accessions representing geographical adaptation within the species, with the aim to better characterize the nucleotide variations (SNPs and InDel) in the quinoa genome and develop markers to be used for population structure and genetic diversity analysis within a panel of 129 quinoa accessions. These data were used for variant calling against the de novo genome assembly of the accession Riobamba, and allowed the identification of 8,441,022 bi-allelic SNPs and 842,783 InDels, corresponding to a density of 5.81 and 0.58 variants per Kb, respectively. InDel length showed a negative correlation with InDel number, with a majority (~72%) of InDels having a short size (1–2 bp), 20% of InDels having a size between 3 and 8 bp, and almost 3% of InDels having a size >8 bp. Out of these InDels, 85 makers were developed and added to a set of 14 genomic screen-derived SSRs (gSSRs) (Jarvis et al. 2008) and 48 EST-derived SSRs (Zhang et al. 2017) and these 147 markers (362 alleles) were used for genotyping 129 quinoas accessions. Allelic information at the loci of these 147 markers was used for population structure analysis among the 129 accessions of quinoa, revealing two main quinoa groups: a highland type from the Andes and a coastal type from Chile. The Andean highland type was further separating into northern highland and southern highland subgroups. Most Chilean and 'United States, New Mexico' accessions clustered in the coastal group and displayed a higher level of genetic diversity (0.38), than the highland quinoas (0.33) which regrouped most accessions from Peru and Bolivia. Within the highland quinoas, the genetic diversity of the Bolivian southern highland subgroup (0.32), was higher than within northern highland Peruvian accessions (0.27), and according to expectations both subgroups displayed lower levels of genetic diversity than the main highlands and coastal groups. A STRUCTURE analysis additionally revealed gene flow between the northern and southern highland subgroups that was more frequent than between the coastal and highland groups, whereas the gene content of the two highland subgroups both contributed to the genomes of admixed individuals in highlands group. Additionally, the authors identified a set of four accessions from both highland and coastal groups that was shown to capture 88% of the total allelic variation from the whole panel of accessions, as well as a core set of 16 accessions sufficient to capture all 362 alleles. These core sets of quinoas represent an important source of information for germplasm conservation and the development of populations for genetic studies.

6.2.4.4 Cytoplasmic Genomes Sequencing

Chloroplast Genome Sequencing

The slow evolution and highly conserved structure of chloroplast genomes make chloroplast sequences an ideal system for assessing plant phylogeny (Clegg and Zurawski 1992). Within chloroplast genomes, *rbc*L (Ribulose-1,5-bisphosphate carboxylase/oxygenase) and *mat*K (Maturase K) regions are among the most rapidly evolving protein-coding genes, providing a high degree of information on phylogeny through the analysis of polymorphism at specific loci.

With the aim to produce molecular profiles to discriminate between quinoa and *C. album* species, Devi and Chrungoo (2017) generated amplicons of the *rbc*L and *mat*K chloroplast regions from 11 quinoa and 8 *C. album* accessions. Sequencing and comparison of the 19 *rbcL* amplicons revealed 1.26% parsimony-informative sites with a level of sequence diversity between species of 0.68%, whereas the nucleotide sequence comparison of amplicons of *mat*K revealed 4.97% parsimony-informative sites with 2.81% interspecific sequence diversity. The SNPs variation within the *rbc*L and *mat*K chloroplast regions was further validated by allele-specific PCR across 36 quinoa and *C. album* accessions. The evolutionary tree of age separated quinoa and *C. album* in two different clusters, with an estimated divergence of 2.5–6.2 million years ago (MYA) from other angiosperms. Furthermore, Himalayan chenopods, including *C. album* accessions, were found to be evolutionarily younger (10.5–4.1 MY) than the Andean chenopods (13.1–7.2 MY), including quinoa. The results established the paraphyletic origin of the genus *Chenopodium*.

Together with the draft sequence assembly of the nuclear genome, Zou et al. (2017) also produced the sequencing of the chloroplast genome of quinoa cv. Real using single-molecule long reads sequencing, resulting in a single contig assembly of a 152,282 bp chloroplast sequence that was further annotated. This sequence provided the first insight into quinoa chloroplast genome, composed of four conserved regions: two inverted repeats (IRA and IRB) that separate a short single copy (SSC) region and a long single copy (LSC) region, characteristic of plant chloroplast genomes.

In 2017, (Hong et al.) sequenced and assembled both quinoa and *C. album* chloroplast genomes from Illumina sequencing, resulting in single circular molecules of 152,099 and 152,167 bp sequences, respectively, composed of 119 genes (78 protein-coding, 37 tRNA and 4 rRNA) each. The gene content, sequence, order, and orientation were highly conserved between the chloroplast genomes of the two species, except in the region of the *mat*K gene which showed only 98.2% homology at the amino acid level. Tandem repeats (TR) were in similar number, length, and repeat unit between both species, except for the copy number variation for which one extra species-specific TR was found in each species. Additionally, 44 and 53 SSRs were identified in the chloroplast genome of quinoa and *C. album*, respectively, out of which 28 SSRs were shared by both species and were mostly detected in the LSC region, inter-genic sequences and mononucleotides. Mononucleotides were the most abundant SSRs motifs, accounting for about 62% of the SRRs in quinoa and 66% in *C. album*, with the majority repeat sequence being A/T. This study allowed the identification of suitable regions to be considered as markers for elucidating the phylogenetic relationship within *Chenopodium*.

Wang et al. (2017) produced an additional chloroplast genome sequence of a quinoa isolated from Jilin Agricultural University (China) through the sequencing of PCR fragments, yielding a sequence of 151,169 bp in length composed of large (LSC, 83,576 bp) and small (SSC, 18,107 bp) single copy regions, separated by a pair of inverted repeat regions (IRs, 24,743 bp each). The sequence harbored 120 genes, including 87 protein-coding genes, 29 transfer RNA and 4 ribosomal RNA gene species. The authors used this sequence in comparison with 72 related complete chloroplast genomes to assess the phylogenetic relationship between them, where quinoa was most closely related to *Chenopodium album*.

Another quinoa chloroplast assembly was produced by Rabah et al. (2017) together with chloroplast assemblies of nine other non-model agronomical species including Cinnamon, Fig, Guava, Pomegranate, Cashew, Mango, Lychee, Okra, and Basil. The sequence assembly of quinoa chloroplast sequence was composed of 152,075 bp including 18,372 large and 25,074 small single copy regions where 78 protein-coding genes and 17 genes with introns were identified. In their study, the authors further identified insertions and deletions in the coding sequence of *rpl23* coding region that were shared with other Amaranthaceae and caused a frame shift and the introduction of stop codons. They also identified an inversion between *ndhC* and *accD* coding sequences of quinoa plastome which isn't shared with other Amaranthaceae members.

Following on the nuclear genome reference sequence published in 2017, Maughan et al. (2019) reported the complete sequences of both mitochondrial and chloroplast genomes from the coastal quinoa accession PI 614886. Using the PacBio and Illumina reads generated for the reference genome project (Jarvis et al. 2017), the authors assembled a single circular contig of 152,079 bp length, which was almost identical in length with the other quinoa chloroplast assemblies reported above (152,075 bp in PI 5105506; 152,099 bp in PI 4332327; and 152,282 bp in 'Real') and slightly longer than those reported in other closely related Amaranthaceae species (149,635 bp in *B. vulgaris* (Li et al. 2014); 150,518 bp in *A. hypochondriacus* (Chaney et al. 2016) and 150,725 bp in *S. oleracea* (Schmitz-Linneweber et al. 2001)). The sequence annotation identified 86 protein-coding, 31 tRNA, and 8 rRNA genes, and the comparison of the quinoa chloroplast sequence with related species revealed a high degree of sequence conservation among chloroplast genomes of the *Amaranthaceae* species, with no additional or lacking genes compared to the set of genes common to the *A. hypochondriacus*, *B. vulgaris* and *S. oleracea* chloroplast genomes. No major structural difference between quinoa and the other species was observed except for a small inversion relative to *Amaranthaceae* members already reported for quinoa (Rabah et al. 2017).

With the aim to evaluate the sequence variation within *Chenopodium*, whole genome re-sequencing data from 13 additional quinoa accessions, seven tetraploid accessions including five *C. berlandieri* and two *C. hircinum*, and one accession each of the A-genome diploid *C. pallidicaule* and the B-genome diploid *C. suecicum* were mapped onto the reference PI 614886 chloroplast assembly. SNPs and InDels calling in each re-sequencing accession relative to the PI 614886 reference assembly revealed no major structural variations among the chloroplast genome assemblies of quinoa, whereas more variants were identified from the diploid species *C. pallidicaule* (709 SNPs and 145 InDels) and *C. suecicum* (1181 SNPs and 221 InDels), than in the chloroplast genome assemblies of the tetraploid species (with an average of 36 SNPs and 22 InDels). Moreover, no SNPs and/or very few InDels were identified in the chloroplast sequences of the three closely related coastal quinoa accessions (Maughan et al. 2019).

Mitochondrial Genome Sequencing

In the same paper, Maughan et al. (2019) report the first assembled mitochondrial genome for the genus *Chenopodium*, which was obtained through the assembly of previously generated Illumina sequencing reads (Jarvis et al. 2017) guided by the reference mitochondrial genome from *Beta vulgaris* L.12 (sugar beet) and gap-filled with long sequencing PacBio reads. The resulting assembly consisted in a 315,003 bp sequence, which is shorter than the reported mitochondrial sequences of *B. vulgaris* (368,801 bp; Kubo et al. (2000), *B. vulgaris* ssp. *maritima* (sea beet; 364,950 bp), and *Spinacia oleracea* L. (spinach; 329,613 bp; Cai et al. (2017))—the only other species in the Amaranthaceae with reported mitochondrial sequences. The annotation of quinoa mitochondrial sequence revealed 30 protein-coding, 21 tRNA, and 3 rRNA genes, which is similar to the number of genes reported in the mitochondrial genomes of *B. vulgaris* (Dohm et al. 2014) and *S. oleracea* (Xu et al. 2017), and is not lacking any genes shared by the other *Amaranthaceae* mitochondrial genome sequences, suggesting that the sequence is likely complete. The annotation revealed an additional copy of the 26S rRNA gene in quinoa compared to related species, which is possibly fragmented but still expressed. In addition, two copies of tRNACys genes have been identified in quinoa mitonchondrial genome: one copy homologous to the native gene trnC1-GCA that is shared by higher plants and a second copy homologous to the trnC2-GCA gene from *B. vulgaris*, which had not previously been identified in other higher plants (Kubo et al. 2000) and suggests that this gene is shared among Amaranthaceae species. The mapping of re-sequencing data (described above) against the mitonchondiral genome of quinoa revealed greater SNP and InDel variants identified in the diploid species (491 and 626 SNPs and 188 and 203 InDels, respectively) than in the tetraploid species (with an average of 90 SNPs and 83 InDels).

The variants identified from both mitochondrial and chloroplast genome sequences were used in the study of the phylogenetic relationships among *Chenopodium* species and revealed concordant results with previous analyses based on the nuclear genome sequences (Jarvis et al. 2017; Maughan et al. 2019). First, the ancestral root of the phylogenetic tree from which the AABB tetraploids are derived is formed by the two diploid species, *C. pallidicaule* (A-genome group) and *C. suecicum* (B-genome group), with the A-genome ancestor being closer to the tetraploids. This suggests that the A-sub-genome ancestor of quinoa was likely the maternal parent in the tetraploidization event. Furthermore, North American *C. berlandieri* accessions are positioned more basally in the tree relative to the cultivated South American quinoa accessions and the wild South American taxon *C. hircinum*, which supports the hypothesis that the initial polyploidization event likely occurred in the New World. In this analysis, quinoa accessions generally fall into two main groups: a highland Andean group and a more distant branch consisting of mostly coastal ecotypes, the Altiplano variety 'Pasankalla' and the two *C. hircinum* accessions. The systematic distal placement of *C. hircinum* accessions on the coastal branch in all three trees suggests these are not ancestral to the Andean quinoas and may be representative of either ancestral or introgressive relationships with the coastal quinoa germplasm. Consequently, cytoplasmic DNA sequencing provides additional support for the hypothesis of separate Andean and Pacific coastal domestication events of South American goosefoot/quinoa.

6.3 Functional Analysis of Quinoa

6.3.1 Genes and Loci Characterization

6.3.1.1 Seed Storage Protein Genes

One of the major qualities of quinoa as a crop lies in its exceptionally nutritive seed characteristics, with high protein content and excellent balance of amino acids. The identification and characterization of key genes contributing to the quinoa seed composition such as those encoding seed storage proteins therefore represent an important clue for maintaining and exploiting this trait in quinoa breeding programs. In 2008, Balzotti et al. isolated the genomic and cDNA sequences of two 11S genes by screening the cDNA library developed from quinoa developing seeds (Coles et al. 2005). The sequencing of the clones revealed that these two genes represented orthologous loci, 11SA and 11SB, of the quinoa genome encoding two variants of a 11S legumin-like seed storage protein sharing 74% homology with the 11S globulin protein from *A. hypochondriacus*. The analysis of the 11S quinoa amino acid sequence revealed a conserved sequence with those found in other species, and evidenced a signal peptide of 25 amino acid residues potentially responsible for the sub-localization of the protein to the lumen of the endoplasmic reticulum (ER). Furthermore, the quinoa 11S amino acid sequence shows a high level of essential amino acids that are beneficial for human nutrition, with notably aspartate/asparagine, glutamate/glutamine, arginine, serine, glycine, and leucine in abundance (Brinegar and Goundan 1993). Gene expression analysis revealed a higher expression of 11S mRNA in the late maturation stages of seed development in nine different quinoa genotypes differing by the geographic origin and maturation rate. The authors demonstrated that the peak of 11S mRNA expression level correlated with the accumulation of 11S protein in the seeds and maturation rate, which is consistent with the pattern of storage protein accumulation during seed development in other plant species (Nakamura et al. 2004). Further work is now needed to clone, sequence, and characterize the expression of the second major seed storage protein gene in quinoa encoding the 2S albumin protein, which combined with 11S seed proteins produces the unique quantity and composition of quinoa seeds.

6.3.1.2 Salt Tolerance Genes

While soil salinity affects almost one-third of all arable land worldwide and causes billions of dollars of damages in agriculture every year, crops with enhanced performances under saline conditions represent an important tool for future food safety around the world (Epstein and Bloom 2005; Qadir et al. 2014; Morton et al. 2019). Quinoa is described as a facultative halophyte crop, with a tolerance to salinity levels ranging between 150 and over 300 mM NaCl depending on genotypes, which is much greater than barley, wheat, and corn crops (Adolf et al. 2013; Ruiz et al. 2016; Schmöckel et al. 2017). With the aim to investigate the mechanisms contributing to the considerable salt tolerance of quinoa, Maughan et al. (2009) made use of the available quinoa BAC library and EST sequence resources to clone and characterize the *Salt Overly Sensitive 1* (*SOS1*) gene which encodes a plasma membrane Na^+/H^+ antiporter that plays an important role in germination and growth of plants in saline environments. The authors identified two homoeologous *SOS1* loci (cq*SOS1A* and cq*SOS1B*) in the quinoa genome, with coding sequence of 3,477 bp and 3,486 bp, respectively, each structured in 23 exons. Sequence analysis revealed a high level of similarity with SOS1 homologs of related species and identified two conserved domains, a cation-antiporter domain (Nhap) and a cyclic-nucleotide binding domain. The translation of cq*SOS1A* and cq*SOS1B* coding sequences predicted proteins of 1158 and 1161 amino acids, respectively, which share a high degree of similarity with *SOS1* sequences from species belonging to the Caryophyllales order. The authors further investigated the relative expression of *SOS1* in root and leaf tissues under saline conditions (450 mmol/L), which was consistently 3–fourfold higher in leaves than in roots, as well as more strongly up-regulated in leaves in response to salt stress than in roots, suggesting a constitutive expression of *SOS1* genes in roots and inducible expression in leaves under stress.

Schmockel et al. (2017) used a combined approach integrating RNA-sequencing analyses with comparative genomics and protein topology prediction in order to identify candidate genes involved in salt tolerance in the genome reference sequence of quinoa accession PI 614886. Differential gene expression analysis in response to salinity allowed the authors to identify 219 candidate genes that were specific to quinoa or were overrepresented relative to other *Amaranthaceae* species, and contained at least one predicted transmembrane domain. Further comparison of single nucleotide polymorphisms and copy number variation (CNV) in these 219 candidate genes and the response to salinity within a panel of 21 *Chenopodium* accessions (14 *C. quinoa*, 5 *C. berlandieri* and 2 *C. hircinum*) allowed the identification of 15 genes that could contribute to the differences in salinity tolerance of these *Chenopodium* accessions.

6.3.1.3 Genes Involved in Drought Tolerance

Another major beneficial trait of quinoa is its capacity to grow and produce seeds in semi-arid or even desert conditions such as those encountered in Chile, northwest Argentina, the Altiplano of Peru and Bolivia, as well as more recently in some Mediterranean, African, and Asian regions. As drought events are predicted to expand and increase in frequency and severity as a result of climate changes, it becomes highly desirable to understand the mechanisms of response of quinoa to drought stress as well as their molecular basis in order to develop quinoa cultivars adapted to the changing environments in the Andes and beyond. With this aim, Liu et al. (2018) searched the quinoa genome reference for genes encoding 70-kDa heat-shock protein (Hsp70s) which constitute a group of conserved chaperone proteins that have been characterized in various plants for their role in drought stress tolerance. Sixteen Cq*hsp*70 genes were identified in the quinoa genome, ranging from 412 amino acids (aa) to 891 aa in length. Phylogenetic analysis of these Cq*hsp*70 genes with additional *Hsp70* gene sequences from 16 species allowed their clustering in eight paralogous pairs which evolved from independent origin, and share high degree of similarity at the gene structure and protein motifs levels, indicating that the expansion of Cq*hsp*70 genes resulted from the recent chromosome-doubling event (allotetraploidization). Furthermore, the expression of 13 *Cqhsp*70s genes was evaluated in two-weeks-old seedlings at five time-points (0, 6, 12, 24, and 48 h) after treatment with 25% polyethylene glycol (PEG) *w*/*v* (with the aim to induce a physiological response analogous to drought stress) by qPCR. Results revealed 'drop-climb-drop' expression patterns following drought stress treatment for half of the genes studied, which is similar to the response observed in Arabidopsis homologous genes, indicating the functional conservation of *Hsp70* genes across species. Moreover, while the majority of paralogous gene pairs showed similar patterns of gene expression in response to the stress, some clades revealed different responses either in terms of direction or amplitude of expression, suggesting functional diversification of some genes after polyploidization.

6.3.1.4 Saponins Synthesis Genes

Quinoa seeds generally contain saponins (a mixture of triterpene glycosides). Although saponins represent a potential benefit for plant growth by repelling pests, they are a mostly unwanted characteristic for human consumption due to haemolitic effect and a bitter taste. Current strategies to produce saponin-free quinoa seeds are to remove them through one of the wet or dry methods, requiring water for soaking the seeds or specialized machinery, respectively, both of which are expensive and time-consuming processes. A promising alternative to these methods will be to develop saponin-free quinoa cultivars by exploiting the genetic diversity of quinoa germplasms in which some genotypes present naturally low levels of seed saponins. This strategy, however, requires to first identify and characterize the function of the genes regulating the absence of saponins in this material.

With this aim, Fiallos-Jurado et al. (2016) used a combined physiological and molecular approach in order to identify new genes involved in the biosynthesis of saponins in quinoa from an Ecuadorian sweet and bitter genotypes. They first validated that methyl jasmonate (MeJA) treatment was inducing the synthesis of saponins in quinoa leaves. Then, they performed a de novo transcriptome assembly from two publicly available quinoa RNA-seq datasets which they used as a query for the identification of quinoa ortholog sequences of 22 genes known to be stably expressed in *Arabidopsis thaliana* to be used as a reference for qPCR analysis in quinoa, as well as the screening of quinoa candidate saponin biosynthesis genes. This approach allowed them to identify the quinoa gene ortholog of At*2g28390* (*MonensinSensitivity 1*, *MON1*) as a stably expressed gene for qPCR analysis in quinoa. The authors also cloned three quinoa gene sequences: Cq*bAS1*, Cq*CYP716A78*, and Cq*CYP716A79* which play a role in the synthesis of saponins as confirmed through functionally analysis in yeast.

Later, Jarvis et al. (2017) made use of two populations segregating for the presence of saponins in the seeds (Kurmi (sweet) × 0654 (bitter) and Atlas (sweet) × Carina Red (bitter)) in order to identify loci associated with the presence/absence of seed saponins through linkage mapping and bulk segregant analysis (BSA). The combined linkage and BSA analyses identified a region on chromosome Cq*B16* that distinguishes bitter and sweet lines. Within this region, two transcription factors genes, AUR62017204 and AUR62017206, annotated as basic helix–loop–helix (bHLH), showed similarity with *triterpene saponin biosynthesis activating regulator* (*TSAR*) genes which have been described to play a role in saponin biosynthesis. Expression analysis of these genes in root, flowers and immature seed tissues using RNA-seq revealed tissue-specific expression of AUR62017206 (*TSAR-like 2*, *TSARL2*) in quinoa roots, while AUR62017204 (*TSARL1*) showed higher expression in the seeds of bitter varieties. The authors further identified *TSARL1* as a strong candidate gene for the regulation of the presence and absence of saponins in quinoa seeds as its sequence contains multiple and independent mutations that co-segregate with the sweet phenotype.

6.3.1.5 Flowering Time Genes

Quinoa is generally defined as a facultative short-day crop species owing to the daylength in its environment of origin as well as its capacity to flower under various photoperiod conditions, depending on genotypes. As cultivation of quinoa expands internationally beyond the Andes, a better characterization of the genetic factors determining the response of quinoa to varying daylength is required in order to develop new varieties better adapted to their novel environments. In 2017, Jarvis et al. identified two genes in the quinoa genome that are homologous to the flowering time (*FT*) regulating gene of *A. thaliana*, *FT1* and *FT2*, a situation that is similar to what has been described in *C. rubrum* (Cháb et al. 2008) and *B. vulgaris* (Pin et al. 2010). The *FT1* and *FT2* genes are present in three and two homoeologous copies in quinoa, respectively, with the *FT1* gene triplicate resulting from a tandem duplication. Further functional studies will be needed in order to determine whether duplicated quinoa *FT* genes have specialized in their function as is the case for the second *B. vulgaris FT* gene which was shown to act as a repressor of flowering before vernalization, while the first *FT* gene acts as a promotor of flowering (Pin and Nilsson 2012; Dally et al. 2014).

6.3.1.6 Genes Providing Herbicide Tolerance

In 2015, Mestanza et al. characterized the acetohydroxyacid synthase (*AHAS*) multigene family in the coastal Chilean quinoa cv. Regalona-Baer. The *AHAS* gene encodes the first enzyme of the biosynthetic pathway that produces the branched-chain of valine, leucine, and isoleucine essential amino acids. Because this gene might exist in multiple copies in polyploid species, it might be of particular interest to study this gene family for the purpose of developing crop resistance to herbicides targeting this enzyme in weedy diploid species usually harboring only one *AHAS* gene copy. With this aim, the authors performed the cloning, sequencing and Southern blotting of the *AHAS* gene, revealing six copies in the tetraploid genome of quinoa. Comparison of the mRNA sequences with *AHAS* mRNA sequences from the non-redundant (NR) Database showed that Cq*AHAS1* and Cq*AHAS2* variants were the most conserved gene copies that would have been inherited each from a different diploid parental ancestor, and would be at the origin of the further copies arising from duplication post-polyploidization. On the other hand, a third copy, Cq*AHAS3*, was hypothesized to arise from a duplication through homeologous chromosomes pairing due to the presence of sequence fragments sharing similarity with Cq*AHAS1* and Cq*AHAS2* genes. Further analysis of maximum parsimony phylogenies suggested that Cq*AHAS1* and Cq*AHAS2* are functional, while there was no evidence supporting the expression of Cq*AHAS3* nor the other variants (4, 5, and 6) of the *AHAS* quinoa genes. Additional gene expression studies are needed in order to determine if the other variants are expressed in a tissue-specific manner.

6.3.2 Expression Profiling

6.3.2.1 (Real-Time)-Quantitative PCR

In 2011, Morales et al. developed primers in order to follow real-time expression of several genes potentially involved in the tolerance of quinoa to high salinity, including *SOS1*, *BADH*, *TIP*, and *NHX1*. The transcriptional response of these genes to salt stress was measured in leaf and root tissues of two quinoa cultivars from the Altiplano Salares (Chipaya and Ollague) adapted to saline soils and one lowland Valley quinoa cultivar (CICA-17). Results showed stable expression of these genes in leaves regardless of the degree of salt stress, and with no significant difference in expression between Salares and Valley cultivars. On the other hand, the expression of the same genes showed significant differences between Salares and Valley ecotypes when measured from the root tissues, with higher up-regulation of *SOS1* and *BADH* genes observed for Chipaya and Ollague in response to salt stress. The authors made the further observations of significant differences in compatible solute accumulation between Altiplano and Valley ecotypes as well as differences in recovery after high extreme salt treatment, with superior resilience of Altiplano cultivars compared to Valley quinoa. Taken together, these results suggest that quinoa tolerates salt through a combination of salt exclusion and accumulation mechanisms.

In the same year, Ruiz-Carrasco et al. (2011) reported on the gene expression of two of the same salt-tolerant candidate genes, *SOS1* and *NHX*, using quantitative RT-PCR on four lowland accessions originating from Chile (PRJ, PRP, UDEC9, and BO78). Results revealed differential expression in response to salt stress between shoot and root tissues and between the same tissues of different genotypes. On the whole, *SOS1A* and Cq*SOS1B*, were more strongly up-regulated by salt stress in shoots than in roots. However, no significant changes in expression of these two genes were observed in the more salt-sensitive genotype BO78 (a quinoa from southern Chile), while a strong down-regulation of both genes was observed in the roots of PRP accession (a quinoa from coastal-central Chile) in response to salt stress. The expression of *SOS1* was found to be up-regulated in salt-treated roots of PRJ accession only, a quinoa from coastal-central Chile and the most salt-tolerant genotype of this experiment. The authors concluded that the tolerance of PRJ accession might be associated with reduced Na translocation. On the other hand, the expression of *NHX1* was significantly up-regulated in response to salt stress in shoots of all genotypes analyzed, except for BO78, the most salt-sensitive genotype. At the root level, *NHX1* expression was up-regulated in the most tolerant accession (PRJ), but not in the other tolerant one (PRP), suggesting that these genotypes acquired salt tolerance through different mechanisms of ion (Na and/or K) uptake/exclusion, translocation, and compartmentation. The results of this work indicate that the response to high salinity varies between quinoa lowland genotypes, both at physiological and molecular levels.

6.3.2.2 EST-Based Microarray

With the aim to identify genetic component involved in the synthesis of saponins, Reynolds (2009) annotated EST sequences expressed from maturing quinoa seed tissues. Using Sanger and 454 sequencing technologies, a total of 39,366 unigene sequences were assembled, from which a custom microarray assay of 102,834 oligonucleotide probes was developed to assess transcriptional changes in developing seeds of saponin-containing and saponin-free quinoa lines. The authors identified a list of 198 candidate genes showing significant differential expression between 'sweet' and 'bitter' seed tissues at two developmental stages, from which they established a list of candidate genes by selecting genes known to be associated with identified saponin biosynthetic pathways, including genes that share homology to cytochrome P450s, cytochrome P450 monooxygenases, and glycosyltransferases.

6.3.2.3 RNA-Sequencing

In 2012, Raney investigated the transcriptomic response to drought treatment in root samples from different quinoa ecotypes, the lowland quinoa variety Ingapirca (drought-sensitive) and a highland quinoa variety Ollague from the Altiplano Salares (drought-tolerant), using Illumina RNA-sequencing. De novo assembly of the transcriptome generated 20,337 unique transcripts, of which 462 putative gene products showed differential expression in response to drought stress, and 27 putative gene products were differentially expressed between quinoa varieties in response to drought treatment. Further gene ontology analysis of these differentially expressed genes indicated an overlap with genes involved in other abiotic stress responses.

Morales et al. (2017) characterized the transcriptional response of the drought tolerant Chilean quinoa R49 (Salares ecotype) to drought and irrigated control conditions through Illumina RNA-sequencing. They performed a de novo transcriptome assembly composed of 150,952 contigs, out of which 31,523 contigs were predicted as full-length coding sequences (CDS). Further functional annotation of these putative genes through sequence comparison to NCBI proteins indicated that 15% of the gene sequences lacked homology to known proteins, a proportion that is higher (19% presenting 306 contigs) within drought-induced genes (2,456 transcripts). The results of the gene ontology (GO) annotation were used to compare GO categories that changed in response to drought in the expression qualitative subset, with the aim to identify a potential set of genes underlying this quinoa-specific drought tolerance response. The expression pattern for these drought tolerance candidate genes obtained from RNA-seq analysis were validated on a set of 15 unigenes selected for their canonical response to drought stress.

With the aim to better understand the molecular mechanisms of how salt is directed into epidermal bladder cells (EBCs) as a potential mechanism contributing to salt tolerance in quinoa, Bohm et al. (2018) performed the RNA-sequencing of EBCs collected from quinoa cv. 5020 plants grown under control and salt-treatment conditions and functionally analyzed bladder-expressed transporters. Using the quinoa cv. Real draft assembly as a reference transcriptome (Zou et al. 2017), the authors identified 40,907 genes expressed in their samples, out of which 6,469 (15.8%) were at least two times higher expressed in bladder than in brushed leaf samples. The authors additionally detected 1,659 and 1,702 transcripts that were differentially expressed in response to salt in non-brushed leaves and in EBCs, respectively. The expression of several genes encoding ion transporters known to be involved in salt tolerance was further analyzed through RNA-sequencing data and qPCR. They observed a low expression of *SOS1* gene in bladders compared to leaf, which was consistent with the function of bladders in Na + accumulation rather than export. On the other hand, they observed the constitutive expression in bladders of the quinoa orthologs of the *HKT1* gene encoding sodium-permeable ion channels, the *NHX1* and *ClC-c* genes encoding vacuolar proton-coupled Na + and anion exchangers, as well genes encoding vacuolar proton pumps. The authors hypothesized that the absence of significant changes in expression of most transporter genes in response to salt stress implies that bladder cells act ‘constitutively’ in the sequestration of salt.

6.3.3 Mutagenesis

In 2013, Gomez-Pando and Eguiluz-de la Barra reported the mutation by gamma irradiation (gamma rays - Gy) of dry seeds of quinoa cv. Pasankalla using three different doses of 150, 250, and 350 Gy. In the first generation after mutation (M_1), the mildest mutation phenotypes included delayed seed germination, reduced seedling height, root length, and leaf development until 250 Gy treatment, while a dose of 350 Gy resulted in complete lethality. The spectrum of chlorophyll mutations was greatest for M_2 generation derived from 150 Gy treated plants, while the maximum frequency was reached from 250 Gy treated plants, with chlorine mutation being the predominant phenotype observed. Additional changes in the M_2 generation affected branch number, pedicel length, plant height, life cycle duration, stem and foliage color, and leaf morphology at both 150 and 250 Gy doses, with multiple mutations per plant recorded in M_2 plants derived from 250 Gy dose. The same spectrum of mutations was observed in the M_3 generation, along with a change in grain color that represents a valuable trait for industrial use. Mutant plants showing reduced life cycle represent a valuable germplasm in order to improve agronomic performance through a better management of flowering time and time to maturity of the plants to occur during an earlier season in the year that is less prone to adverse climatic conditions. Similarly, mutants with reduced plant height represent an interesting trait which could contribute to reduce plant lodging and therefore indirectly improve yield. Repeating the same approach, Pando and Deza (2017) selected eight gamma ray mutant lines that showed improved characteristics compared to the original cultivars, including higher yield and increased grain protein content.

In 2018, Imamura et al. performed ethyl methane sulfonate (EMS) mutagenesis on the quinoa variety CQ127 and subsequently identified two green hypocotyl mutant (ghy) lines in the M3 progeny, named ghy1 and ghy2, which both had no accumulation of betacyanin pigments. In order to identify the causative gene of the ghy mutant phenotype, the authors sequenced the DNA of bulks of wild-type and mutant M_3 plants, and then applied the MutMap + method (Fekih et al. 2013) which allows the identification of the causal mutations through the comparison of the frequencies of mutant and wild-type alleles at SNP variant loci. Results of the MutMap + method using the quinoa draft genome allowed the identification of Cq*CYP76AD1-1* as the causative gene of the ghy mutant phenotype. Further expression analysis indicated that Cq*CYP76AD1-1* was expressed in a light-dependent manner. Further functional characterization of the gene through transient expression in *Nicotiana benthamiana* leaves and loss of function associated with sequence polymorphism in the gene sequence suggested that Cq*CYP76AD1-1* is involved in betalain biosynthesis during the hypocotyl pigmentation process in quinoa.

Finally, Mestanza et al. (2018) created a mutant quinoa population with a frequency of one mutation every 203 kilobases through 2% EMS treatment of the quinoa Regalona–Baer variety in order to induce mutations in acetolactate synthase (AHAS) genes, which represent the enzyme target site of several herbicides. The screening of this population through the sequencing of tri-dimensional overlapping pools of M_2 mutant lines by next-generation sequencing (NGS) allowed to identify a mutation in the *AHAS2b* gene fragment that alters the amino acid composition (Val → Ile), but unfortunately this mutation did not result in herbicide resistance.

6.3.4 Transformation and Genome Editing

The capacity to transform plant species is an indispensable technique required to enable molecular analysis and gene function characterization. Yet, there are to our knowledge only few reports describing transformation of the genus *Chenopodium*, and no study reporting the regeneration of transformants into mature plants. Early attempts to develop a transformation method for quinoa using various strains of *Agrobacterium tumefaciens* and a common binary vector were not successful until the use of *A. tumefaciens* strain A281, a 'super-virulent' strain with a wide host range provided by the Ti plasmid pTiBo542. In 1990, Komari described the successful transformation of quinoa cells in a suspension culture using a combination of binary vectors and the fragment from the virulence region of pTiBo542, but attempts to regenerate plants from cultured cells or protoplasts were unsuccessful, despite extensive efforts. The author discussed the need for further improvements in techniques of tissue culture for quinoa, before we can generate transgenic plants.

Jung et al. (1992), used electroporation of quinoa protoplasts with viral RNAs for the study of gene expression, but again, no regeneration of the plants was performed.

In 2003, Solís et al. used sonication-assisted *Agrobacterium*-mediated transformation (SAAT)

(Trick and Finer 1997) to transform *Chenopodium rubrum*, a method which showed increased transformation efficiency in two-days seedling plants; however, no mature plant regeneration was attempted.

Though no transgenic quinoa plants have been reported, successful protocols for transformation and plant regeneration have been reported in close relatives such as *Amaranthus* despite being primarily described as a recalcitrant crop for tissue culture and transformation (Brenner et al. 2000; Joshi et al. 2018). These include a culture-based shoot regeneration protocol from stem internode explants, as well as a protocol of plant regeneration from callus tissue derived from mature embryo and epicotyl (Jofre-Garfias et al. 1997; Swain et al. 2010; Pal et al. 2013) Furthermore, Munusamy et al. (2013) have developed a tissue culture-independent genetic transformation system for Amaranths using mature inflorescence as an explant, which although showing a low transformation efficiency of 1.8%, represents a promising starting point for further improvement of transformation in amaranth as well as in quinoa.

6.4 Perspectives

As quinoa is gaining international importance, both in terms of demand and production, a better knowledge of its genome structure and function will become increasingly needed in order to support and fasten breeding efforts in various countries to make quinoa more productive and better adapted to its novel cultivation environments. Within this context, we identify four main genomics areas that must develop in order to meet the expectations of the breeding programs.

The first area to be developed is the production of genotype specific sequence resources for the development of molecular markers associated to agronomical important traits. Indeed, quinoa disposes of a huge phenotypic and genetic diversity for domestication and adaption to a wide range of environments that needs to be better characterized in order to identify the genomic loci associated with important agronomic traits as well as the allelic variation available in the natural quinoa diversity at these important loci. Currently, only low throughput genomic markers were developed and characterization of genetic and phenotypic diversity of quinoa was performed at the ecotype scale only. Moreover, genome sequences are available for only a small number of accessions which all belong to the most cultivated quinoa varieties, and therefore a large fraction of the beneficial allelic variation available in the quinoa diversity is likely undiscovered. In order to change this paradigm, high-throughput technologies such as whole genome re-sequencing or target genotyping (e.g., exome capture and sequencing, SNP array…) must be leveraged in order to enable a comprehensive genotyping of the quinoa diversity and the identification of loci/genes associated with important agronomic traits.

The second key area will be to build comparative genomics knowledge between quinoa and other crop plant genomes. Indeed, quinoa breeding and genomics are still at its early ages and genetic components of important traits such as flowering time, grain size, and plant architecture for example are still very poorly characterized. However, these traits have been extensively studied in older crops such as wheat, rice, and maize, as well as in other economically important quinoa relatives such as Amaranths, beet, and spinach. Therefore, quinoa breeding would largely benefit from a transfer of knowledge gained from older crops to identify and characterize such pathways in quinoa genome. An important component of this work is the definition of syntenic regions between quinoa and related species genomes which would enable to provide a better support for the identification of orthologous genes between quinoa and related species and refining the sequence homology search to sequences supported by a shared evolutionary relationship.

The third area of development is the need for a better characterization of the different genomics levels that might impact the expression of quinoa genome's function. With the recent access to a quinoa reference genome sequence and its annotation (Jarvis et al. 2017), it would be interesting to investigate different levels of genomics functionality such as transcriptomics, DNA methylation, and chromatin accessibility, as well as transposable elements mobility in the different tissues and developmental stages of quinoa in control conditions and in response to various biotic and abiotic stresses. Indeed, various studies in plants have shown that alternative splicing, for example, was shown to play an important role in the regulation of flowering time, environmental stress response, and resistance to biotic stress, among others (Egawa et al. 2006; Lee et al. 2013; Pose et al. 2013; Staiger and Brown 2013; Bennetzen and Wang 2014). On the other hand, transposable elements (TEs) have been shown to be key players in generating genomic and phenotypic novelty through a combination of chromosome rearrangements and altered gene regulation (Bennetzen and Wang 2014). However, still very little attention has been paid to the repetitive sequences of quinoa, accounting for most of the size ($\sim$65%) of its genome. Further studies are therefore needed to characterize the structural and functional organization of quinoa genome in order to better understand their potential implications for quinoa stress resilience and adaptation to various environments.

Finally, with the identification of genetic loci associated with phenotypic variation for traits of interest, we will need the development of forward genetics toolbox for the functional characterization of quinoa agronomical important genes such as transformation, gene editing, and plant regeneration protocols for quinoa. In the past decade, genome editing technology has revolutionized basic research and trait development in crop plants by modifying genomes in a rapid yet precise and predictable manner (Bortesi and Fischer 2015). Genome editing allows the direct transformation of favorable alleles or gene complexes into an elite genetic background, even if these allele combinations are not occurring in the population. Different methods have been used for editing plants genomes including Zinc finger nucleases (ZFNs), transcription activator-like effector nucleases (TALENs)-based genome editing, and the simpler and most widely used CRISPR/Cas9 system (Ma et al. 2016). In order to enable the use of these techniques for editing genes in quinoa, efforts must be applied to the development of efficient transformation and plant regeneration protocols in which success rate is still limited as quinoa is a recalcitrant plant. Advancements in transformation of quinoa will open opportunities to identify and improve causative genes under different agronomic traits related to domestication and yield, such as grain size, panicle shape, and plant architecture through genome editing in near future.

References

Adolf VI, Jacobsen S-E, Shabala S (2013) Salt tolerance mechanisms in quinoa (Chenopodium quinoa Willd.). Environ Exp Bot 92:43–54

Aellen P, Just T (1943) Key and synopsis of the American species of the genus Chenopodium L. Am Midland Nat 30:47–76

Balzotti MRB, Thornton JN, Maughan PJ, McClellan DA, Stevens MR, Jellen EN, Fairbanks DJ, Coleman CE (2008) Expression and evolutionary relationships of the *Chenopodium quinoa* 11S seed storage protein gene. Int J Plant Sci 169:281–291

Bennetzen JL, Wang H (2014) The contributions of transposable elements to the structure, function, and evolution of plant genomes. Annu Rev Plant Biol 65:505–530

Bhargava A, Shukla S, Ohri D (2006) Chenopodium quinoa—An Indian perspective. Ind Crop Prod 23:73–87

Bohm J, Messerer M, Muller HM, Scholz-Starke J, Gradogna A, Scherzer S, Maierhofer T, Bazihizina N, Zhang H, Stigloher C, Ache P, Al-Rasheid KAS, Mayer KFX, Shabala S, Carpaneto A, Haberer G, Zhu JK, Hedrich R (2018) Understanding the molecular basis of salt sequestration in epidermal bladder cells of Chenopodium quinoa. Curr Biol 28:3075+

Bonifacio A (1995) Interspecific and intergeneric hybridization in chenopod species. Brigham Young University. Department of Botany and Range Sciences

Bortesi L, Fischer R (2015) The CRISPR/Cas9 system for plant genome editing and beyond. Biotechnol Adv 33:41–52

Brenner D, Baltensperger D, Kulakow P, Lehmann J, Myers R, Slabbert M, Sleugh B (2000) Genetic resources and breeding of Amaranthus. Plant Breed Rev 19:227–285

Brinegar C, Goundan S (1993) Isolation and characterization of Chenopodin, the 11 s seed storage protein of Quinoa (Chenopodium-Quinoa). J Agric Food Chem 41:182–185

Cai XF, Jiao C, Sun HH, Wang XL, Xu CX, Fei ZJ, Wang QH (2017) The complete mitochondrial genome sequence of spinach, Spinacia oleracea L. Mitochondrial DNA B 2:339–340

Cháb D, Kolář J, Olson MS, Štorchová H (2008) Two flowering locus T (FT) homologs in Chenopodium rubrum differ in expression patterns. Planta 228:929

Chaney L, Mangelson R, Ramaraj T, Jellen EN, Maughan PJ (2016) The complete chloroplast genome sequences for four Amaranthus species (Amaranthaceae). Appl Plant Sci 4

Christensen S, Pratt DB, Pratt C, Nelson P, Stevens M, Jellen EN, Coleman CE, Fairbanks DJ, Bonifacio A, Maughan PJ (2007) Assessment of genetic diversity in the USDA and CIP-FAO international nursery collections of quinoa (Chenopodium quinoa Willd.) using microsatellite markers. Plant Genet Resour 5:82–95

Clegg MT, Zurawski G (1992) Chloroplast DNA and the study of plant phylogeny: present status and future prospects. In: Molecular systematics of plants. Springer, pp 1–13

Coles ND, Coleman CE, Christensen SA, Jellen EN, Stevens MR, Bonifacio A, Rojas-Beltran JA, Fairbanks DJ, Maughan PJ (2005) Development and use of an expressed sequenced tag library in quinoa (Chenopodium quinoa Willd.) for the discovery of single nucleotide polymorphisms. Plant Sci 168:439–447

Dally N, Xiao K, Holtgräwe D, Jung C (2014) The B2 flowering time locus of beet encodes a zinc finger transcription factor. Proc Natl Acad Sci 111:10365–10370

Danial D, Parlevliet J, Almekinders C, Thiele G (2007) Farmers' participation and breeding for durable disease resistance in the Andean region. Euphytica 153:385–396

del Castillo C, Winkel T, Mahy G, Bizoux J-P (2007) Genetic structure of quinoa (Chenopodium quinoa Willd.) from the Bolivian altiplano as revealed by RAPD markers. Genet Resour Crop Ev 54:897–905

Devi RJ, Chrungoo NK (2017) Evolutionary divergence in Chenopodium and validation of SNPs in chloroplast rbcL and matk genes by allele-specific PCR for development of Chenopodium quinoa-specific markers. Crop J 5:32–42

Dillehay TD, Quivira MP, Bonzani R, Silva C, Wallner J, Le Quesne C (2007) Cultivated wetlands and emerging complexity in south-central Chile and long distance effects of climate change. Antiquity 81:949–960

Dodsworth S, Leitch AR, Leitch IJ (2015) Genome size diversity in angiosperms and its influence on gene space. Curr Opin Genet Dev 35:73–78

Dohm JC, Minoche AE, Holtgrawe D, Capella-Gutierrez S, Zakrzewski F, Tafer H, Rupp O, Sorensen T, Stracke R, Reinhardt R, Goesmann A, Kraft T, Schulz B, Stadler PF, Schmidt T, Gabaldon T, Lehrach H, Weisshaar B, Himmelbauer H (2014) The genome of the recently domesticated crop plant sugar beet (Beta vulgaris). Nature 505:546+

Egawa C, Kobayashi F, Ishibashi M, Nakamura T, Nakamura C, Takumi S (2006) Differential regulation of transcript accumulation and alternative splicing of a DREB2 homolog under abiotic stress conditions in common wheat. Genes Genet Syst 81:77–91

Epstein E, Bloom A (2005) Mineral nutrition of plants: principles and perspectives, 2nd edn. Sinauer Association Inc., Sunderland, UK

Fairbanks DJ, Burgener KW, Robison LR, Andersen WR, Ballon E (1990) Electrophoretic characterization of Quinoa seed proteins. Plant Breeding 104:190–195

Fekih R, Takagi H, Tamiru M, Abe A, Natsume S, Yaegashi H, Sharma S, Sharma S, Kanzaki H, Matsumura H, Saitoh H, Mitsuoka C, Utsushi H, Uemura A, Kanzaki E, Kosugi S, Yoshida K, Cano L, Kamoun S, Terauchi R (2013) MutMap plus: genetic mapping and mutant identification without crossing in rice. Plos One 8

Fiallos-Jurado J, Pollier J, Moses T, Arendt P, Barriga-Medina N, Morilloi E, Arahana V, Torres MD, Goossens A, Leon-Reyes A (2016) Saponin determination, expression analysis and functional characterization of saponin biosynthetic genes in Chenopodium quinoa leaves. Plant Sci 250:188–197

Fuentes FF, Bazile D, Bhargava A, Martinez EA (2012) Implications of farmers' seed exchanges for on-farm conservation of quinoa, as revealed by its genetic diversity in Chile. J Agric Sci 150:702–716

Fuentes FF, Martinez EA, Hinrichsen PV, Jellen EN, Maughan PJ (2009) Assessment of genetic diversity patterns in Chilean quinoa (Chenopodium quinoa Willd.) germplasm using multiplex fluorescent microsatellite markers. Conserv Genet 10:369–377

Gandarillas H (1968) Caracteres botánicos más importantes para la clasificación de la quinua. Anales de la Primera convención de Quenopodiáceas quinoa–cañahua. Puno, Perú: Universidad Nacional Técnica del Altiplano, pp 41–49

Gill BS, Friebe B, Endo TR (1991) Standard karyotype and nomenclature system for description of chromosome bands and structural-aberrations in wheat (Triticum-Aestivum). Genome 34:830–839

Gomez-Pando LR, Eguiluz-de la Barra A (2013) Developing genetic variability of quinoa (Chenopodium quinoa Willd.) with gamma radiation for use in breeding programs. Am J Plant Sci 4:349

Hong SY, Cheon KS, Yoo KO, Lee HO, Cho KS, Suh JT, Kim SJ, Nam JH, Sohn HB, Kim YH (2017) Complete chloroplast genome sequences and comparative analysis of Chenopodium quinoa and C. album. Front Plant Sci 8

Imamura T, Takagi H, Miyazato A, Ohki S, Mizukoshi H, Mori M (2018) Isolation and characterization of the betalain biosynthesis gene involved in hypocotyl pigmentation of the allotetraploid Chenopodium quinoa. Biochem Bioph Res Co 496:280–286

Jacobsen SE (2003) The worldwide potential for quinoa (Chenopodium quinoa Willd.). Food Rev Int 19:167–177

Jarvis DE, Ho YS, Lightfoot DJ, Schmockel SM, Li B, Borm TJA, Ohyanagi H, Mineta K, Michell CT, Saber N, Kharbatia NM, Rupper RR, Sharp AR, Dally N, Boughton BA, Woo YH, Gao G, Schijlen EGWM, Guo XJ, Momin AA, Negrao S, Al-Babili S, Gehring C, Roessner U, Jung C, Murphy K, Arold ST, Gojobori T, van der Linden CG, van Loo EN, Jellen EN, Maughan PJ, Tester M (2017) The genome of Chenopodium quinoa (vol 542, pg 307, 2017). Nature 545:510–510

Jarvis DE, Kopp OR, Jellen EN, Mallory MA, Pattee J, Bonifacio A, Coleman CE, Stevens MR, Fairbanks DJ, Maughan PJ (2008) Simple sequence repeat marker development and genetic mapping in quinoa (Chenopodium quinoa Willd.). J Genet 87:39–51

Jellen EN, Kolano BA, Sederberg MC, Bonifacio A, Maughan PJ (2011) Chenopodium. In: Wild crop relatives: genomic and breeding resources. Springer, pp 35–61

Jofre-Garfias A, Villegas-Sepúlveda N, Cabrera-Ponce J, Adame-Alvarez R, Herrera-Estrella L, Simpson J (1997) Agrobacterium-mediated transformation of Amaranthus hypochondriacus: light-and tissue-specific expression of a pea chlorophyll a/b-binding protein promoter. Plant Cell Rep 16:847–852

Joshi D, Sood S, Hosahatti R, Kant L, Pattanayak A, Kumar A, Yadav D, Stetter MG (2018) From zero to hero: the past, present and future of grain amaranth breeding. Theor Appl Genet 131:1807–1823

Jung JL, Bouzoubaa S, Gilmer D, Hahne G (1992) Visualization of transgene expression at the single protoplast level. Plant Cell Rep 11:346–350

Karcz J, Kolano B, Maluszynska J (2005) SEM studies on fruit and seed of some Chenopodium L. species (Chenopodiaceae). Acta Biol Cracov Bot 47:61–61

Kolano B, Siwinska D, Pando LG, Szymanowska-Pulka J, Maluszynska J (2012a) Genome size variation in Chenopodium quinoa (Chenopodiaceae). Plant Syst Evol 298:251–255

Kolano B, Tomczak H, Molewska R, Jellen EN, Maluszynska J (2012b) Distribution of 5S and 35S rRNA gene sites in 34 Chenopodium species (Amaranthaceae). Bot J Linn Soc 170:220–231

Komari T (1990) Transformation of cultured-cells of Chenopodium-Quinoa by binary vectors that carry a fragment of DNA from the virulence region of Ptibo542. Plant Cell Rep 9:303–306

Kubo T, Nishizawa S, Sugawara A, Itchoda N, Estiati A, Mikami T (2000) The complete nucleotide sequence of the mitochondrial genome of sugar beet (Beta vulgaris L.) reveals a novel gene for tRNA(Cys) (GCA). Nucleic Acids Res 28:2571–2576

Lee JH, Ryu HS, Chung KS, Pose D, Kim S, Schmid M, Ahn JH (2013) Regulation of temperature-responsive flowering by MADS-box transcription factor repressors. Science 342:628–632

Li H, Cao H, Cai YF, Wang JH, Qu SP, Huang XQ (2014) The complete chloroplast genome sequence of sugar beet (Beta vulgaris ssp vulgaris). Mitochondr DNA 25:209–211

Liu JX, Wang RM, Liu WY, Zhang HL, Guo YD, Wen RY (2018) Genome-wide characterization of heat-shock protein 70s from Chenopodium quinoa and expression analyses of Cqhsp70s in response to drought stress. Genes-Basel 9

Ma XL, Zhu QL, Chen YL, Liu YG (2016) CRISPR/Cas9 platforms for genome editing in plants: developments and applications. Mol Plant 9:961–974

Mason SL, Stevens MR, Jellen EN, Bonifacio A, Fairbanks DJ, Coleman CE, McCarty RR, Rasmussen AG, Maughan PJ (2005) Development and use of microsatellite markers for germplasm characterization in quinoa (Chenopodium quinoa Willd.). Crop Sci 45:1618–1630

Maughan PJ, Bonifacio A, Coleman CE, Jellen EN, Stevens MR, Fairbanks DJ (2007) Quinoa (Chenopodium quinoa). In: Pulses, sugar and tuber crops. Springer, pp 147–158

Maughan PJ, Bonifacio A, Jellen EN, Stevens MR, Coleman CE, Ricks M, Mason SL, Jarvis DE, Gardunia BW, Fairbanks DJ (2004) A genetic linkage map of quinoa (Chenopodium quinoa) based on AFLP, RAPD, and SSR markers. Theor Appl Genet 109:1188–1195

Maughan PJ, Chaney L, Lightfoot DJ, Cox BJ, Tester M, Jellen EN, Jarvis DE (2019) Mitochondrial and chloroplast genomes provide insights into the evolutionary origins of quinoa (Chenopodium quinoa Willd.). Sci Rep-Uk 9

Maughan PJ, Kolano BA, Maluszynska J, Coles ND, Bonifacio A, Rojas J, Coleman CE, Stevens MR, Fairbanks DJ, Parkinson SE, Jellen EN (2006) Molecular and cytological characterization of ribosomal RNA genes in Chenopodium quinoa and Chenopodium berlandieri. Genome 49:825–839

Maughan PJ, Smith SM, Rojas-Beltran JA, Elzinga D, Raney JA, Jellen EN, Bonifacio A, Udall JA, Fairbanks DJ (2012) Single nucleotide polymorphism identification, characterization, and linkage mapping in Quinoa. Plant Genome-Us 5:114–125

Maughan PJ, Turner TB, Coleman CE, Elzinga DB, Jellen EN, Morales JA, Udall JA, Fairbanks DJ, Bonifacio A (2009) Characterization of Salt Overly Sensitive 1 (SOS 1) gene homoeologs in quinoa (Chenopodium quinoa Wilid.). Genome 52:647–657

McElhinny E, Peralta E, Mazon N, Danial DL, Thiele G, Lindhout P (2007) Aspects of participatory plant breeding for quinoa in imarginal areas of Ecuador. Euphytica 153:373–384

Mestanza C, Riegel R, Silva H, Vasquez SC (2015) Characterization of the acetohydroxyacid synthase multigene family in the tetraploide plant Chenopodium quinoa. Electron J Biotechnol 18:393–398

Mestanza C, Riegel R, Vásquez SC, Veliz D, Cruz-Rosero N, Canchignia H, Silva H (2018) Discovery of mutations in Chenopodium quinoa Willd through EMS mutagenesis and mutation screening using pre-selection phenotypic data and next-generation sequencing. J Agric Sci 156:1196–1204

Morales A, Zurita-Silva A, Herman JM, Silva H (2017) Transcriptional responses of Chilean Quinoa (Chenopodium quinoa Willd.) under water deficit conditions uncovers ABA-independent expression patterns. Front Plant Sci 8

Morales AJ, Bajgain P, Garver Z, Maughan PJ, Udall JA (2011) Physiological responses of Chenopodium quinoa to salt stress. Int J Plant Physiol Biochem 3:219–232

Morton MJ, Awlia M, Al-Tamimi N, Saade S, Pailles Y, Negrão S, Tester M (2019) Salt stress under the scalpel–dissecting the genetics of salt tolerance. Plant J 97:148–163

Mujica A, Jacobsen S-E (2006) La quinua (Chenopodium quinoa Willd.) y sus parientes silvestres. Bótanica económica de los Andes Centrales 32:449–457

Munusamy U, Abdullah SNA, Aziz MA, Khazaai H (2013) Female reproductive system of Amaranthus as the target for Agrobacterium-mediated transformation. Adv Biosci Biotechnol 4:188

Nakamura S, Ikegami A, Mizuno M, Yagi F, Nomura K (2004) The expression profile of lectin differs from that of seed storage proteins in Castanea crenata trees. Biosci Biotech Biochem 68:1698–1705

Orsini F, Accorsi M, Gianquinto G, Dinelli G, Antognoni F, Carrasco KBR, Martinez EA, Alnayef M, Marotti I, Bosi S, Biondi S (2011) Beyond the ionic and osmotic response to salinity in Chenopodium quinoa: functional elements of successful halophytism. Funct Plant Biol 38:818–831

Pal A, Swain SS, Das AB, Mukherjee AK, Chand PK (2013) Stable germ line transformation of a leafy vegetable crop amaranth (Amaranthus tricolor L.) mediated by Agrobacterium tumefaciens. Vitro Cell Dev Biol-Plant 49:114–128

Palomino G, Hernandez LT, Torres ED (2008) Nuclear genome size and chromosome analysis in Chenopodium quinoa and C-berlandieri subsp nuttalliae. Euphytica 164:221–230

Palomino G, Segura M, Bye R, Mercado P (1990) Cytogenetic distinction between Teloxys and Chenopodium (Chenopodiaceae). Southwest Nat 35:351–353

Pando G, Deza P (2017) Development of Advanced Mutant Lines of Native Grains through Radiation-induced Mutagenesis in Peru. Horticult Int J 1:15–19

Pin PA, Benlloch R, Bonnet D, Wremerth-Weich E, Kraft T, Gielen JJL, Nilsson O (2010) An antagonistic pair of FT homologs mediates the control of flowering time in sugar beet. Science 330:1397–1400

Pin PA, Nilsson O (2012) The multifaceted roles of FLOWERING LOCUS T in plant development. Plant, Cell Environ 35:1742–1755

Pose D, Verhage L, Ott F, Yant L, Mathieu J, Angenent GC, Immink RG, Schmid M (2013) Temperature-dependent regulation of flowering by antagonistic FLM variants. Nature 503:414–417

Pulvento C, Riccardi M, Lavini A, Iafelice G, Marconi E, d'Andria R (2012) Yield and quality characteristics of quinoa grown in open field under different saline and non-saline irrigation regimes. J Agron Crop Sci 198:254–263

Qadir M, Quillérou E, Nangia V, Murtaza G, Singh M, Thomas RJ, Drechsel P, Noble AD (2014) Economics of salt-induced land degradation and restoration. In: Natural resources forum. Wiley Online Library, pp 282–295

Rabah SO, Lee C, Hajrah NH, Makki RM, Alharby HF, Alhebshi AM, Sabir JSM, Jansen RK, Ruhlman TA (2017) Plastome sequencing of ten nonmodel crop species uncovers a large insertion of mitochondrial DNA in cashew. Plant Genome-Us 10

Rana T, Narzary D, Ohri D (2010) Genetic diversity and relationships among some wild and cultivated species of Chenopodium L. (Amaranthaceae) using RAPD and DAMD methods. Curr Sci:840–846

Raney JA (2012) Transcriptome analysis of drought induced stress in Chenopodium quinoa

Renny-Byfield S, Chester M, Kovarik A, Le Comber SC, Grandbastien MA, Deloger M, Nichols RA, Macas J, Novak P, Chase MW, Leitch AR (2011) Next generation sequencing reveals genome downsizing in allotetraploid Nicotiana tabacum, predominantly through the elimination of paternally derived repetitive DNAs. Mol Biol Evol 28:2843–2854

Reynolds DJ (2009) Genetic dissection of triterpenoid saponin production in Chenopodium quinoa using microarray analysis

Ruas PM, Bonifacio A, Ruas CF, Fairbanks DJ, Andersen WR (1999) Genetic relationship among 19 accessions of six species of Chenopodium L., by random amplified polymorphic DNA fragments (RAPD). Euphytica 105:25–32

Ruiz K, Biondi S, Martínez E, Orsini F, Antognoni F, Jacobsen S-E (2016) Quinoa–a model crop for understanding salt-tolerance mechanisms in halophytes. Plant Biosyst-Int J Deal Aspects Plant Biol 150:357–371

Ruiz KB, Biondi S, Oses R, Acuña-Rodríguez IS, Antognoni F, Martinez-Mosqueira EA, Coulibaly A, Canahua-Murillo A, Pinto M, Zurita-Silva A, Bazile D, Jacobsen S-E, Molina-Montenegro MA (2014) Quinoa biodiversity and sustainability for food security under climate change. A review. Agron Sustain Dev 34:349–359

Ruiz-Carrasco K, Antognoni F, Coulibaly AK, Lizardi S, Covarrubias A, Martinez EA, Molina-Montenegro MA, Biondi S, Zurita-Silva A (2011) Variation in salinity tolerance of four lowland genotypes of quinoa (Chenopodium quinoa Willd.) as assessed by growth, physiological traits, and sodium transporter gene expression. Plant Physiol Biochem 49:1333–1341

Salazar J, Torres MD, Gutierrez B, Torres AF (2019) Molecular characterization of Ecuadorian quinoa (Chenopodium quinoa Willd.) diversity: implications for conservation and breeding. Euphytica 215

Schmitz-Linneweber C, Maier RM, Alcaraz JP, Cottet A, Herrmann RG, Mache R (2001) The plastid chromosome of spinach (Spinacia oleracea): complete nucleotide sequence and gene organization. Plant Mol Biol 45:307–315

Schmockel SM, Lightfoot DJ, Razali R, Tester M, Jarvis DE (2017) Identification of putative transmembrane proteins involved in salinity tolerance in Chenopodium quinoa by integrating physiological data, RNAseq, and SNP analyses. Front Plant Sci 8

Shabala S, Hariadi Y, Jacobsen SE (2013) Genotypic difference in salinity tolerance in quinoa is determined by differential control of xylem Na + loading and stomatal density. J Plant Physiol 170:906–914

Simmonds NW (1971) Breeding System of Chenopodium-Quinoa. 1. Male sterility. Heredity 27:73–000

Solís JF, Mlejnek P, Studená K, Procházka S (2003) Application of sonication-assisted Agrobacterium-mediated transformation in Chenopodium rubrum L. Plant Soil Environ 49:255–260

Staiger D, Brown JWS (2013) Alternative splicing at the intersection of biological timing, development, and stress responses. Plant Cell 25:3640–3656

Stevens MR, Coleman CE, Parkinson SE, Maughan PJ, Zhang HB, Balzotti MR, Kooyman DL, Arumuganathan K, Bonifacio A, Fairbanks DJ, Jellen EN, Stevens JJ (2006) Construction of a quinoa (Chenopodium quinoa Willd.) BAC library and its use in identifying genes encoding seed storage proteins. Theor Appl Genet 112:1593–1600

Štorchová H, Drabesova J, Chab D, Kolar J, Jellen EN (2015) The introns in FLOWERING LOCUS T-LIKE (FTL) genes are useful markers for tracking paternity in tetraploid Chenopodium quinoa Willd. Genet Resour Crop Evol 62:913–925

Swain SS, Sahu L, Barik DP, Chand PK (2010) Agrobacterium × plant factors influencing transformation of 'Joseph's coat'(Amaranthus tricolor L.). Sci Hortic-Amsterdam 125:461–468

Tartara SMC, Manifesto MM, Bramardi SJ, Bertero HD (2012) Genetic structure in cultivated quinoa (Chenopodium quinoa Willd.), a reflection of landscape structure in Northwest Argentina. Conserv Genet 13:1027–1038

Trick HN, Finer JJ (1997) SAAT: sonication-assisted Agrobacterium-mediated transformation. Transgenic Res 6:329–336

Vega-Galvez A, Miranda M, Vergara J, Uribe E, Puente L, Martinez EA (2010) Nutrition facts and functional potential of quinoa (Chenopodium quinoa willd.), an ancient Andean grain: a review. J Sci Food Agr 90:2541–2547

Vidueiros SM, Curti RN, Dyner LM, Binaghi MJ, Peterson G, Bertero HD, Pallaro AN (2015) Diversity and interrelationships in nutritional traits in cultivated quinoa (Chenopodium quinoa Willd.) from Northwest Argentina. J Cereal Sci 62:87–93

Wang KY, Li L, Li SK, Sun HH, Zhao MZ, Zhang MP, Wang Y (2017) Characterization of the complete chloroplast genome of Chenopodium quinoa Willd. Mitochondr DNA B 2:812–813

Ward SM (2000) Allotetraploid segregation for single-gene morphological characters in quinoa (Chenopodium quinoa Willd.). Euphytica 116:11–16

Wilson H, Manhart J (1993) Crop-weed gene flow - Chenopodium-Quinoa Willd and C-Berlandieri Moq. Theor Appl Genet 86:642–648

Wilson HD (1981) Genetic-variation among south-american populations of tetraploid Chenopodium-Sect Chenopodium-Subsect Cellulata. Syst Bot 6:380–398

Wilson HD (1988a) Allozyme variation and morphological relationships of Chenopodium hircinum (sl). Syst Bot:215–228

Wilson HD (1988b) Quinua biosystematics I: domesticated populations. Econ Bot 42:461–477

Wilson HD (1988c) Quinua biosystematics II: free-living populations. Econ Bot 42:478–494

Wilson HD (1990) Quinua and relatives (Chenopodium Sect Chenopodium Subsect Cellulata). Econ Bot 44:92–110

Xu C, Jiao C, Sun HH, Cai XF, Wang XL, Ge CH, Zheng Y, Liu WL, Sun XP, Xu YM, Deng J, Zhang ZH, Huang SW, Dai SJ, Mou BQ, Wang QX, Fei ZJ, Wang QH (2017) Draft genome of spinach and transcriptome diversity of 120 Spinacia accessions. Nat Commun 8

Yasui Y, Hirakawa H, Oikawa T, Toyoshima M, Matsuzaki C, Ueno M, Mizuno N, Nagatoshi Y, Imamura T, Miyago M, Tanaka K, Mise K, Tanaka T, Mizukoshi H, Mori M, Fujita Y (2016) Draft genome sequence of an inbred line of Chenopodium quinoa, an allotetraploid crop with great environmental adaptability and outstanding nutritional properties. DNA Res 23:535–546

Zhang TF, Gu MF, Liu YH, Lv YD, Zhou L, Lu HY, Liang SQ, Bao HB, Zhao H (2017) Development of novel InDel markers and genetic diversity in Chenopodium quinoa through whole-genome re-sequencing. BMC Genom 18

Zou CS, Chen AJ, Xiao LH, Muller HM, Ache P, Haberer G, Zhang ML, Jia W, Deng P, Huang R, Lang D, Li F, Zhan DL, Wu XY, Zhang H, Bohm J, Liu RY, Shabala S, Hedrich R, Zhu JK, Zhang H (2017) A high-quality genome assembly of quinoa provides insights into the molecular basis of salt bladder-based salinity tolerance and the exceptional nutritional value. Cell Res 27:1327–1340

Zurita-Silva A, Fuentes F, Zamora P, Jacobsen SE, Schwember AR (2014) Breeding quinoa (Chenopodium quinoa Willd.): potential and perspectives. Mol Breed 34:13–30

7 Quinoa Diversity and Its Implications for Breeding

Katharina B. Böndel and Karl J. Schmid

Abstract

Genetic and phenotypic diversity constitutes the raw material for natural selection and artificial selection by breeders. The diversity observed in many crop species reflects adaptation to cultivation in different regions or selection for different end uses. Therefore, evolutionary and population genetic approaches are highly useful to characterize trait variation and select suitable genotypes for breeding programs. Quinoa (*Chenopodium quinoa*) gained a lot of interest in the recent past due to its high nutritious value and stress tolerance. It has been originally cultivated in harsh climatic regions stretching from the high altitudes in Bolivia and Peru (altiplano) to the lowlands in Chile (e.g., Atacama Desert), where it is confronted with diverse extreme abiotic environmental conditions. It therefore is an ideal crop for using an evolutionary framework to characterize its diversity. We provide a review of current studies on quinoa diversity in different ranges of the cultivation range with a focus on abiotic stress tolerance and on other traits relevant for breeding, such as nutritional content or germination rates. We also outline how a better understanding of genotypic diversity and differences between the different ecotypes of quinoa can improve breeding programs, particularly during early stages when the selection of suitable parents with beneficial traits is critical. Although quinoa is currently a minor crop with limited funding resources available for its improvement, recent trends in genomic and phenotypic analysis will facilitate the characterization and utilization of the abundant diversity present in quinoa genetic resources and contribute to its development as a crop of global importance for future food security.

7.1 Why Does Diversity Matter in Breeding Populations?

In his seminal work "On the origin of species" (1859) Charles Darwin outlined the importance of intraspecific variation for evolution, which he introduced by the example of plant and animal breeding. Only if individuals show phenotypic variation, if this variation is heritable and if it causes fitness differences among individuals, evolution occurs. Advantageous phenotypes will have more offspring on average and increase in frequency. Likewise, disadvantageous phenotypes will have less offspring and eventually disappear. Darwin used this theory to explain the

K. B. Böndel · K. J. Schmid (✉)
Institute for Plant Breeding, Seed Sciences and Population Genetics, University of Hohenheim, Stuttgart, Germany
e-mail: Karl.Schmid@uni-hohenheim.de

K. B. Böndel
e-mail: Katharina.Boendel@uni-hohenheim.de

S. M. Schmöckel (ed.), *The Quinoa Genome*, Compendium of Plant Genomes,
https://doi.org/10.1007/978-3-030-65237-1_7

diversity of domesticated plants and animals resulting from artificial selection for cultivation in different climatic regions or various uses by humans. New variation may originate spontaneously as "sporting plants" and "assume […] a new and sometimes very different character" (Darwin 1859). Variation can also persist in populations for long periods of time and may eventually become advantageous if conditions change, which is referred to as adaptation from standing genetic variation (Barrett and Schluter 2007). Standing genetic variation may become advantageous when individuals move to new territories that differ from the home territory (Innan & Kim 2008) or when the environment or climate change rapidly. In contrast, adaptation from new mutations arising in the population is slower and it may take a long time before a new favorable mutation arises. In both wild and domesticated species, a high degree of genetic diversity provides an advantage under a rapid environmental change and strong selection due to natural or artificial selection. For example, disease resistance (R) genes in plants are among the most polymorphic genes in the genome (Friedman and Baker 2007; Monteiro and Nishimura 2018), because a high diversity increases the chance that a new pathogen invading a population is recognized by an allele segregating at an R locus to trigger the immune response and react against the pathogen. Self-incompatibility (SI) genes are also very diverse (Schierup and Vekemans 2008). They prevent inbreeding and counteract the loss of genetic diversity. High levels of diversity are frequently maintained by balancing selection in natural populations (Delph and Kelly 2014) or reflect differential adaptation in subpopulations. Darwin (1859) proposed that natural selection and artificial selection largely follow the same rules. In artificial selection, or breeding, individuals with a specific characteristic are selected from a larger pool of variable individuals. If selection occurs over multiple generations the selected phenotype is ultimately fixed in a population. Such an (admittedly simple) Darwinian view of plant breeding has experienced a renaissance recently and an evolutionary approach to describing the phenotypic and genetic diversity of crops is becoming common (Turner-Hissong et al. 2020).

7.2 Diversity and Population Structure in Quinoa

Quinoa is an ideal crop for applying an evolutionary approach to characterize and utilize phenotypic and genetic diversity for breeding purposes. Since its domestication several thousand years ago, quinoa is today cultivated predominantly in many South American countries but increasingly also worldwide (Bazile et al. 2016). Several thousand accessions of different geographic origins are listed in various germplasm banks (Christensen et al. 2007; FAO 2010), and several studies investigated the diversity and population structure of quinoa accessions to better understand similarities and differences between accessions.

7.2.1 Phenotypic Diversity

Early studies investigated diversity and population structure on the phenotypic level. Multiple accessions were cultivated in common garden experiments, phenotyped and analyzed with multivariate statistical methods. In 1983, Risi and Galway (1989a, b) investigated 294 accessions at Cambridge, UK. This Cambridge quinoa germplasm collection is expected to represent the existing quinoa diversity as it contains accessions from Columbia to southern Chile and from the entire altitudinal range. The collection was phenotyped for 19 continuous and discrete traits. Accessions differed significantly for all traits and showed correlations between some traits. Clustering methods identified seven groups, which largely reflect the different sources of the germplasm material (Risi and Galway 1989a), but further multivariate analysis (Risi and Galwey 1989b) did not reveal distinct groups. A lack of distinct phenotypic differentiation among genebank accessions may be promising for breeding quinoa because various combinations of traits that are advantageous for particular cultivation regions or end uses may already exist in genebank accessions. Rojas (2003) phenotyped a collection of over 1,500 accessions of which most originate from Bolivia and Peru, with additional accessions from Chile and Argentina, for 15 traits. All traits showed considerable variation between accessions. Cluster analysis revealed seven clusters that largely matched the geographic origin of the accessions.

7.2.2 Genetic Diversity

7.2.2.1 SNPs, InDels, Microsatellites, and Other Marker

Various studies investigated the genetic diversity and population structure in quinoa. Although they applied different methods and used different genetic markers and accession sets, they were consistent in their results. Quinoa accessions cluster genetically in two distinct groups: an Andean highland and a Chilean coastal group (e.g., Christensen et al. 2007; Maughan et al. 2012; Zhang et al. 2017). Furthermore, subgroups within these major groups (e.g., Christensen et al. 2007) or substructure within regions of the putative geographic range of quinoa domestication (e.g., Costa Tártara et al. 2012; Del Castillo et al. 2007) were found.

Christensen et al. (2007) genotyped 151 accessions with various origins predominantly from the USDA germplasm bank and the CIP-FAO international nursery collection with 36 microsatellites. They applied two methods to assess the population structure: a dendrogram constructed with the unweighted pair-group method with arithmetic averages (UPGMA) and Jaccard's similarity coefficients, and a principal component analysis (PCA). Both methods identified two major groups. A highland group with accessions primarily from Argentina, Bolivia, Ecuador, and Peru, and a coastal lowland group of all Chilean accessions in the sample. The dendrogram further revealed a southern highland group nested within the highland group with accessions primarily from the area around Lake Titicaca. Maughan et al. (2012) first generated a set of Single Nucleotide Polymorphisms (SNPs) and then genotyped a diversity panel of 113 accessions from various origins for these SNPs. A combination of three different methods to infer population structure revealed again two main groups: Andean highlands and Chilean coastal. Zhang et al. (2017) first developed a set of 85 InDel markers and then used them together with 62 simple sequence repeat (SSR) markers to genotype 129 quinoa accessions from various origins. A STRUCTURE analysis and a neighbor joining (NJ) tree clearly supported two genetic groups, Andean highland and Chilean coastal. A PCA also supported these two groups but further showed two subgroups within the highland group. In the context of the reference genome of quinoa, Jarvis et al. (2017) also constructed a phylogeny of 20 re-sequenced quinoa accessions. Their phylogeny also supports the previous findings of two distinct highland and coastal groups.

While these studies used germplasm from the whole cultivation range in South America, others focused on specific geographic regions. Del Castillo et al. (2007) investigated populations from the Bolivian altiplano. They used two populations from each of four defined ecological sectors: the Northern lake area, the Central altiplano, the Salar region, and the Inter-Andean valleys. A UPGMA dendrogram constructed from random amplified polymorphic DNA (RAPD) markers revealed two major clades. The first contains all samples from the northern and central altiplano. The second is split in two clades, one with the interandean valley samples and the other with the southern altiplano samples. Fuentes et al. (2009) focused on accessions from Chile. Using 20 microsatellite markers they investigated 59 accessions from the Chilean altiplano and coastal regions. Their combined analysis of UPGMA and PCA shows a distinct grouping in altiplano and coastal accessions. The coastal accessions further show more diversity and possibly even some further substructure. Costa Tártara et al. (2012) used microsatellite markers to investigate 35 accessions mainly from the Jujuy and Salta provinces in northwest Argentina. They found a remarkably high diversity and also some substructure with

Table 7.1 Selection of studies investigating genetic diversity in multiple *C. quinoa* accessions

Study	Sample size *C. quinoa*	Marker type	Marker count	Estimator for genetic diversity	Genetic diversity[a]
Christensen et al. (2007)	152	Microsatellite marker	36	Heterozygosity	0.75 (0.45–0.94)
Mason et al. (2005)	31	Microsatellite marker	208	Heterozygosity	0.57 (0.20–0.90)
Fuentes et al. (2012)	34	Microsatellite marker (SSR)	20	Heterozygosity	0.64 (0.12–0.87)
Jarvis et al. (2008)	22	SSR	216	Heterozygosity	0.57 (0.12–0.90)
Maughan et al. (2012)	113	SNPs	511	Minor allele frequency	0.28 (0.02–0.50)
Coles et al. (2005)	5	EST loci/SNPs	51	SNP frequency	0.0022 (1 in 462 bp)
Costa Tártara et al. (2012)	35 (Argentina only)	Microsatellite marker (SSR)	22	Heterozygosity	0.82 (0.62–0.93)
Del Castillo et al. (2007)	87 (8 populations, Bolivia only)	RAPD marker		Heterozygosity	Across populations 0.165 (0.104–0.215)
Fuentes et al. (2009)	59 (Chile only)	Microsatellite marker (SSR)	20	Heterozygosity	0.65 (0.30–0.90); highland 0.42 (0.14–0.95); coastal 0.51 (0.12–1.00)

[a]If not stated otherwise: mean and range of genetic diversity across marker

accessions clustering according to their origin. Using 15 SSR markers Salazar et al. (2019) investigated 84 accessions sampled across seven provinces in Ecuador. They also reported a high diversity in this sample and three distinct groups. But contrary to the other studies focusing on specific regions they could not find any correlation between group and origin of the accessions as each group contained accessions from multiple provinces. They rather concluded that these three groups represent ancestral lineages that were used throughout Ecuador.

Several studies also estimated the genetic diversity from marker data (for an overview see Table 7.1). Since they differ in types and numbers of markers as well as in the choice of accessions, these estimates may not be directly comparable. However all studies suggest that quinoa shows a fairly high level of genetic diversity. This suggests a high potential for quinoa breeding and highlights the necessity for a more thorough large-scale investigation of quinoa genetic diversity.

7.2.2.2 Genetic Diversity Beyond Markers

Genetic diversity can not only be measured from DNA sequence data, i.e., markers, but also as gene expression variation. It has been a long standing debate among evolutionary biologists whether mutations in coding or regulatory regions were more important for evolution (e.g., Carroll 2008; Hoekstra and Coyne 2007). Since changes in both coding and regulatory regions in genes important for domestication were identified in various crop species (reviewed in Doebley et al. 2006), the same may be true for breeding (Turner-Hissong et al. 2020). So far, only few studies investigated gene expression in quinoa and even fewer used different accessions. A study on salinity stress responses measured the expression of four salt response genes in three salt stressed accessions (Morales et al. 2011). There was no significant expression difference for any of the genes in leaves, but some differences were observed in roots between the Salares and valley ecotypes. Ruiz-Carrasco et al. (2011) also looked at expression of genes involved in the salt stress response in four Chilean coastal accessions after salt treatment of seedlings. They found differences in both root and shoot tissue between the accessions. Raney et al. (2014) sequenced transcriptomes of drought stressed plants. They used two accessions, one Inter-Andean valley and one Altiplano Salares ecotype, and reported 6,170 differentially expressed genes between them. Although there is so far only little evidence for expression variation between accessions and none on the whole genome scale due to the lack of studies on the respective topic, the existing genetic diversity may suggest that also gene expression differs between accessions.

The biggest scale on which individuals of the same species can vary genetically is their genome size. The main source of intraspecific genome size differences is likely due to transposable elements (Hawkins et al. 2008). Genome size differences were detected in quinoa. Kolano et al. (2012) used flow cytometry to measure the genome sizes of 20 accessions from different regions of the cultivation range. They found significant genome size differences between the accessions. The difference between the smallest (a Peruvian accession from the southern highlands) and the biggest (a Chilean accession from the lowlands) genome was 5.9%. A significant correlation between genome size and geographic origin was not detected. However, a tendency for larger genomes in the Chilean lowland accessions was noted, consistent with similar observations in maize and teosinte (Diez et al. 2013). Despite these substantial differences, it is not clear at this stage, however, whether variation in genome size affects phenotypic variation and therefore is relevant for breeding.

7.3 Diversity for Specific Phenotypic Traits

The clustering of quinoa accessions according to their geographic origin in multivariate analyses of phenotypic traits or genome-wide marker raises the question to the extent of phenotypic variation of individual traits and the roles of local adaptation to the cultivation area or artificial selection by humans in the distribution of variation among accessions. The following section presents a selection of studies, which focused on specific traits with substantial variation among quinoa accessions. A better understanding of how accessions differ for a specific trait and which accessions express a particular phenotype from a specific geographic range will ultimately help to decide upon which accessions to use for selected breeding or cultivation in specific environments.

7.3.1 Abiotic Stress Response

As a native species from extreme environments in South America, quinoa is tolerant to a variety of abiotic stresses like drought, salinity, or low temperatures (e.g., Jacobsen et al. 2003; Hinojosa et al. 2018). Therefore, stress responses to various types of stresses have been extensively investigated. Several studies used accessions originating from different areas of the cultivation range. Such an experimental set-up allows not only to understand the stress response itself but also to study differences between accessions. Possible associations between the degree of tolerance and the environment of origin can then be subject to further investigation.

7.3.1.1 Variation in Salinity Stress Response

Several experiments were conducted on salinity stress. Adolf et al. (2012) found significant differences in biomass and height after three weeks of salt treatment between 14 accessions from various origins. They further focused on a Bolivian accession from the altiplano and a Danish accession with Chilean coastal origin for a more thorough investigation of the physiological response to salinity stress. These two accessions showed differences in many tests, including net photosynthetic rate and chlorophyll content index. The analyses identified accessions with increased salt tolerance, but no correlation with origin was observed. Morales et al. (2011) found differences in compatible solute accumulation, recovery after salt stress and root gene expression between two altiplano Salares accessions and a Peruvian valley accession with the Peruvian valley accession performing worse, i.e., it was less tolerant. Ruiz-Carrasco et al. (2011) investigated the responses to salt stress at early developmental stages of four Chilean lowland accessions. Overall, the southernmost accession was most sensitive to salt stress and the performance of the four accessions followed their origin from South to North. This study suggests that differences in trait variation may not only be distributed between the major genetic groups of quinoa but also on a much smaller scale, i.e., within a geographic region. Ruiz et al. (2016) investigated the physiological response to salinity in three cultivars: one Salares and two coastal lowlands. Across experiments they found differences between the cultivars. Although the Salares ecotype did not show any better adaptation to salinity in form of height or yield, it performed better in physiological responses indicating tolerance mechanisms like sodium exclusion from leaves or longer roots. In a recent study, Kiani-Pouya et al. (2019) subjected 114 accessions to six weeks of salt stress. They found differences in fresh and dry weight between the accessions and also found a positive correlation with epidermal bladder cell density and size among the tolerant accessions.

7.3.1.2 Variation in Drought Stress Response

Raney et al. (2014) investigated drought stress responses in the valley ecotype variety Ingapirca from Ecuador and the Salares ecotype variety Ollague from Chile. They examined their phenotypic responses and found the Salares ecotype to be more tolerant to dry conditions. They also investigated the transcriptomic responses under drought stress and identified several genes significantly different expressed under stress between the two accessions including genes involved in the response to abiotic stresses. Morales et al. (2017) also reported a Salares ecotype the most tolerant to drought stress. Their study focused on three accessions from Chile, a Salares and two coastal ecotypes. Compared to the coastal ecotypes the Salares ecotype did not show a significant increase in electrolyte leakage or decrease for photosystem functionality.

7.3.1.3 Variation in Low Temperature Stress Response

Jacobsen et al. (2005) performed a set of experiments to investigate the effect of frost on the phenotype and solute content. They used one valley and four altiplano cultivars from Peru, but different subsets were used for each experiment. Across the different experiments the valley cultivar was affected stronger or earlier compared to the altiplano cultivars. For example, its death rate was twice as high and it accumulated less soluble sugars. A follow-up study focusing on the mechanisms behind quinoa's frost resistance confirmed that an altiplano cultivar was more frost resistant than a valley cultivar (Jacobsen et al. 2007). Bois et al. (2006) investigated the phenotypic responses of ten accessions in several experiments. They also used subsets of accessions for some of the experiments. Altogether they found differences between the accessions but could not clearly relate them to origin or ecotype.

7.3.2 Biotic Stress Response

Not only abiotic but also biotic stresses can affect plant growth and survival. It is therefore of great importance for plant breeding to investigate plant responses to pathogens and to identify resistant or less susceptible genotypes for breeding. Downy mildew (*Peronospora farinosa*) is the most damaging pathogen for quinoa leading to substantial yield losses (Danielsen et al. 2003). Several studies screened multiple accessions for their resistance to downy mildew in field trials in different countries (Kumar et al. 2006; Mhada et al. 2014; Khalifa and Thabet 2018). They discovered variation in resistance and could also identify resistant genotypes, but since the resistant genotypes originated from both Chilean coastal lowlands and Peru (Kumar et al. 2006; Khalifa and Thabet 2018) a clear correlation between geographic origin and resistance does not seem to exist. Ochoa et al. (1999) investigated the interaction of 60 quinoa accessions and 24 downy mildew isolates from Ecuador. They identified different virulence groups within downy mildew and resistance factors within quinoa, which allowed them to classify type and extent of disease resistance. They identified accessions resistant to some isolates and found that no accession was resistant to all isolates. This study shows that not only quinoa but also pathogens are highly variable and a resistant accession likely is not resistant against all strains of a pathogen.

7.3.3 Germination

Germination experiments including multiple accessions at different temperatures were conducted. Bois et al. (2006) investigated ten accessions at temperatures ranging von 2–20 °C. The accessions varied for the time until 50% germination was reached and this variation increased with lower temperatures. Across temperatures the cultivars largely followed the same ranking from fastest to slowest with the Surumi variety always being the fastest. Although accessions from different origins were used, no clear correlation with origin was observed. In a similar experiment conducted by Gonzalez et al. (2017), where germination of ten different genotypes at nine different temperatures between 8 and 50 °C was investigated, total germination varied between the accessions and ranged between 72.5 and 90.8%. They also noticed differences in seed abortion and lack of germination among accessions.

7.3.4 Saponins

Saponins are secondary metabolites found in the seed coat of quinoa (Kuljanabhagavad et al. 2008). Since they have harmful properties they need to be removed before the seeds can be consumed. On the other hand, saponins are useful as biopesticides or for therapeutic purposes (Ruiz et al. 2017). Various studies measured saponin content (De Santis et al. 2016; Miranda et al. 2012) and composition (Ruiz et al. 2017) in multiple accessions. They found differences between the accessions but no correlation with the region of origin suggesting either multiple independent origins of saponin-free genotypes or seed exchange and crosses between accessions across the cultivation range. Recently, a possible candidate variant in a basic helix-loop-helix (bHLH) transcription factor possibly responsible for the sweet saponin-free genotype was identified (Jarvis et al. 2017). A more thorough investigation of multiple accessions will show whether this variant is the causal mutation and where it originated.

7.3.5 Nutritional Aspects

Quinoa seeds are highly nutritious and have antioxidant properties. They are further characterized by high protein and vitamin contents and also contain important minerals (Vega-Gálvez et al. 2010). It is therefore of interest to investigate whether accessions differ in their nutritional quality and to investigate the genetic basis of these differences. In a field experiment in India 18 accessions were screened for mineral content in leaves (Bhargava et al. 2008). These accessions differed for all minerals investigated. Among these, sodium (15 fold), chromium (18 fold), cadmium (13 fold) showed the greatest variation. Differences in amino acid composition and content were found between ten accessions grown at two different sites in South America (Gonzalez et al. 2012). Most amino acids varied around 1.5 to twofold and histidine varied almost two to threefold. A field trial conducted in Southwestern Germany also revealed differences in amino acid composition and content between four accessions from different origins (Präger et al. 2018).

7.4 Implications for Breeding

The available studies of genetic and phenotypic diversity demonstrate the high potential of plant breeding to improve quinoa varieties. Multiple breeding programs were established in the native countries of quinoa cultivation, e.g., Bolivia, Peru, Ecuador, Chile, and Argentina and, more recently, in the United States, China, and Europe. The history and current state of quinoa breeding as well as the breeding methods used were recently described in several review articles (Zurita-Silva et al. 2014; Gomez-Pando 2015; Murphy et al. 2018; Gomez-Pando et al. 2019). Due to the increasing demand for quinoa, its improvement by classical breeding is gaining momentum and expected to lead to multiple new and improved varieties. These efforts will benefit strongly from multiple resources available for quinoa. They include in particular high-quality genome sequences of two different genotypes (Jarvis et al. 2017; Zou et al. 2017), re-sequenced genomes and gene expression data, all of which facilitate genetic mapping and genome-enabled breeding. High throughput and precision phenotyping methods specifically developed for quinoa or other crops, and new low-cost genotyping technologies will provide strong benefits for minor crops with limited funds for research and breeding. Examples of emerging technologies are speed breeding for rapid advancement of breeding populations (Ghosh et al. 2018; Jähne et al. 2020), low coverage whole genome sequencing and graph-based imputation for cost-efficient genotyping (Jensen et al. 2020), or image analysis for the rapid characterization of particular phenotypes (Tovar et al. 2020).

The increasingly low costs of sequencing and the ability for high throughput phenotyping of multiple traits will also have a major positive impact on research of minor crops, like quinoa for which available resources and commercial interest is currently limited. The reference genome sequences provide a framework for the subsequent analysis of genetic and phenotypic variation. This includes genome-wide association studies after genotyping or resequencing of multiple accessions from genebanks or of segregating populations produced by crossing genetically diverse parents as shown with the pseudocereal amaranth (Stetter et al. 2020). Although currently only few quinoa accessions have been re-sequenced, several projects are underway to re-sequence hundreds of quinoa accessions to use these data, which attempt to identify genes relevant for agronomic traits.

The improvement of minor crops like quinoa will also greatly benefit from exploiting evolutionary relationships to model plants like *Arabidopsis thaliana* or to major crops like wheat. In the latter groups, many genes that are involved in domestication or agronomically relevant traits have been characterized. An evolutionary comparison (i.e., a phylogenomic approach) of genes that control various phenotypes like seed size, seed shattering, plant height, or flowering time revealed that many of them are conserved in the quinoa genome (López-Marqués et al. 2020). These homologs are therefore potential targets for targeted mutagenesis using either genome editing or tilling. Although genome editing of quinoa using CRISPR/Cas9 or related methods has not yet been demonstrated, the individual steps such as transformation, callus culture, and somatic embryogenesis are possible with this crop. However, recent genome editing methods combine the application of plant hormones and genome editing vectors to induce stem meristems in somatic tissues (Maher et al. 2020). Such protocols allow to bypass tissue regeneration after transformation and overcome a frequent limitation in using genetic engineering in crops for which no robust tissue culture and regeneration protocols have been developed.

Although genome editing is considered a key technology for future plant improvement, one needs to consider that many traits are controlled by multiple genes. For this reason, genome editing may contribute to improved varieties if the major genes influencing quantitative traits can be identified, but approaches like marker assisted selection or genomic selection, which utilizes total genetic variation influencing a trait, will continue to be important for future crop improvement. Genomic prediction can be applied to different steps of a breeding program from the characterization of crossing parents in genetic resources to the advancement of best genotypes in later stages of breeding programs (Crossa et al. 2017). Since most quinoa varieties are still closely related to landrace varieties and not as advanced as elite breeding material in other crops, using genomic selection to predict the quantitative traits in genetically diverse genebank material (Yu et al. 2016) allows to maintain both an overall high diversity of in quinoa breeding populations and achieve rapid genetic gain.

The future availability of large numbers of sequenced quinoa genebank accessions will enable population genomics approaches to identify genomic regions that control adaptation to the diverse and extremely harsh environmental conditions to which quinoa became adapted since its domestication. Genome-wide analysis of selective sweeps ("Selective sweep mapping") identified candidate genes for environmental adaptation in multiple crop species during or after domestication (e.g., Hufford et al. 2020). Favorable genetic variation at adaptive genes can be subsequently introgressed into current cultivars. Combining genomic and phenotypic diversity together with environmental data for sites of origin identifies candidate genes for environmental adaptation without requiring large-scale multi location field trials. This approach helped to define a small core collection from tens of thousands soybean genebank accessions for a targeted production environment in Central Europe, which can then be used for subsequent phenotypic evaluation to identify new genetic resources for breeding programs (Haupt and Schmid 2020). A related approach is to use measures of genetic differentiation and genomic signatures of past selection to identify genomic regions, which contribute to adaptation to different ecogeographic zones (Stetter et al. 2020). Such an approach can be particularly powerful in quinoa because well characterized quinoa ecotypes are adapted to different environments (Murphy et al. 2019) and can be expected to exhibit genetic differentiation resulting from local adaptation.

The high level of phenotypic and genetic diversity of quinoa and its wild relatives provides the potential to develop this minor crop into a high-yielding and high-quality crop of global importance with the help of modern phenotyping and genome-enabled breeding technologies. In contrast to major crops, however, some important issues need to be resolved. First, the > 10,000 genebank accessions of quinoa are distributed over different genebanks (FAO 2010), which makes a joint and standardized analysis of diversity more difficult. Second, quinoa is currently not included in Annex 1 of the International Treaty on Plant Genetic Resources for Food and Agriculture. For this reason, the exchange of seeds and other plant materials is not possible using the Standard Material Transfer Agreement (SMTA) of the multilateral system that was established together with the International Treaty. Instead, individual mutual deliberations and agreements between countries according to the Nagoya Protocol are required, which makes an efficient characterization of quinoa more difficult. Finally, most quinoa breeding is mostly currently out by public institutions with limited funding. For this reason, continued support by public organizations and private enterprises will be required to efficiently characterize and utilize the diversity of the species using state of the art scientific methods for plant breeding. Only then quinoa will be able to fulfill the promise as a future crop of importance for food security in a rapidly changing world.

References

Adolf VI, Shabala S, Andersen MN, Razzaghi F, Jacobsen SE (2012) Varietal differences of quinoa's tolerance to saline conditions. Plant Soil 357:117–129

Barrett RDH, Schluter D (2007) Adaptation from standing genetic variation. Trends Ecol Evol 23:38–44

Bazile D, Jacobsen SE, Verniau A (2016) The global expansion of Quinoa: trends and limits. Front Plant Sci 7:622

Bhargava A, Shukla S, Srivastava J, Singh N, Ohri D (2008) Genetic diversity for mineral accumulation in the foliage of *Chenopodium* ssp. Sci Hortic 118:338–346

Bois JF, Winkel T, Lhomme JP, Raffaillac JP, Rocheteau A (2006) Response of some Andean cultivars of quinoa (Chenopodium quinoa Willd.) to temperature: effects on germination, phenology, growth and freezing. Europ J Agron 25:299–308

Carroll SB (2008) Evo-Devo and an expanding evolutionary synthesis: a genetic theory of morphological evolution. Cell 134:25–36

Christensen SA, Pratt DB, Pratt C, Nelson PT, Stevens MR, Jellen EN, Coleman CE, Fairbanks DJ, Bonifacio A, Maughan PJ (2007) Assessment of genetic diversity in the USDA and CIP-FAO international nursery collections of quinoa (*Chenopodium quinoa* Willd.) using microsatellite markers. Plant Genet Resour: Char Util 5(2):82–95

Coles ND, Coleman CE, Christensen SA, Jellen EN, Stevens MR, Bonifacio A, Rojas-Beltran JA, Fairbanks DJ, Maughan PJ (2005) Development and use of an expressed sequenced tag library in quinoa (*Chenopodium quinoa* Willd.) for the discovery of single nucleotide polymorphisms. Plant Sci 168:439–447

Costa Tártara SM, Manifesto MM, Bramardi SJ, Bertero HD (2012) Genetic structure in cultivated quinoa (*Chenopodium quinoa* Willd.), a reflection of landscape structure in Northwest Argentina. Conserv Genet 13:1027–1038

Crossa J, Pérez-Rodríguez P, Cuevas J, Montesinos-López O, Jarquín D, de los Campos G, Burgueño J, González-Camacho JM, Pérez-Elizalde S, Beyene Y et al (2017). Genomic selection in plant breeding: methods, models, and perspectives. Trends Plant Sci 22:961–975

Danielsen S, Bonifacio A, Ames T (2003) Diseases of quinoa (*Chenopodium quinoa*). Food Rev Int 19:43–59

Darwin C (1859) On the origin of species by means of natural selection, or, the preservation of favoured races in the struggle for life, 6th edn. J Murray, London, UK

De Santis G, Maddaluno C, D'Ambrosio T, Rascio A, Rinaldi M, Troisi J (2016) Characterisation of quinoa (*Chenopodium quinoa* Willd.) accessions for the saponin content in Mediterranean environment. Italian J Agron 11:277–281

Del Castillo C, Winkel T, Mahy G, Bizoux JP (2007) Genetic structure of quinoa (*Chenopodium quinoa* Willd.) from the Bolivian altiplano revealed by RAPD markers. Genet Resour Crop Evol 54:897–905

Delph LF, Kelly JK (2014) On the importance of balancing selection in plants. New Phytol 201:45–56

Díez CM, Gaut BS, Meca E, Scheinvar E, Montes-Hernandez S, Eguiarte LE, Tenaillon MI (2013) Genome size variation in wild and cultivated maize along altitudinal gradients. New Phytol 199:264–276

Doebley JF, Gaut BS, Smith BD (2006) The molecular genetics of crop domestication. Cell 127:1309–1321

FAO (2010) The second report on the State of World's plant genetic resources for food and agriculture. http://www.fao.org/3/i1500e/i1500e00.htm

Friedman AR, Baker BJ (2007) The evolution of resistance genes in multi-protein plant resistance systems. Curr Opin Genet Dev 17:493–499

Fuentes FF, Martinez EA, Hinrichsen PV, Jellen EN, Maughan PJ (2009) Assessment of genetic diversity patterns in Chilean quinoa (*Chenopodium quinoa* Willd.) germplasm using multiplex fluorescent microsatellite markers. Conserv Genet 10:369–377

Fuentes FF, Bazile D, Bhargava A, Martínez EA (2012) Implications of farmers' seed exchanges for on-farm conservation of quinoa, as revealed by its genetic diversity in Chile. J Agric Sci 150:702–716

Ghosh S, Watson A, Gonzalez-Navarro OE, Ramirez-Gonzalez RH, Yanes L, Mendoza-Suárez M, Simmonds J, Wells R, Rayner T, Green P et al (2018) Speed breeding in growth chambers and glasshouses for crop breeding and model plant research. Nat Protoc 13:2944–2963

Gomez-Pando L (2015) Quinoa breeding. In: Quinoa: Improvement and sustainable production, 1st edn. Wiley

Gomez-Pando LR, Aguilar-Castellanos E, Ibañez-Tremolada M (2019) Quinoa (*Chenopodium quinoa* Willd.) breeding. In: Al-Khayri JM, Jain SM, Johnson DV (eds) Advances in plant breeding strategies: cereals. Springer International Publishing, Cham, pp 259–316

González JA, Konishi Y, Bruno M, Valoy M, Prado FE (2012) Interrelationships among seed yield, total protein and amino acid composition of ten quinoa (*Chenopodium quinoa*) cultivars from two different agroecological regions. J Sci Food Agric 92:1222–1229

González JA, Buedo SE, Bruno M, Prado FE (2017) Quantifying cardinal temperatures in Quinoa (*Chenopodium quinoa*) cultivars. Lilloa 54(2):179–194

Haupt M, Schmid K (2020) Combining focused identification of germplasm and core collection strategies to identify genebank accessions for central European soybean breeding. Plant Cell Environ 43:1421–1436

Hawkins JS, Grover CE, Wendel JF (2008) Repeated big bangs and the expanding universe: Directionality in plant genome size variation. Plant Sci 174:557–562

Hinojosa L, González JA, Barrios-Masias FH, Fuentes F, Murphy KM (2018) Quinoa abiotic stress responses: a review. Plants 7:106

Hoekstra HE, Coyne JA (2007) The locus of evolution: Evo Devo and the genetics of adaptation. Evolution 61:995–1016

Hufford MB, Xu X, van Heerwaarden J, Pyhäjärvi T, Chia J-M, Cartwright RA, Elshire RJ, Glaubitz JC, Guill KE, Kaeppler SM et al (2012) Comparative population genomics of maize domestication and improvement. Nat Genet 44:808–811

Innan H, Kim Y (2008) Detecting local adaptation using the joint sampling of polymorphism data in the parental and derived populations. Genetics 179:1713–1720

Jacobsen SE, Mujica A, Jensen CR (2003) The resistance of Quinoa (*Chenopodium quinoa* Willd.) to adverse abiotic factors. Food Rev Int 19:99–109

Jacobsen SE, Monteros C, Christiansen JL, Bravo LA, Corcuera LJ, Mujica A (2005) Plant responses of quinoa (*Chenopodium quinoa* Willd.) to frost at various phenological stages. Europ J Agron 22:131–139

Jacobsen SE, Monteros C, Corcuera LJ, Bravo LA, Christiansen JL, Mujica A (2007) Frost resistance mechanisms in quinoa (*Chenopodium quinoa* Willd.). Europ J Agron 26:471–475

Jähne F, Hahn V, Würschum T, Leiser WL (2020) Speed breeding short-day crops by LED-controlled light schemes. Theor Appl Genet. https://doi.org/10.1007/s00122-020-03601-4

Jarvis DE, Kopp OR, Jellen EN, Mallory MA, Pettee J, Bonifacio A, Coleman CE, Stevens MR, Fairbanks DJ, Maughan PJ (2008) Simple sequence repeat marker development and genetic mapping in quinoa (*Chenopodium quinoa* Willd.). J Genet 87:39–51

Jarvis DE, Ho YS, Lightfoot DJ, Schmöckel SM, Li B, Borm TJA, Ohyanagi H, Mineta K, Michell CT, Saber N, Kharbatia NM, Rupper RR, Sharp AR, Dally N, Boughton BA, Woo YH, Gao G, Schijlen EGWM, Guo X, Momin AA, Negrão S, Al-Babili S, Gehring C, Roessner U, Jung C, Murphy K, Arold ST, Gojobori T, van der Linden CG, van Loo EN, Jellen EN, Maughan PJ, Tester M (2017) The genome of *Chenopodium quinoa*. Nature 542:307–312

Jensen SE, Charles JR, Muleta K, Bradbury PJ, Casstevens T, Deshpande SP, Gore MA, Gupta R, Ilut DC, Johnson L et al (2020) A sorghum practical haplotype graph facilitates genome-wide imputation and cost-effective genomic prediction. Plant Genome: e20009

Khalifa W, Thabet M (2018) Variation in downy mildew (*Peronospora variabilis* Gäum) resistance of some quinoa (*Chenopodium quinoa* Willd) cultivars under Egyptian conditions. Middle East J Agric 7:671–682

Kiani-Pouya A, Rasouli F, Bazihizina N, Zhang H, Hedrich R, Shabala S (2019) A large-scale screening of quinoa accessions reveals an important role of epidermal bladder cells and stomatal patterning in salinity tolerance. Environ Exp Bot 168:103885

Kolano B, Siwinska D, Gomez Pando L, Szymanowska-Pulka J, Maluszynska J (2012) Genome size variation in *Chenopodium quinoa* (Chenopodiaceae). Plant Syst Evol 298:251–255

Kuljanabhagavad T, Thongphasuk P, Chamulitrat W, Wink M (2008) Triterpene saponins from *Chenopodium quinoa* Willd. Phytochemistry 69:1919–1926

Kumar A, Bhargava A, Shukla S, Singh HB, Ohri D (2006) Screening of exotic *Chenopodium quinoa* accessions for downy mildew resistance under mid-eastern conditions in India. Crop Protect 25:879–889

López-Marqués RL, Nørrevang AF, Ache P, Moog M, Visintainer D, Wendt T, Østerberg JT, Dockter C, Jørgensen ME, Salvador AT, Hedrich R, Gao C, Jacobsen SE, Shabala S, Palmgren M (2020) Prospects for the accelerated improvement of the resilient crop quinoa. J Exp Bot. https://doi.org/10.1093/jxb/eraa285

Maher MF, Nasti RA, Vollbrecht M, Starker CG, Clark MD, Voytas DF (2020) Plant gene editing through de novo induction of meristems. Nat Biotechnol 38:84–89

Mason SL, Stevens MR, Jellen EN, Bonifacio A, Fairbanks DJ, Coleman CE, McCarty RR, Rasmussen AG, Maughan PJ (2005) Development and Use of Microsatellite Markers for Germplasm Characterization in Quinoa (*Chenopodium quinoa* Willd.). Crop Sci 45:1618–1630

Maughan PJ, Smith SM, Rojas-Beltrán JA, Elzinga D, Raney JA, Jellen EN, Bonifacio A, Udall JA, Fairbanks DJ (2012) Single nucleotide polymorphism identification, characterization, and linkage mapping in Quinoa. Plant Genome 5(3):114–125

Mhada M, Ezzahiri B, Benlhabib O (2014) Assessment of Downy mildew Resistance (*Peronospora farinosa*) in a Quinoa (*Chenopodium quinoa* Willd.) Germplasm. Int J Agric Biosyst Eng 8:277–280

Miranda M, Vega-Gálvez A, Quispe-Fuentes I, Rodríguez MJ, Maureira H, Martínez EA (2012) Nutritional aspects of six quinoa (*Chenopodium quinoa* Willd.) exotypes from three geographical areas of Chile. Chilean J Agric Res 72(2):175–181

Monteiro F, Nishimura MT (2018) Structural, functional, and genomic diversity of plant NLR proteins: an evolved resource for rational engineering of plant immunity. Ann Rev Phytopathol 56:243–267

Morales AJ, Bajgain P, Zackary Garver, Maughan PJ, Udall JA (2011) Physiological responses of *Chenopodium quinoa* to salt stress. Int J Plant Physiol Biochem 3:219–232

Morales A, Zurita-Silva A, Maldonado J, Silva H (2017) Transcriptional responses of Chilean Quinoa (*Chenopodium quinoa* Willd.) under water deficit conditions uncovers ABA-independent expression patterns. Front Plant Sci 8:216

Murphy KM, Matanguihan JB, Fuentes FF, Gomez-Pando LR, Jellen EN, Maughan PJ, Jarvis DE (2018) Quinoa breeding and genomics. In: Goldman I (ed) Plant breeding reviews. Wiley, Hoboken, NJ, USA, pp 257–320

Ochoa J, Frinking HD, Jacobs T (1999) Postulation of virulence groups and resistance factors in the quinoa/downy mildew pathosystem using material from Ecuador. Plant Pathol 48:425–430

Präger A, Munz S, Nkebiwe PM, Mast B, Graeff-Hönninger S (2018) Yield Quality Characteristics of Different Quinoa (*Chenopodium quinoa* Willd.) Cultivars Grown under Field Conditions in Southwestern Germany. *Agronomy* **8**: 197

Raney JA, Reynolds DJ, Elzinga DB, Page J, Udall JA, Jellen EN, Bonifacio A, Fairbanks DJ, Maughan PJ (2014) Transcriptome Analysis of Drought Induced Stress in *Chenopodium quinoa*. American Journal of Plant Sciences 5:338–357

Risi CJ, Galwey NW (1989a) The pattern of genetic diversity in the Andean grain crop quinoa (*Chenopodium quinoa* Willd.). I. Associations between characteristics. Euphytica 41:147–162

Risi CJ, Galwey NW (1989b) The pattern of genetic diversity in the Andean grain crop quinoa (*Chenopodium quinoa* Willd.) II Multivariate methods. Euphytica 41:135–145

Rojas W (2003) Multivariate analysis of genetic diversity of Bolivian Quinoa germplasm. Food Rev Int 19:9–23

Ruiz KB, Aloisi I, Del Duca S, Canelo V, Torrigiani P, Silva H, Biondi S (2016) *Salares* versus coastal ecotypes of quinoa: salinity responses in Chilean landraces from contrasting habitats. Plant Physiol Biochem 101:1–13

Ruiz KB, Khakimov B, Engelsen SB, Bak S, Biondi S, Jacobsen SE (2017) Quinoa seed coats as an expanding and sustainable source of bioactive compounds: An investigation of genotypic diversity in saponin profiles. Ind Crops Prod 104:156–163

Ruiz-Carrasco K, Antognoni F, Coulibaly AK, Lizzardi S, Covarrubias A, Martínez EA, Molina-Montenegro MA, Biondi S, Zurita-Silva A (2011) Variation in salinity tolerance of four lowland genotypes of quinoa (*Chenopodium quinoa* Willd.) as assessed by growth, physiological traits, and sodium transporter gene expression. Plant Physiol Biochem 49:1333–1341

Salazar J, de Lourdes Torres M, Gutierrez B, Torres AF (2019) Molecular characterization of Ecuadorian quinoa (*Chenopodium quinoa* Willd.) diversity: implications for conservation and breeding. Euphytica 215:60

Schierup MH, Vekemans X (2008) Genomic consequences of selection on self-incompatibility genes. Curr Opin Plant Biol 11:116–122

Stetter MG, Vidal-Villarejo M, Schmid KJ (2020) Parallel seed color adaptation during multiple domestication attempts of an ancient new world grain. Mol Biol Evol 37:1407–1419

Tovar JC, Quillatupa C, Callen ST, Castillo SE, Pearson P, Shamin A, Schuhl H, Fahlgren N, Gehan MA (2020) Heating quinoa shoots results in yield loss by inhibiting fruit production and delaying maturity. Plant J. https://doi.org/10.1111/tpj.14699

Turner-Hissong SD, Mabry ME, Beissinger TM, Ross-Ibarra J, Pires JC (2020) Evolutionary insights into plant breeding. Curr Opin Plant Biol 54:93–100

Vega-Gálvez A, Miranda M, Vergara J, Uribe E, Puente L, Martínez EA (2010) Nutrition facts and functional potential of quinoa (*Chenopodium quinoa* willd.), an ancient Andean grain: a review. J Sci Food Agric 90:2541–2547

Yu X, Li X, Guo T, Zhu C, Wu Y, Mitchell SE, Roozeboom KL, Wang D, Wang ML, Pederson GA et al (2016) Genomic prediction contributing to a promising global strategy to turbocharge gene banks. Nat Plants 2:16150

Zhang T, Gu M, Liu Y, Lv Y, Zhou L, Lu H, Liang S, Bao H, Zhao H (2017) Development of novel InDel markers and genetic diversity in *Chenopodium quinoa* through whole-genome re-sequencing. BMC Genom 18:685

Zou C, Chen A, Xiao L, Muller HM, Ache P, Haberer G, Zhang M, Jia W, Deng P, Huang R et al (2017) A high-quality genome assembly of quinoa provides insights into the molecular basis of salt bladder-based salinity tolerance and the exceptional nutritional value. Cell Res 27:1327–1340

Zurita-Silva A, Fuentes F, Zamora P, Jacobsen S-E, Schwember AE (2014) Breeding quinoa (*Chenopodium quinoa* Willd.): potential and perspectives. Mol Breed 34:13–30

Saponins of Quinoa: Structure, Function and Opportunities

8

Sophie Otterbach, Gordon Wellman, and Sandra M. Schmöckel

Abstract

In quinoa, saponins are found predominantly on the outside of the seeds. Saponins are triterpenoid glucosides that have diverse structures. Over 90 different saponins have been identified in quinoa seed hulls. They are bitter in taste and produce foam, which make them undesirable for human consumption. Their function in quinoa seeds is poorly understood. In this chapter, we provide an overview of these diverse compounds, their structure, and biosynthesis. We provide an overview of techniques for detection and quantification of saponins. We discuss their potential biological roles, from antifungal and anti-herbivory activity to their impact on germination and stress tolerance. Finally, we explore traditional and commercial methods for saponin removal from grain prior to consumption, breeding programs to reduce or alter saponin content, and the use of saponin waste as a novel bioproduct. While there is still much more to investigate regarding quinoa saponins, this work summarises the current knowledge and provides a basis for future studies.

S. Otterbach · S. M. Schmöckel (✉)
Department Physiology of Yield Stability, Institute of Crop Science, Faculty of Agriculture, University of Hohenheim, Stuttgart, Germany
e-mail: sandra.schmoeckel@uni-hohenheim.de

G. Wellman
Division of Biological and Environmental Sciences and Engineering (BESE), King Abdullah University of Science and Technology (KAUST), Thuwal, Kingdom of Saudi Arabia

8.1 *Chenopodium Quinoa* Contains Saponins

Saponins are a broad class of bitter-tasting and foaming compounds that are found in many species including *Chenopodium quinoa* Willd. (quinoa) seeds. They are amphipathic glycosides occurring in two major classes of the plant kingdom, Magnoliopsida (dicotyledon) and Liliopsida (monocotyledon) (Vincken et al. 2007), as well a number in marine animals such as sea cucumbers (Holothuroidea) (van Dyck et al. 2010) and starfish (Asteroidea) (Liu et al. 2008). Most known saponins are plant-derived secondary metabolites, displaying a wide range of activities, like defense against fungi, microbes, insects and molluscs (Abdel-Gawad et al. 1999; Sindambiwe et al. 1998; Sparg et al. 2004; Woldemichael and Wink 2001). Recent studies on the biological activity of saponins suggest saponins have beneficial properties to health, such as anticarcinogenic and anti-inflammatory properties (Ismail et al. 2018; Man et al. 2010). The name saponin originates from the Latin word "sapo", which reflects their ability to produce soap-like foams in aqueous solutions. This is due to their structure of one or more

S. M. Schmöckel (ed.), *The Quinoa Genome*, Compendium of Plant Genomes,
https://doi.org/10.1007/978-3-030-65237-1_8

Fig. 8.1 Triterpene aglycone backbone

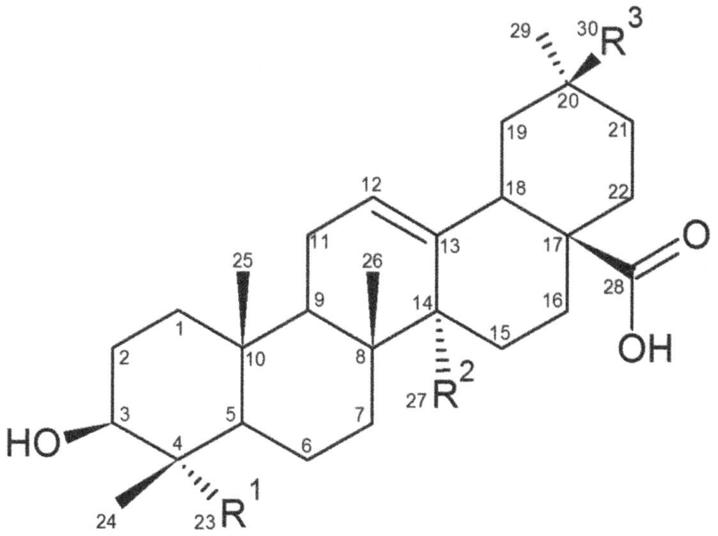

	R^1	R^2	R^3
Oleanolic acid (OA) (**I**)	CH_3	CH_3	CH_3
Hederagenin (HED) (**II**)	CH_2OH	CH_3	CH_3
Phytolaccagenic acid (PA) (**III**)	CH_2OH	CH_3	$COOCH_3$
Serjanic acid (SA) (**IV**)	CH_3	CH_3	$COOCH_3$
3*β*-Hydroxy-23-oxo-olean-12-en-28-oic acid (**V**)	CHO	CH_3	CH_3
3*β*-Hydroxy-27-oxo-olean-12-en-28-oic acid (**VI**)	CH_3	CHO	CH_3
3*β*, 23α, 30*β*-Trihydroxy-olean-12-en-28-oic acid (**VII**)	CH_2OH	CH_3	CH_2OH

hydrophilic glycoside moieties combined with a lipophilic aglycone. Saponins are divided into two major classes depending on the aglycone backbone: triterpenoid or steroid glycosides (Abe et al. 1993).

Saponins in quinoa are mainly triterpene glycosides (Kuljanabhagavad et al. 2008; Madl et al. 2006; Woldemichael and Wink 2001), consisting of a pentacyclic C30 skeleton (sapogenin, Fig. 8.1). They exist as a mixture of derivatives of oleanolic acid (OA), hederagenin (HED), phytolaccagenic acid (PA), serjanic acid (SA) as the main aglycones, and others like 3β-hydroxy-23-oxo-olean-12-en-28-oic acid, 3β-hydroxy-27-oxo-olean-12-en-28-oic acid, and 3β, 23α,30β-trihydroxy-olean-12-en-28-oic acid, decorated with hydroxyl- and carboxyl groups at C3 and C28. The major sugars are arabinose, glucose, and galactose; less common are glucuronic acid and xylose (Dini et al. 2001a, b, 2002; Kuljanabhagavad et al. 2008; Madl et al. 2006; Mastebroek et al. 2000; Mizui et al. 1988, 1990; Woldemichael and Wink 2001; Zhu et al. 2002). Further, the number of sugar chains on the aglycone characterizes the saponins as mono-, bi- or tri-desmosidic (Meyer et al. 1990). The triterpene saponins in quinoa contain mainly three carbohydrate units; one at C28 and two units at C3, where they can be linear or branched (Fig. 8.1.) (Kuljanabhagavad and Wink 2009).

8.2 Saponins Accumulate Predominantly in the Seeds of Quinoa

Saponins have been found in different organs and tissues of plants (Fig. 8.2). In many plants, roots are the major storage and synthesis organ, e.g. *Beta vulgaris* (beet), *Glycyrrhiza glabra* (liquorice), and *Panax ginseng* (ginseng) (Cheok et al. 2014).

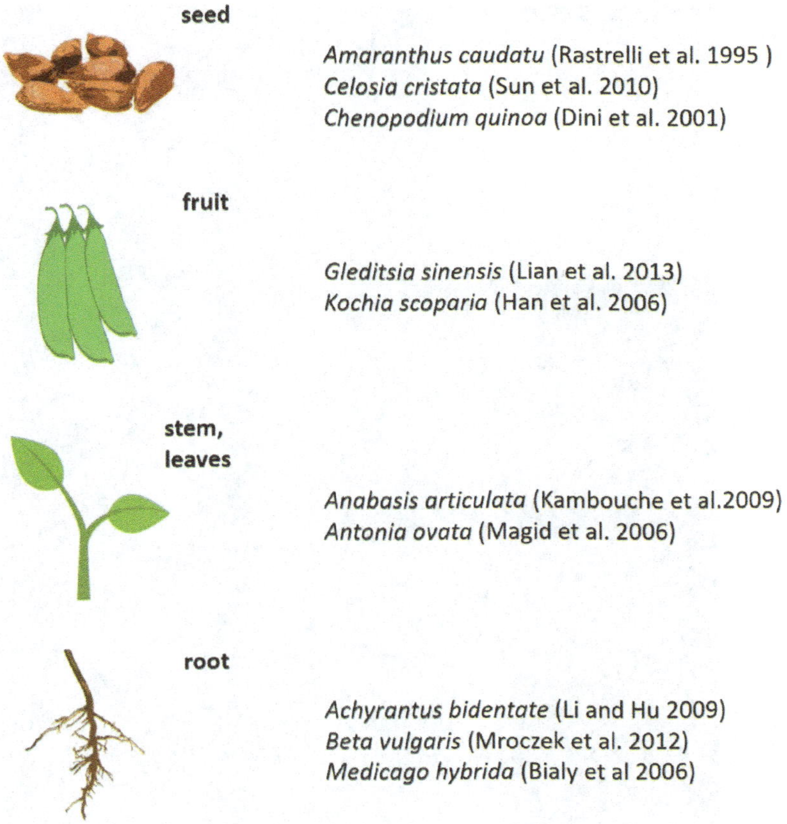

Fig. 8.2 Plant tissues containing triterpene saponins in other plant species. The pictures in the graphic are all licensed under CC BY-SA 4.0 or CC0 and modified by Sophie Otterbach

In the medicinal plant *Achyrantus bidentate* it has been demonstrated that the root is the main storage tissue of triterpene saponins, but the leaves seem to be the biosynthetic active site and the vascular bundle the transport organ (Li and Hu 2009). Additionally, saponins can also be stored at multiple storage sites, as in the case of *Medicago truncatula*, where the highest amounts of triterpene saponins were detected in the roots, followed by leaves and seeds (Huhman et al. 2005).

In quinoa, the triterpene saponins accumulate predominantly in the seeds. Matrix-Assisted Laser Desorption/Ionization-Mass Spectrometry (MALDI-MS) analyses suggest that saponins are localized in the outer layers of the seeds, more specifically in the external layer, the pericarp and the two seed coat layers (Figs. 8.3 and 8.4). The pericarp appears as a friable layer consisting mostly of saponins (Fig. 8.4) (Jarvis et al. 2017; Koyro and Eisa 2008; Prego et al. 1998; Varriano-Marston and Defrancisco 1984). Moreover, saponins have been identified in different tissues of quinoa, including flowers and fruits, and to a lesser extent in leaves (Kuljanabhagavad et al. 2008; Mastebroek et al. 2000). In leaves the saponin content is produced from the onset of flowering and increases until the end of the seed ripening, appearing in the later stages of the plant development (Mastebroek et al. 2000). Measurements in shoots of quinoa showed a similar peak in saponin concentration coinciding with the beginning of the blooming, while the content was lowest in the branching stage and during grain filling (Solíz et al. 2002). Saponins make up to 4% (w/w) of the seed mass (Jarvis et al. 2017).

In quinoa, accessions with naturally low amounts of saponins can be found (Mastebroek et al. 2000). These accessions are classified as 'sweet' if the saponin levels are under 0.11% of the dry weight and as 'bitter' if above (Koziol

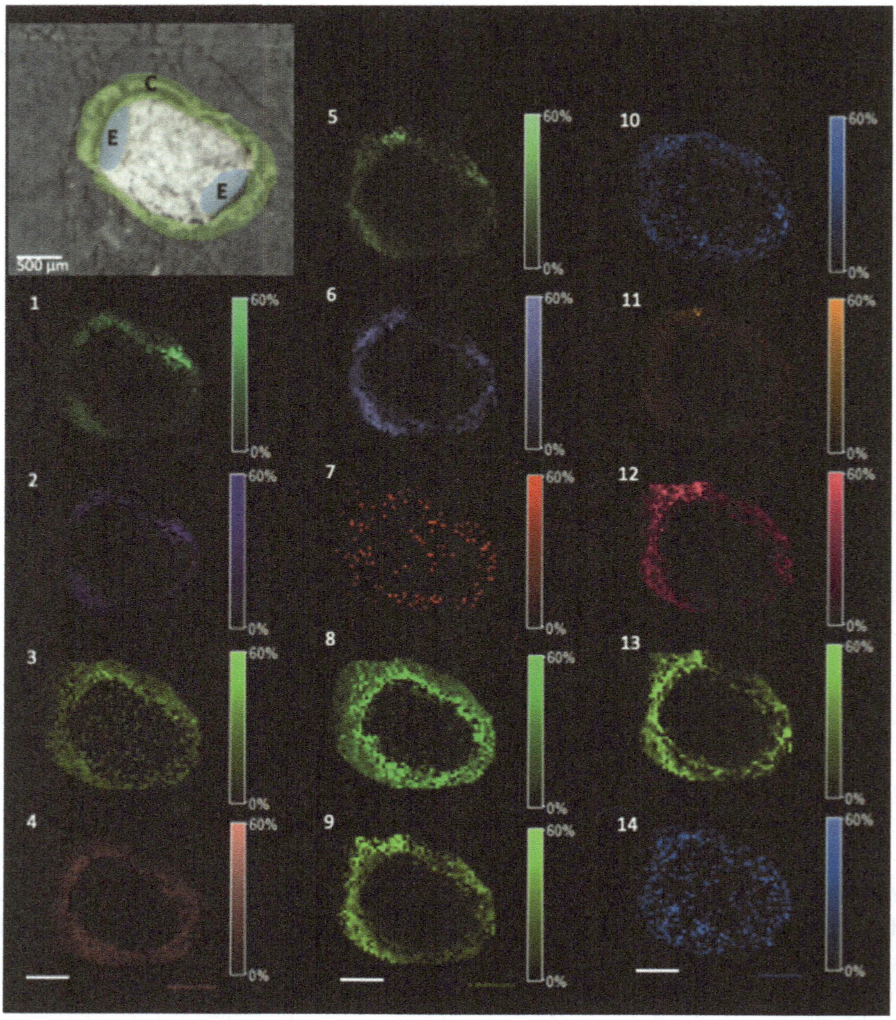

ID	Saponin	Formula	Calculated	Observed	Error (ppm)
1	PA(unknown) +Na	$[C_{36}H_{56}O_{10}+Na]^+$	671.37657	671.37080	-8.6
2	PA(unknown) +K	$[C_{36}H_{56}O_{10}+K]^+$	687.35051	687.34390	-9.6
3	AG487(Pent) +Na	$[C_{41}H_{64}O_{14}+Na]^+$	803.41883	803.41090	-9.9
4	AG487(Pent) +K	$[C_{41}H_{64}O_{14}+K]^+$	819.39277	819.38620	-8.0
5	PA(Pent) +Na	$[C_{42}H_{66}O_{15}+Na]^+$	833.42939	833.42941	0.0
6	PA +K	$[C_{42}H_{66}O_{15}+K]^+$	849.40333	849.40366	0.4
7	OA(Pent-Hex)b +Na	$[C_{47}H_{76}O_{17}+Na]^+$	935.49747	935.49550	-2.1
8	OA(Pent-Hex)b +K	$[C_{47}H_{76}O_{17}+K]^+$	951.47141	951.47800	6.9
9	PA(Hex-Pent)x2 +Na	$[C_{48}H_{76}O_{20}+Na]^+$	995.48222	995.48150	-0.7
10	AG515(HexA-Hex-Pent) +Na	$[C_{47}H_{68}O_{22}+Na]^+$	1007.40945	1007.40000	-9.4
11	Hed(Hex-Hex-Pent) +K	$[C_{53}H_{86}O_{23}+K]^+$	1129.51915	1129.51100	-7.2
12	PA((Hex-Pent-Hex)/(Hex-Hex-Pent)b +Na	$[C_{54}H_{86}O_{25}+Na]^+$	1157.53504	1157.53340	-1.4
13	PA((Hex-Pent-Hex)/(Hex-Hex-Pent)b +K	$[C_{54}H_{86}O_{25}+K]^+$	1173.50898	1173.50820	-0.7
14	AG489(Hex-Hex-HexA-) +K	$[C_{54}H_{86}O_{26}+K]^+$	1173.52995	1173.52580	-3.5

Fig. 8.3 Saponins are predominantly located in the pericarp in quinoa.
Imaging mass spectrometry analyses (analyses performed exactly as described in Jarvis et al. (2017)) detected 14 different saponins, which are predominantly located in the seed coat (C) and not in the embryo (E). Scale bar indicates 500 µm

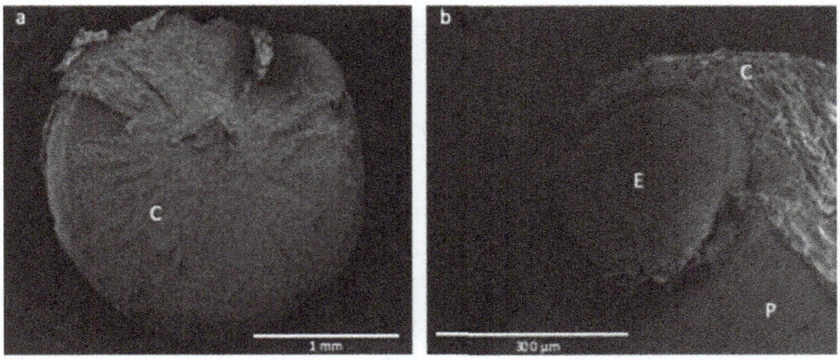

Fig. 8.4 Scanning Electron Microscopy picture of quinoa seed.
Scanning Electron Microscopy (SEM) picture of (a) a bitter seed containing saponins and (b) a cross-section. Perisperm (P), embryo (E) and seed coat (C) are indicated

1992). It appears that bitter and sweet accessions show a similar saponin content in the leaves, with hederagenin as the major saponin. Oleanolic acid is the main sapongenin in seeds (Mastebroek et al. 2000).

Congruent with the notion that saponins predominantly accumulate in the seed coat, it was found that the seed coat of bitter seeds is significantly thicker than that of sweet seeds. However, the origin of the saponins in quinoa seeds remains unclear and has not been addressed in the literature thus far.

8.3 Methods to Detect and Quantify Saponins

Saponins are a large group of compounds with a great natural diversity of their structures. The aglycone varieties are multiplied by their composition of sugar chains, a number of sugar moieties, and the branching patterns. In plant extracts, saponins occur in a mixture of structurally related forms with similar polarities, which complicates the separation.

Detection methods for saponins can be classified into quantitative and qualitative methods. The first semi-quantitative determination methods of saponins in plant materials were predominantly based on gravimetry or on the exploitation of some of their chemical and biological features (Van Atta et al. 1961). The well-known foaming property of saponins in water and the building of stable froth makes it easy to measure saponin content (Fig. 8.5). A standardized method to determine saponin content in the seeds is the afrosimetric method (Koziol 1991), which is fast but with some considerable disadvantages. The method underestimates high levels of grain saponin and may be affected by other surfactants in the plant material. Additionally, saponins with one or two-branched sugar chains do not form stable froth and, conversely, some plants may not contain saponins and yet produce a froth (Oleszek 2002).

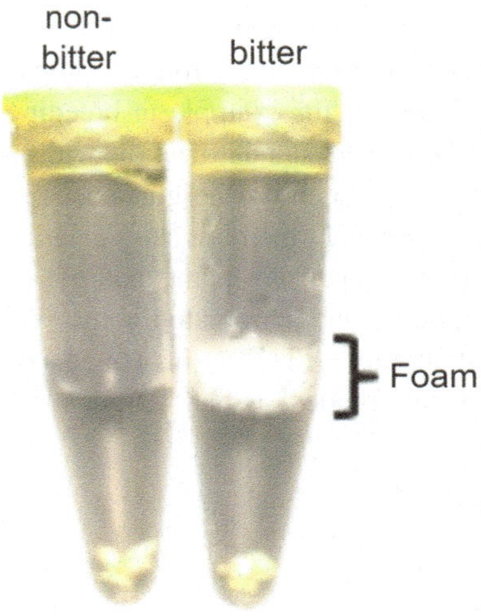

Fig. 8.5 Afrosimetric method (foaming test) to estimate saponin content in quinoa.
Here, five seeds are placed in 1.5 mL microcentrifuge tube with 0.5 mL of water and shaken vigorously until a stable foam can be measured

A biological feature of the saponins is its ability to cause membrane perturbation, also referred to as hemolytic activity. This property can be used in a semi-quantitative erythrocyte-assay, where saponins cause lysis of the cells. It is thought that saponins can form complexes with membrane sterols of erythrocytes, causing an increase in permeability and a subsequent loss of hemoglobin (Baumann et al. 2000). Saponins differ in their hemolytic activity, depending on the structure and on the hemolytic assay used. Furthermore, quantitation of saponins can also be performed with other biological methods including the growth of sensitive fungus colonies on saponin containing plants (Shany et al. 1970; Zimmer et al. 2010).

These tests are best to be seen as an approximate of the total saponin content, they are simple tests and good for a rough comparison of saponin content, for example in breeding experiments. However, standardization with saponins is necessary to quantify the glucosides.

Non-biological methods to determine the quantity of saponins in plant material are spectrophotometry and chromatography. Thin-layer chromatography (TLC) in one- or two-dimensional modes is an efficient tool for separation of saponins and has been used to determine saponin content of quinoa seeds in breeding programs (Dini et al. 2002; Ng et al. 1994). However, the simultaneous analysis of saponin standards on a plate is complicated but necessary to minimize variations of the method. Nowadays, TLC is only used as a supportive separation technique for saponin analysis.

TLC-colorimetry uses colorimetric determination of saponins using crude plant extracts. The TLC is the means of compound separation: the bands are scraped off the plate, extracted with alcohol, and treated with a colorant. The typically used reagents, ehrlich and vanillin, are prone to react with the hydroxyl group at C3 of bile acids or sterols (Nakajima 1976). This means that also sterols are detected, which can result in misleading information of the saponin character in crude plant extracts.

Gas chromatography (GC) is a semi-quantitative method, which separates the saponins into their aglycones and sugar moieties through acid hydrolysis (Fig. 8.6). This means that the large mixture of saponins present in the

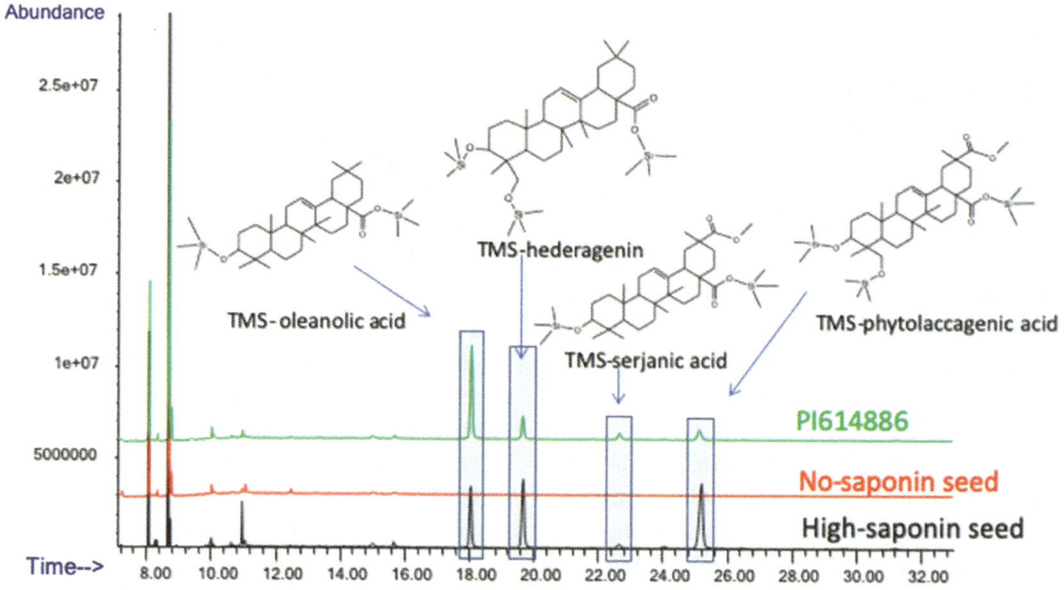

Fig. 8.6 Detection of saponins using Gas Chromatography-Mass Spectrometry (GC-MS). Overlaid chromatograms from extracts of seeds with no detectable saponins (red), high amount of saponins (green), and accessions PI614886 (QQ74) also with high saponin content. Four main sapogenins were detected. Procedure as described in Jarvis et al. (2017). All samples were extracted in methanol-water, hydrolyzed, and derivatized with trimethylsilyl (TMS) agent (removing the sugar groups).

seeds is broken down to only their aglycone backbones. This will allow estimation of the total amount of saponins based on the backbone, but not identify the individual saponins. In addition, under hydrolysis conditions, several artifacts can be formed which makes GC-analysis complicated. However, the method has been successfully employed in combination with mass spectrometry (MS) or flame ionizing detector (FID) to profile saponins in quinoa (Madl et al. 2006; Medina-Meza et al. 2016).

Separation of individual saponins is complicated and time-consuming, due to their similar polarities in plant extracts. A cascade of separation techniques (such as TLC, column chromatography, flash chromatography, Sephadex chromatography, high-pressure LC (HPLC)) is used to isolate individual saponins. Recent studies used liquid chromatography (LC) methods with mass spectrometry (LC-MS) to identify and quantify individual saponins.

The lack of chromophores prevents UV detection, therefore, wavelengths of 200-210 nm are applied for HPLC methods. These wavelengths are not ideal, as other components in the plant sample may overlap. The development of new methods like liquid chromatography-electrospray mass-spectrometry (LC-ESI-MS) has improved saponin determination. Additions like tandem mass spectrometry (MS/MS) support identification of structural information from complex triterpene saponin samples as (Madl et al. 2006) showed using nano-HPLC electrospray ionization multistage tandem mass spectrometry (nLC-ESI-MS/MS). This method allowed detection of over 80 saponins in crude extracts of quinoa seed hull. However, new technologies, such as nuclear magnetic resonance (NMR) should enable to elucidate saponin structures in the future.

Those spectroscopic methods are quite time-consuming and resource expensive, a simple and non-invasive method is needed to detect saponin content in seeds. Near-infrared spectroscopy (NIRS) is frequently used for grain quality analysis due to its speed, low cost, and lack of requirement for sample preparation. NIRS has been optimized to detect protein, carbohydrate, lipid, ash, moisture content, and amino acids in quinoa seeds (Ferreira et al. 2015). The method may also be applied to measure concentrations of fiber, fatty acids, vitamins, and minerals, though the determination of secondary metabolites has not been successfully used in quinoa seeds yet (Bittner et al. 2017; Czekus et al. 2019; Graf et al. 2015).

8.4 Role of Saponins in Quinoa

Many plant species, from multiple plant families, have developed the ability to produce saponins, regardless of their geographical region or climate zone and growth habit. The fact that saponins have different concentrations in plant organs and tissues, and that these concentrations are affected by developmental factors, suggests that there are differential synthesis and regulation depending on biological activities. Due to the often toxic nature of saponins, especially to fungi, insects, and molluscs, saponins are generally considered to be plant defense compounds; however, their role is not yet fully understood.

8.4.1 Toxicity and Biological Activity

The biological activity of saponins is closely linked to their chemical structure, which determines polarity, hydrophobicity, and acidity. The cytotoxicity and hemolytic activity of saponins are well known and based on the aglycone and the linked sugar chains (Oda et al. 2000). Several studies suggest that mono-desmosidic saponins (C3) are more active hemolytically toward erythrocytes than bi-desmosidic saponins (Takagi et al. 1986; Voutquenne et al. 2002; Wang et al. 2007; Woldemichael and Wink 2001). However, reports show that the hemolytic activities and the cytotoxicity of saponins are not necessarily linked to each other, as the underlying mechanism seems to be different (Gauthier et al. 2009).

Although the exact mechanism of the hemolytic property is not yet clearly understood, it was found to be correlated with their amphiphilic properties. One hypothesis is that saponins interact with plasmatic membrane sterols and form insoluble complexes, those may cause curvature and evoke pore-like structures or result in sterol extraction via vesiculation, which reduces the integrity of the membrane (Baumann et al. 2000). An alternative model describes a prior step to complex formation with sterols, saponins may migrate toward sphingolipid/sterol-enriched membrane domains and interfere with specific domains (Lin and Wang 2010). However, these properties differ for each individual saponin.

8.4.1.1 Antifungal Activity of Saponins

In recent years, there has been an increasing amount of research on the antifungal activity of various saponins (De Lucca et al. 2006; Kuljanabhagavad et al. 2008; Ribeiro et al. 2013; Sindambiwe et al. 1998; Stuardo and San Martín 2008), which could be used as natural fungicides in organic agriculture.

The main triterpene saponin avenacin A-1 in *Avena spp.* (oats) is predominantly localized in the epidermal cells of the root tips and released into the surrounding soil, where it shows its antifungal potential by providing resistance to phytopathogenic fungi (Carter et al. 1999; Crombie and Crombie 1986).

Total saponin content extracted from quinoa showed inhibition of the growth of the fungal pathogen *Candida albicans* at a concentration of 50 μg/mL. However, no individually tested saponins displayed a similar effect toward the fungi, suggesting a possible synergistic effect between all saponins or a different compound exhibiting the antifungal activity (Woldemichael and Wink 2001).

In another study, alkali-treated saponins showed the best results against the fungus *Botrytis cinerea*, an important disease of grapes. The mycelial growth and conidial germination were significantly inhibited at 5 mg saponins/mL, whereas untreated quinoa extract showed only minimal activity against *B. cinera.* The antifungal activity in alkali-treated saponins arises from membrane disruption, probably due to the higher number of hydrophobic saponin derivatives, which may have a higher affinity toward the sterols molecules in the membrane (Stuardo and San Martín 2008). San Martín et al. (2008) obtained similar results testing alkali-treated saponins from quinoa hulls of their molluscicidal activity and argues that the newly formed hydrophobic compounds are responsible for the membrane disturbance.

The first commercial biochemical pesticide/fungicide based on quinoa saponins was released in the Early twenty first century based on the patent of Dutcheshen and Danyluk (2004). "Heads Up"® plant protectant is based on the triterpene bi-desmosidic glycosides of quinoa seeds extract (HeadsUp Plant Protectant Ltd., Kamsack, Canada). The product is applied as suspension directly on tuber (e.g. potato seed pieces), legume (e.g. bean, pea), and cereal (e.g. wheat) seeds, as a root dip or foliar spray on tomato seedlings to prevent fungal or bacterial growth, and viral plant diseases (USEPA 2002). Although, the company campaigns with proof of vigour improvement of plants (applied to seeds) by initiating the plants own defense mechanism, little research is available on the efficacy of the product. Research on potato brown leaf spot disease shows no effect of "Heads Up"®; however, a tuber yield improvement was observed compared to other commercial products (Soleimani and Kirk 2012).

8.4.1.2 Anti-herbivory and Anti-nutritional Activity of Quinoa Saponins

Due to the potentially toxic/hemolytic properties and perceived bitterness of saponins, these compounds are generally assumed to play a role in plant defense, particularly as antifeedants and predation deterrents. Interestingly, despite the hemolytic properties of saponins, there is little evidence to support for direct toxicity of saponins when consumed normally by non-aquatic animals, although saponins can be effective piscicidal agents by damage respiratory epithelia in fish (Francis et al. 2002; Roy et al. 1990).

Animal feed produced from high saponin species such as alfalfa (Sen et al. 1998), soya, and legumes (Cheeke and Carlsson 1978) can impact animal growth; however, the mode of action is still unclear. The antimicrobial activity of saponins may affect gut microbial communities, especially in ruminants (Francis et al. 2002). Additional antinutritional effects may be due to saponins reducing protein digestibility, through the formation of stable protein-saponin complexes (Potter et al. 1993) and reducing the absorption of nutrients (Gee et al. 1997). However, the differing saponin compositions and concentrations in different plant species make comparisons related to the role of quinoa saponins difficult. It is suggested that the major antinutritional activity of saponins is the result of food aversion due to the perceived bitterness (Fleming and Galwey 1995).

As for quinoa specifically, there has been limited animal feeding trials to investigate the effect of saponins (Martens et al. 2012). In rats, quinoa meal had an expected negative effect on growth (Cheeke and Carlsson 1978), while sweet (low-saponin) quinoa was shown to have little effect (Grant et al. 1995). Similarly, quinoa containing meal showed reduced growth performance in poultry and pigs (Carlson et al. 2012; Jeroch et al. 1993). In addition, in feeding trials of broiler chickens with feed containing: unwashed, polished, or washed quinoa grain, chicks fed unwashed grains had significantly lower survival (Improta and Kellems 2001). Due to the unpalatability of unwashed grain, quinoa seed saponins are traditionally thought to be a bird deterrent, although evidence for this in control studies is limited. In field trials in Saudi Arabia with saponin containing bitter and low saponin sweet accessions of quinoa, a strong preference for birds feeding on the sweet accessions was observed (personal observation).

Similarly, although saponins are suggested to be involved in deterring insect predation (Dowd et al. 2011; Städler et al. 1992), few studies have examined this in quinoa, with one study showing increased insect activity in high-saponin accessions compared to low-saponin accessions (Yábar et al. 2002). Further studies are needed to confirm the anti-herbivory and anti-nutritional role of saponins in quinoa directly.

8.4.2 Germination

The localization of saponins on the outside of the seeds may play a role in plant development and seed germination. *Avena spp.* (*oats*) release saponins into the rhizosphere where they might serve as allelopathic agents, suppressing the growth of adjoining plants (Carter et al. 1999). In crop rotation of *Medicago sativa* (alfalfa) with cotton, a 50% reduced emergence of cotton seeds was observed (Mishustin 1955). This effect was verified with in vitro assays of extracted alfalfa saponins, which showed a detrimental effect on cottonseed germination (Pedersen 1965). However, on wheat seeds, the alfalfa saponins had no effect, but the growth of the seedlings was indirectly influenced by inhibition of the root growth (Oleszek and Jurzysta 1987). Stimulated germination was observed of barley seeds (*Hordeum vulgare*) and broomrape (*Orobanche minor*) when treated with the triterpene saponin soyasapogenol B (Evidente et al. 2011; Macías et al. 1997). However, similar effects have not been reported for quinoa and their saponins and allow only room for speculations on their role in germination processes.

8.4.3 Stress Tolerance

Abiotic stresses often negatively influence plant growth; drought and salinity are two major factors limiting crop growth and production. Quinoa is considered a drought and salt-tolerant plant (Jacobsen et al. 2003). Whether saponins play a role in conveying resistance to abiotic stresses is not well understood and only a few studies focus on secondary metabolites.

The influence of different irrigation levels was evaluated on bioactive compounds in quinoa seeds. The plants were tested under different saline-water and non-saline-water irrigation treatments (Gómez-Caravaca et al. 2012). Interestingly, the saponin content in the seeds decreased by 45% under drought conditions compared with non-saline watered conditions and by 50% under drought conditions compared with saline-water irrigation. The total saponin content is 35% higher in plants watered with saline-water than with non-saline water. This could be interpreted as a stress response, but further studies are needed to understand saponin production under saline conditions (Gómez-Caravaca et al. 2012; Pulvento et al. 2012). Quinoa seeds contain higher levels of saponins when watered normally. This observation suggests, that reduced irrigation for quinoa plants could be beneficial to reduce the saponin levels. As to why saponin production is reduced under drought stress is still unclear; however, we can speculate that increased saponins under normal watering conditions may protect seeds from fungal pathogens that may not be present in water-limited conditions during grain filling; or possibly that the synthesis of saponins is too energetically expensive to produce under stress conditions.

Seed priming has been shown to be beneficial in adapting plants to abiotic stresses. This strategy can invoke cross-tolerance in seeds when previously exposed to abiotic stresses. (Bradford et al. 1990). In a priming experiment with quinoa seeds *cv. Titicaca* showed that salt stress tolerance was positively influenced when treated with saponins. A concentration of 10-25% saponins alleviated the negative effect of the salt stress during seed germination. Despite, the salt tolerance of mature quinoa plants, at the germination stage, they are prone to the salty surroundings and less likely to establish. Priming with saponins (15%) seems to improve the germination rate under salt stress by shortening the metabolic phase and weakening the pericarp, acting as biostimulant (Yang et al. 2018). These results demonstrate a fruitful avenue for future research in salt stress tolerance for other crops.

8.4.4 Uses of Saponins

Saponins derived from a large variety of plants have been traditionally used for a wide range of purposes, from soaps and detergents for washing (such as from *Yucca schidigera* (Oleszek and Hamed 2010), toxins for fishing (such as from *Serjania lethalis* (Teixeira et al. 1984)), and are present in many traditional medicines. Industrially, the foaming properties of saponins have been used for the production of detergents, shampoos, foaming agents for beer and fire-extinguishers (Güçlü-Üstündağ and Mazza 2007; Zurita-Silva et al. 2014).

The biochemical activity of saponins, especially the hemolytic activities make them potential organic fungicides, insecticides, and antimicrobial agents. Saponins from quinoa have been registered for use as a biochemical pesticide and fungicide for treatment of tuber, legume, and cereal seeds, for root and for foliar application of tomato seedlings (USEPA 2002) due to their apparent antifungal and anti-insect properties as previously discussed.

The increased intestinal permeability mediated by saponins may also aid drug absorption and assist with reduction of cholesterol (Jacobsen 2003); however, the role of saponins from quinoa for cholesterol reduction is still under debate as the hypocholesterolemic effect was seen with desaponified grain meal in rat studies (Takao et al. 2005).

Additionally, a wide range of plant-derived saponins have been linked to anti-cancer activity (Man et al. 2013; Rao and Sung 1995), with some quinoa derived saponins were shown to induce apoptosis of cancer cells in cell-cultures (Kuljanabhagavad et al. 2008; Kuljanabhagavad and Wink 2009). Saponins may also serve as immunological adjuvants or immunostimulators (Campbell and Peerbaye 1992; Kenarova et al. 1990; reviewed in Rajput et al. 2007), such as from *Quillaja saponaria* (Kensil et al. 1991). Similar uses have been suggested for quinoa derived saponins and shown promise in mice immunization studies, where bulk quinoa saponins used as an adjunct increased systemic and mucosal antibodies, possibly by increasing permeability of antigens (Estrada et al. 1998).

For these reasons, there are exciting new prospects for the use of quinoa saponins. However, further study is required to determine the safety and beneficial effects, either nutritionally or medicinally, if any. More investigation into the use of saponins from quinoa is needed to exploit these interesting compounds.

8.5 Removal of Saponins from Quinoa Seeds

The bitter taste of unprocessed quinoa seeds and potential anti-nutritional properties of saponins necessitates the removal of these compounds before consumption. As a large proportion of saponins are located in the seed coat (34%) (Chauhan et al. 1992; Jarvis et al. 2017) the majority of saponins can be removed through milling to reduce this layer. Dehulling and washing decreases saponin content by up to 72% (Chauhan et al. 1999; Gee et al. 1993; Ruales and Nair 1993) although even following these processes some saponins remain in the seed. Other forms of processing such as alkaline washing and heat treatment can also help in removal or debittering of quinoa seeds by degradation of saponins (Gee et al. 1993; Zhu et al. 2002); however, these processes individually reduce saponin content to a lesser extent than with dehulling (Gee et al. 1993).

For traditional consumers, quinoa seeds are generally stored unprocessed, likely to help improve storage life due to the anti-herbivory and antifungal properties of saponins. Directly prior to consumption, an intensive process of roasting, hand milling, and repeated washing is used to remove saponins (reviewed in Quiroga Ledezma 2015) (Fig. 8.7

Fig. 8.7 Traditional quinoa processing by hand in Bolivia.
Traditionally, harvested quinoa seeds are threshed and separated from chaff by hand for storage. Prior to consumption, several laborious rounds of washing and hand milling are required to remove saponins before cooking. Photo credit: G. Fiene, KAUST. Community of Aroma, Boliva (2019)

As quinoa has rapidly gained in popularity in the last decades, worldwide production has increased from approximately 80 mT in 2010 to nearly 200 mT at the peak in 2015 (FAO 2019). In response, improvements in processing grain to dehull and remove saponins have been made to allow increased availability to consumers.

Depending on the variety (and starting grain saponin content), either washing (wet) or mechanical milling (dry) methods can be used (Bacigalupo and Tapia 1990; Quiroga Ledezma 2015). Generally, a combined wet-dry method for saponin removal, which removes saponins and maintains seed quality, is used to comply with market expectations. This process is similar to traditional methods where grain is first milled to remove the hull and majority of saponins, several washing steps to extract remaining saponins, followed by centrifugation and a final drying heat treatment to reduce the water content of the grain prior to storage. The use of water required for processing is estimated at >5 m^3/t of grain (Quiroga Ledezma 2015) and is contaminated with potentially bio-active saponins which may pose an environmental hazard if not disposed of responsibly (Jiang et al. 2018).

Although progress has been made to increase the efficiency of quinoa grain processing and reducing water usage, further improvements can be made. Some "water-less" quinoa processing machines are available on the market (such as from Schule Mühlenbau GmbH, Germany), which operate primarily by dehulling to remove the majority of saponins, or through non-aqueous extractions (Muir et al. 2008), although their uptake of these methods is still limited.

Further research into the efficient and inexpensive removal of saponins is required to make quinoa a more widely consumed crop. Additionally, the purification of interesting bioactive compounds from saponins mixtures is required before these compounds can be used in widespread industrial and medicinal products.

8.6 Regulation of Synthesis

Triterpene saponins in quinoa derive from the precursor isopentyl pyrophosphate (IPP, isoprenoid), resulting from the cytosolic mevalonate (MVA) pathway (Fig. 8.8). These precursors are shared with the phytosterol and steroidal saponin synthesis pathway. The oxidation of oxidosqualene is the first diversifying step in the triterpenoid biosynthesis (Abe et al. 1993; Kalinowska et al. 2005). Oxidosqualene is formed through the condensation of two units of IPP with one unit of its allylic isomer dimethylallyl pyrophosphate (DMAPP) yielding in farnesyl pyrophosphate (FPP). Subsequently, two units of FPP are condensated to squalene (30 carbon atoms) (Moses et al. 2013) and further epoxidized to 2,3-oxidosqualene catalyzed by the 2,3-oxidosqualene cyclase (OSC) to a tetra or pentacyclic structure to form dammarenes, tirucallanes and phytosterols, or the oleananes, ursanes, lupanes, and taraxasteranes, respectively (Augustin et al. 2011; Phillips et al. 2006). In quinoa, the oleane types are the common sapogenins, those result from the precursor ß-amyrin which is cyclized by a specific ß-amyrin synthase (*CqbAS1*) from 2,3-oxidosqualene. The following oxidation step facilitated by the quinoa-specific cytochrome P450-dependent monooxygenases CYP716A78 and CYP716A79 oxidize ß-amyrin at C28 to form sapogenin. In the subsequent reactions, the sapogenins are further modified by acyl-, malonyl- or methyltransferases (Fiallos-Jurado et al. 2016; Thimmappa et al. 2014). The probable last step to form saponins is promoted by enzymes belonging to the family of uridine diphosphate glycosyltransferases (UGTs) to glycosylate the sapogenin at C3 and/or C28 (ester linkage) (Augustin et al. 2011).

The seed saponin content in quinoa seeds varies with accession, can be influenced by environmental factors and is therefore considered

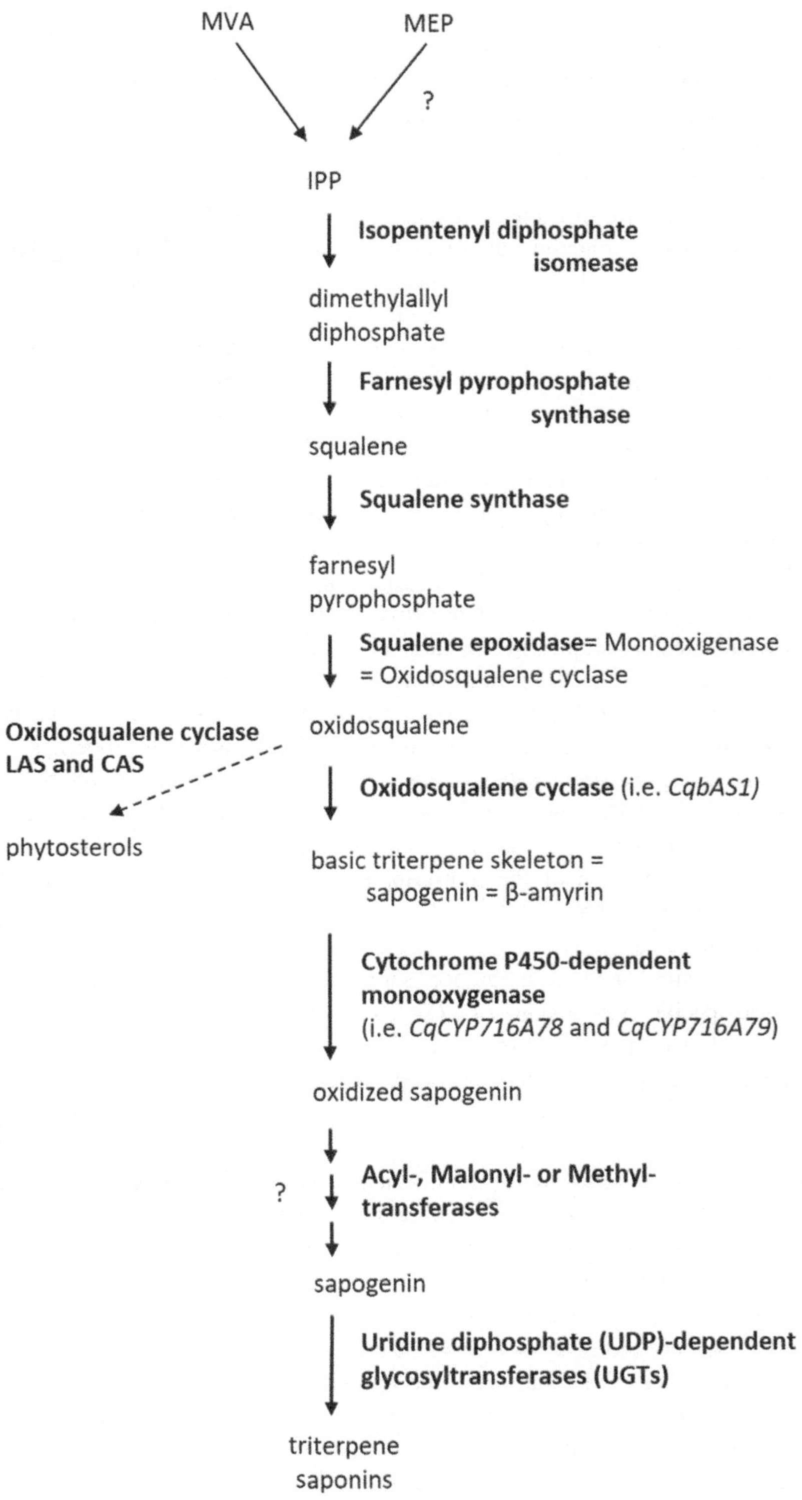

Fig. 8.8 Pathway for saponin biosynthesis in quinoa. The triterpenoids are built from isopentyl diphosphate (IPP), which is supplied from the cytosolic mevalonic acid (MVA) pathway and/or potentially the plastidial methylerythritol phosphate pathway (MEP). The contribution of the MEP pathway to the IPP pool leading to triterpenoid saponins is still being investigated, as denoted by the "?". The cyclization of oxidosqualene is the first diversifying step in triterpenoid biosynthesis. The biodiversity of oxidosqualene cyclases commits the production of phytohormones or triterpenoid saponins. For instance, the β-amyrin synthase in quinoa produces (i.e. *CqbAS1*) β-amyrin, which is the sapogenin for oleanolic saponins. The quinoa specific cytochrome P450 monooxygenases (i.e. *CYP716A78 and CYP716A79*) oxidize ß-amyrin to sapogenin (Fiallos-Jurado *et al.* 2016). A diverse number of acyl-, malonyl, and/or methyltransferases are involved for further decorations, however, the degree of involvement and order is unknown and indicated by "?". The probable last step is promoted by Uridine diphosphate (UDP)-dependent glycosyltransferases (UGTs) to produce a large variety of saponins

a quantitative trait (Kenwright 1989; Risi 1986). However, crossing experiments of bitter and sweet quinoa varieties showed an F2-segregation of 3:1, suggesting that bitterness is determined by a single dominant locus (Gandarillas 1948; Jarvis et al. 2017; Ward 2002). Recently, a reference genome sequence of quinoa has been published offering new insights into the regulation mechanism of saponins. By means of linkage mapping and bulk segregant analysis (BSA) in two populations of crossing events of sweet and bitter varieties (Kurmi (sweet) × 0654 (bitter), and Atlas (sweet) × Carina Red (bitter)) the same region on chromosome *Cq*B16 was identified to contain the locus conferring bitterness. In this region, two neighboring genes were striking: *AUR62017204* (*TSAR-like 1, TSARL1*) and *AUR62017206* (*TSAR-like 2, TSARL2*), annotated as a basic helix–loop–helix (bHLH) transcription factors similar to the class IVa bHLH genes (Jarvis et al. 2017). In *Medicago truncatula* were two transcription factors of the same class reported to regulate triterpenoid biosynthesis, *triterpene saponin biosynthesis activating regulator 1* (*TSAR1*) and *TSAR2*. Overexpression of these transcription factors in hairy roots showed an increase of the transcript levels of triterpene biosynthetic genes and accumulation of triterpene saponins (Mertens et al. 2016).

Expression analysis in quinoa showed root tissue-specific expression of *TSARL2*, while *TSARL1* was mainly expressed in flowers and immature seeds. The DNA binding motif for TSAR of *M. truncatula* was found upstream of several saponin biosynthetic pathway genes in quinoa. Those genes were significantly downregulated in sweet accessions, correlating with the lower expression levels of *TSARL1*. Comparing the transcripts of *TSARL1* in sweet and bitter accessions revealed a spliced alternative in the sweet accessions, showing a single nucleotide polymorphism (SNP) that alters the intron/exon splicing boundary. The spliced alternative of *TSARL1* in sweet varieties results in a truncated protein predicted to be compromised to form homodimers and/or to bind DNA and is therefore likely restricted in transcription regulation. Further analysis of *TSARL1* in sweet varieties showed multiple, independent gene mutations which co-segregate with the sweet phenotype, suggesting that this gene regulates the presence and absence of saponins in quinoa seeds.

8.7 Perspectives in Quinoa Breeding for Altered Saponin Content

Although increased yield remains the highest priority for the majority of quinoa breeding programs, both commercial and indigenous communities also target "sweetness" or reduced seed saponin content (McElhinny et al. 2007). Reduction of saponin content in quinoa grain would greatly reduce the labor and resources used for processing grain for consumption. The use of aqueous methods for saponin removal produces a significant amount of saponin contaminated wastewater which is potentially hazardous to the waterways due to the general piscicidal properties of saponins, if not properly disposed of. Additionally, the quantities of water required for saponin removal limits industrial grain processing, especially in traditional quinoa growing areas in South America where access to good quality water is scarce. The drought tolerance of quinoa is contrasted by the water usage required for processing. Some saponin-free or at least very low saponin varieties do exist (Koziol 1991, 1992) so there is scope for development of reduced saponin lines (Fig. 8.9).

Complete removal of beneficial saponins in quinoa, however, may result in increased insecticide usage, bird predation, and possibly reduce shelf-life of unprocessed quinoa due to fungal contamination.

Further study is required to identify particular toxic, anti-nutritional, or bitter saponins to target breeding removal of these compounds, as well as simple, high-throughput methods for assessing saponin diversity in quinoa grain. Removal of particularly pH- or heat-stable saponins may also improve processability, reducing the need for heat-treatment or alkaline washing processes, while retaining some of the beneficial saponins.

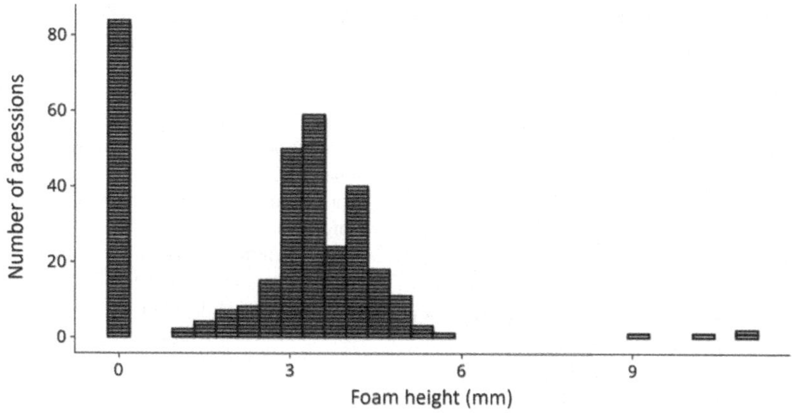

Fig. 8.9 Histogram indicating saponin content in 334 accessions of *Chenopodium quinoa*. Foam height generated by vigorous shaking of 5 seeds in 0.5 mL water in a 1.5 mL microcentrifuge tube was measured as an indication for saponin content in 334 *Chenopodium quinoa* accessions

Additionally, the wide-spread industrial, food additive and medicinal uses of saponins from other species warrant further investigation of the uses for quinoa saponins for these purposes. Development of products using quinoa saponins would turn current waste into valuable by-products.

With recent advances in genome editing in plants, such as via CRISPR technology (Li et al. 2015), there is scope for selectively altering saponin biosynthesis in quinoa. Targeting of key saponin biosynthesis and regulatory genes, previously described in this chapter, such as replicating the SNP leading to inactivated or altered expression of *TSARL1* in sweet varieties would be one such approach for rapid breeding for reduced saponin content.

However, to our knowledge, there is yet no established protocols for the stable transformation or whole plant regeneration of quinoa, although there are several works describing callus induction and transient transformation (Eisa et al. 2005; Hesami and Daneshvar 2016; Komari 1990). The transformation of related species such as amaranth (Li et al. 2012; Yaroshko et al. 2019), sugar beet (Dhalluin et al. 1992; Yang et al. 2005), and spinach (Al-Khayri 1995; Naderi et al. 2012) indicates that transformation of quinoa is similarly possible. With some effort and the large diversity of quinoa germplasm available, it is likely that transformation tractable varieties can be identified.

In the short term, the use of virus-induced gene silencing (VIGS), such as recently done by Adhikary and colleagues (Adhikary et al. 2019) in *Amaranth spp.*, may be of use to rapidly screen for gene candidates useful for quinoa breeding, however, how amenable the VIGS system is to altering gene activity in developing seeds is yet to be tested.

8.8 Conclusion

Saponins are a diverse range of bitter-tasting, biologically active secondary compounds, generally considered to be bio-protectants. Natively, the role of saponins in quinoa is likely to serve a similar role, particularly due to their concentration in seed coats as a deterrent to bird predation and limiting fungal growth. However, further research into the role of saponins in quinoa, especially into their impact on abiotic stress tolerance is required. With recent advances in quinoa genomics, there are good prospects for molecular plant breeding and gene editing for altering saponin content and diversity. Such developments will benefit both basic research into the role saponins play in quinoa and related *Chenopodium* species, as well as further the development of this currently underutilized crop, particularly in the production of sweet varieties that require less post-harvest processing.

Investigation into potential uses of quinoa saponins, currently being disposed of as a waste-byproduct of quinoa processing is also required, to help maximize the value and limit the impacts of quinoa processing.

Acknowledgements We thank Berin Boughton and Dinaiz Thinagaran, University of Melbourne (Australia) for providing the Imaging-MS pictures; Gabriele Fiene, KAUST (Saudi Arabia), for images of quinoa cleaning; David Jarvis, Brigham Young University (USA), and Ohoud Mohammed Eid Alharbi, KAUST (Saudi Arabia), for providing SEM pictures; and Najeh Kharbatia, KAUST (Saudi Arabia), for providing the GC-MS chromatogram. SMS received funding from the Ministry for Science, Research and Art of Baden-Wuerttemberg, Germany (Az: 75533-30-20/1).

References

Abdel-Gawad MM, El-Sayed MM, Abdel-Hameed ES (1999) Molluscicidal steroidal saponins and lipid content of Agave decipiens. Fitoterapia 70(4):371–381. https://doi.org/10.1016/S0367-326X(99)00057-X

Abe I, Rohmer M, Prestwich GD (1993) Enzymatic cyclization of squalene and oxidosqualene to sterols and triterpenes. Chem Rev 93(6):2189–2206. https://doi.org/10.1021/cr00022a009

Adhikary D, Khatri-Chhetri U, Tymm FJM, Murch S, Deyholos M (2019) A virus-induced gene-silencing system for functional genetics in a betalainic species, Amaranthus tricolor (Amaranthaceae). Appl Plant Sci 7:e01221. doi:10.1002/aps3.1221

Al-Khayri JM (1995). Genetic Transformation in Spinacia oleracea L. (Spinach). In: Bajaj YPS (ed) Plant protoplasts and genetic engineering, vol VI. Springer, Berlin Heidelberg, pp 279–288

Augustin JM, Kuzina V, Andersen SB, Bak S (2011) Molecular activities, biosynthesis and evolution of triterpenoid saponins. Phytochemistry 72(6):435–457. https://doi.org/10.1016/j.phytochem.2011.01.015

Bacigalupo A, Tapia M (1990) Potencial agroindustrial de los cultivos andinos subexplotados. Cultivos Andinos subexplotados y su aporte a la alimentación. FAO. Ediciones Gegra SA Santiago, Chile, pp 136–163

Baumann E, Stoya G, Völkner A, Richter W, Lemke C, Linss W (2000) Hemolysis of human erythrocytes with saponin affects the membrane structure. Acta Histochem 102(1):21–35. https://doi.org/10.1078/0065-1281-00534

Bittner M, Krähmer A, Schenk R, Springer A, Gudi G, Melzig MF (2017) NIR spectroscopy of Actaea racemosa L rhizome—en route to fast and low-cost quality assessment. Planta Med 83(12–13):1085–1096. https://doi.org/10.1055/s-0043-108122

Bradford KJ, Steiner JJ, Trawatha SE (1990) Seed priming influence on germination and emergence of pepper seed lots. Crop Sci 30(3):718. https://doi.org/10.2135/cropsci1990.0011183x003000030049x

Campbell JB, Peerbaye YA (1992) Saponin. Res Immunol 143(5):526–530 (discussion 577–578)

Carlson D, Fernandez JA, Poulsen HD, Nielsen B, Jacobsen SE (2012) Effects of quinoa hull meal on piglet performance and intestinal epithelial physiology. J Anim Physiol Anim Nutr (Berl) 96(2):198–205. https://doi.org/10.1111/j.1439-0396.2011.01138.x

Carter JP, Spink J, Cannon PF, Daniels MJ, Osbourn AE (1999) Isolation, characterization, and avenacin sensitivity of a diverse collection of cereal-root-colonizing fungi. Appl Environ Microbiol 65(8):3364–3372

Chauhan GS, Cui W, Eskin NAM (1999) Effect of saponin on the surface properties of quinoa proteins. Int J Food Prop 2(1):13–22. https://doi.org/10.1080/10942919909524586

Chauhan GS, Eskin NAM, Tkachuk R (1992) Nutrients and antinutrients in quinoa seed. Cereal Chem 69 (1):85–88

Cheeke PR, Carlsson R (1978). Evaluation of several crops as sources of leaf meal: composition, effect of drying procedure, and rat growth response. Nutr Rep Int (USA)

Cheok CY, Salman HAK, Sulaiman R (2014) Extraction and quantification of saponins: A review. Food Res Int 59:16–40. https://doi.org/10.1016/j.foodres.2014.01.057

Crombie W, Crombie L (1986) Distribution of avenacins A-1, A-2, B-1 and B-2 in oat roots: Their fungicidal activity towards 'take-all' fungus. Phytochemistry 25 (9):2069–2073. https://doi.org/10.1016/0031-9422(86)80068-1

Czekus B, Pećinar I, Petrović I, Paunović N, Savić S, Jovanović Z, Stikić R (2019) Raman and Fourier transform infrared spectroscopy application to the Puno and Titicaca cvs. of quinoa seed microstructure and perisperm characterization. J Cereal Sci 87:25–30. https://doi.org/10.1016/j.jcs.2019.02.011

De Lucca AJ, Boue S, Palmgren MS, Maskos K, Cleveland TE (2006) Fungicidal properties of two saponins from Capsicum frutescens and the relationship of structure and fungicidal activity. Can J Microbiol 52 (4):336–342. https://doi.org/10.1139/w05-137

Dhalluin K, Bossut M, Bonne E, Mazur B, Leemans J, Botterman J (1992) Transformation of sugar-beet (beta-Vulgaris L) and evaluation of herbicide resistance in transgenic plants. Bio-Technology 10(3):309–314. https://doi.org/10.1038/nbt0392-309

Dini I, Schettino O, Simioli T, Dini A (2001a) New oleanane saponins in Chenopodium quinoa. J Agric Food Chem 49(8):3976–3981. https://doi.org/10.1021/jf010361d

Dini I, Schettino O, Simioli T, Dini A (2001b) Studies on the constituents of Chenopodium quinoa seeds: isolation and characterization of new Triterpene Saponins. J Agric Food Chem 49(2):741–746. https://doi.org/10.1021/jf000971y

Dini I, Tenore GC, Dini A (2002) Oleanane saponins in "kancolla", a sweet variety of Chenopodium quinoa. J Nat Prod 65(7):1023–1026. https://doi.org/10.1021/np010625q

Dowd PF, Berhow MA, Johnson ET (2011) Differential activity of multiple saponins against omnivorous insects with varying feeding preferences. J Chem Ecol 37 (5):443–449. https://doi.org/10.1007/s10886-011-9950-3

Dutcheshen JM, Danyluk TA (2004) WO Patent No. WO2001060153A2. World Intellectual Property Organization

Eisa S, Koyro HW, Kogel KH, Imani J (2005) Induction of somatic embryogenesis in cultured cells of Chenopodium quinoa. Plant Cell Tissue Organ Cult 81(2):243–246. https://doi.org/10.1007/s11240-004-4793-z

Estrada A, Li B, Laarveld B (1998) Adjuvant action of Chenopodium quinoa saponins on the induction of antibody responses to intragastric and intranasal administered antigens in mice. Comp Immunol Microbiol Infect Dis 21(3):225–236. https://doi.org/10.1016/S0147-9571(97)00030-1

Evidente A, Cimmino A, Fernández-Aparicio M, Rubiales D, Andolfi A, Melck D (2011) Soyasapogenol B and trans-22-dehydrocam- pesterol from common vetch (Vicia sativa L.) root exudates stimulate broomrape seed germination. Pest Manag Sci 67 (8):1015–1022. https://doi.org/10.1002/ps.2153

FAO (2019) Quinoa production worldwide from 2010 to 2017 (in metric tons)* [Graph]. Statista. https://www.statista.com/statistics/486442/global-quinoa-production/. Accessed 1 Feb 2019

Ferreira DS, Pallone JAL, Poppi RJ (2015) Direct analysis of the main chemical constituents in Chenopodium quinoa grain using Fourier transform near-infrared spectroscopy. Food Control 48:91–95. https://doi.org/10.1016/j.foodcont.2014.04.016

Fiallos-Jurado J, Pollier J, Moses T, Arendt P, Barriga-Medina N, Morillo E, Leon-Reyes A (2016) Saponin determination, expression analysis and functional characterization of saponin biosynthetic genes in Chenopodium quinoa leaves. Plant Sci 250:188–197. https://doi.org/10.1016/j.plantsci.2016.05.015

Fleming JE, Galwey NW (1995) Quinoa (Chenopodium quinoa) In: Williams JT (ed) Cereals and Pseudocereals. Chapman & Hall, London

Francis G, Kerem Z, Makkar HPS, Becker K (2002) The biological action of saponins in animal systems: a review. Br J Nutr 88(6):587–605. https://doi.org/10.1079/BJN2002725

Gandarillas H (1948) Efecto fisiológico de la saponina de la quinua en los animales. Rev Agric 4:52–56

Gauthier C, Legault J, Girard-Lalancette K, Mshvildadze V, Pichette A (2009) Haemolytic activity, cytotoxicity and membrane cell permeabilization of semi-synthetic and natural lupane- and oleanane-type saponins. Bioorg Med Chem 17(5):2002–2008. https://doi.org/10.1016/j.bmc.2009.01.022

Gee JM, Price KR, Ridout CL, Wortley GM, Hurrell RF, Johnson IT (1993) Saponins of quinoa (Chenopodium quinoa): Effects of processing on their abundance in quinoa products and their biological effects on intestinal mucosal tissue. J Sci Food Agric 63 (2):201–209. https://doi.org/10.1002/jsfa.2740630206

Gee JM, Wal JM, Miller K, Atkinson H, Grigoriadou F, Wijnands MVW, Johnson IT (1997) Effect of saponin on the transmucosal passage of β-lactoglobulin across the proximal small intestine of normal and β-lactoglobulin-sensitised rats. Toxicology 117(2–3):219–228

Gómez-Caravaca AM, Iafelice G, Lavini A, Pulvento C, Caboni MF, Marconi E (2012) Phenolic compounds and saponins in quinoa samples (Chenopodium quinoa Willd.) grown under different saline and nonsaline irrigation regimens. J Agric Food Chem 60(18):4620–4627. https://doi.org/10.1021/jf3002125

Graf BL, Rojo LE, Delatorre-Herrera J, Poulev A, Calfio C, Raskin I (2015) Phytoecdysteroids and flavonoid glycosides among Chilean and commercial sources of Chenopodium quinoa: variation and correlation to physico-chemical characteristics. J Sci Food Agric 96 (2):633–643. https://doi.org/10.1002/jsfa.7134

Grant G, More LJ, McKenzie NH, Dorward PM, Buchan WC, Telek L, Pusztai A (1995) Nutritional and haemagglutination properties of several tropical seeds. J Agric Sci 124(3):437–445. https://doi.org/10.1017/S0021859600073391

Güçlü-Üstündağ Ö, Mazza G (2007) Saponins: properties, applications and processing. Crit Rev Food Sci Nutr 47(3):231–258

Hesami M, Daneshvar MH (2016) Development of a regeneration protocol through indirect organogenesis in chenopodium quinoa willd. Indo Am J Agric Vet Sci 4(2):10

Huhman DV, Berhow MA, Sumner LW (2005) Quantification of saponins in aerial and subterranean tissues of Medicago truncatula. J Agric Food Chem 53(6):1914–1920. https://doi.org/10.1021/jf0482663

Improta F, Kellems RO (2001) Comparison of raw, washed and polished quinoa (Chenopodium quinoa Willd.) to wheat, sorghum or maize based diets on growth and survival of broiler chicks. Livestock Res Rural Dev 13(1):10

Ismail H, Dragišic Maksimovic J, Maksimovic V, Shabala L, Živanovic BD, Tian Y, Shabala S (2018) Rutin, a flavonoid with antioxidant activity, improves plant salinity tolerance by regulating K + retention and Na + exclusion from leaf mesophyll in quinoa and broad beans. Funct Plant Biol 43(1):75. https://doi.org/10.1071/fp15312

Jacobsen S-E (2003) The worldwide potential for Quinoa (Chenopodium quinoa Willd.). Food Rev Int 19(1–2):167–177. https://doi.org/10.1081/fri-120018883

Jacobsen SE, Mujica A, Jensen CR (2003) The resistance of quinoa (Chenopodium quinoa Willd.) to adverse abiotic factors. Food Rev Int 19(1–2):99–109. https://doi.org/10.1081/fri-120018872

Jarvis DE, Ho YS, Lightfoot DJ, Schmöckel SM, Li B, Borm TJA, Ohyanagi H, Mineta K, Mitchell CT, Saber N, Kharbatia NM, Rupper RR, Sharp AR,

Dally N, Boughton BA, Woo YH, Gao G, Schijlen EGWM, Guo X, Momin AA, Negrão S, Al-Babili S, Gehring C, Roessner U, Jung C, Murphy K, Arold ST, Gojobori T, Linden CG, van der Loo EN, van Jellen EN, Maughan PJ, Tester M (2017). The genome of Chenopodium quinoa. Nature 542 (7641):307–312. https://doi.org/10.1038/nature21370 http://www.nature.com/nature/journal/v542/n7641/abs/nature21370.html#supplementary-information

Jeroch H, Flachowsky G, Weissbach F (1993) Futtermittelkunde Gustav Fischer Verlag Jena. Stuttgart, Germany

Jiang X, Hansen HCB, Strobel BW, Cedergreen N (2018) What is the aquatic toxicity of saponin-rich plant extracts used as biopesticides? Environ Pollut 236:416–424. https://doi.org/10.1016/j.envpol.2018.01.058

Kalinowska M, Zimowski J, Pączkowski C, Wojciechowski ZA (2005) The formation of sugar chains in triterpenoid saponins and glycoalkaloids. Phytochem Rev 4(2–3):237–257. https://doi.org/10.1007/s11101-005-1422-3

Kenarova B, Neychev H, Hadjiivanova C, Petkov VD (1990) Immunomodulating activity of ginsenoside Rg1 from Panax ginseng. Jpnese J Pharmacol 54(4):447–454

Kensil CR, Patel U, Lennick M, Marciani D (1991) Separation and characterization of saponins with adjuvant activity from Quillaja saponaria Molina cortex. J Immunol 146(2):431–437

Kenwright PA (1989) Breeding the Andean grain crop Quinoa (Chenopodium quinoa) for cultivation in Britain. PhD thesis, University of Cambridge

Komari T (1990) Transformation of cultured cells of Chenopodium quinoa by binary vectors that carry a fragment of DNA from the virulence region of pTiBo542. Plant Cell Rep 9(6):303–306. https://doi.org/10.1007/BF00232856

Koyro HW, Eisa SS (2008) Effect of salinity on composition, viability and germination of seeds of Chenopodium quinoa Willd. Plant Soil 302(1–2):79–90. https://doi.org/10.1007/s11104-007-9457-4

Koziol MJ (1991) Afrosimetric estimation of threshold saponin concentration for bitterness in quinoa (Chenopodium quinoa Willd). J Sci Food Agric 54(2):211–219. https://doi.org/10.1002/jsfa.2740540206

Koziol MJ (1992) Chemical Composition and Nutritional Evaluation of Quinoa (Chenopodium quinoa Willd.). J Food Compos Anal 5:35–68. https://doi.org/10.1016/0889-1575(92)90006-6

Kuljanabhagavad T, Thongphasuk P, Chamulitrat W, Wink M (2008) Triterpene saponins from Chenopodium quinoa Willd. Phytochemistry 69(9):1919–1926. https://doi.org/10.1016/j.phytochem.2008.03.001

Kuljanabhagavad T, Wink M (2009) Biological activities and chemistry of saponins from Chenopodium quinoa Willd. Phytochem Rev 8(2):473–490. https://doi.org/10.1007/s11101-009-9121-0

Li JF, Zhang D, Sheen J (2015) Targeted plant genome editing via the CRISPR/Cas9 technology. Methods Mol Biol (Clifton, N.J.) 1284:239–255. https://doi.org/10.1007/978-1-4939-2444-8_12

Li J, Hu Z (2009) Accumulation and dynamic trends of triterpenoid saponin in vegetative organs of Achyranthus bidentata. J Integr Plant Biol 51(2):122–129. https://doi.org/10.1111/j.1744-7909.2008.00764.x

Li T, Liu B, Spalding MH, Weeks DP, Yang B (2012) High-efficiency TALEN-based gene editing produces disease-resistant rice. Nat Biotechnol 30(5):390–392. https://doi.org/10.1038/nbt.2199

Lin F, Wang R (2010) Hemolytic mechanism of dioscin proposed by molecular dynamics simulations. J Mol Model 16(1):107–118. https://doi.org/10.1007/s00894-009-0523-0

Liu HW, Li JK, Zhang DW, Zhang JC, Wang NL, Cai GP, Yao XS (2008) Two new steroidal compounds from starfish Asterias amurensis Lutken. J Asian Nat Prod Res 10(6):521–529. https://doi.org/10.1080/10286020801966674

Macías FA, Simonet AM, Galindo JCG (1997) Bioactive steroids and triterpenes from Melilotus messanensis and their allelopathic potential. J Chem Ecol 23 (7):1781–1803. https://doi.org/10.1023/B:JOEC.0000006451.19649.a0

Madl T, Sterk H, Mittelbach M, Rechberger GN (2006) Tandem mass spectrometric analysis of a complex triterpene saponin mixture of Chenopodium quinoa. J Am Soc Mass Spectrom 17(6):795–806. https://doi.org/10.1016/j.jasms.2006.02.013

Man S, Gao W, Zhang Y, Huang L, Liu C (2010) Chemical study and medical application of saponins as anti-cancer agents. Fitoterapia 81(7):703–714

Man S, Wang Y, Li Y-Y, Gao W-Y, Huang X-X (2013) Phytochemistry, pharmacology, toxicology, and structure-cytotoxicity relationship of Paridis Rhizome Saponin. Chin Herbal Med 5(1):33–46

Martens SD, Tiemann TT, Bindelle J, Peters M, Lascano CE (2012) Alternative plant protein sources for pigs and chickens in the tropics – nutritional value and constraints: a review. J Agric Rural Dev Trop Subtrop (JARTS) 113(2):101–123

Mastebroek HD, Limburg H, Gilles T, Marvin HJP (2000) Occurrence of sapogenins in leaves and seeds of quinoa (Chenopodium quinoa Willd). J Sci Food Agric 80(1):152–156. https://doi.org/10.1002/(SICI)1097-0010(20000101)80:1%3c152:AID-JSFA503%3e3.0.CO;2-P

McElhinny E, Peralta E, Mazón N, Danial DL, Thiele G, Lindhout P (2007) Aspects of participatory plant breeding for quinoa in␣marginal areas of Ecuador. Euphytica 153(3):373–384. https://doi.org/10.1007/s10681-006-9200-x

Medina-Meza IG, Aluwi NA, Saunders SR, Ganjyal GM (2016) GC-MS profiling of Triterpenoid Saponins from 28 Quinoa varieties (Chenopodium quinoa Willd.) Grown in Washington State. J Agric Food Chem 64(45):8583–8591. https://doi.org/10.1021/acs.jafc.6b02156

Mertens J, Pollier J, Bossche RV, Lopez-Vidriero I, Franco-Zorrilla JM, Goossens A (2016) The bHLH Transcription Factors TSAR1 and TSAR2 regulate Triterpene Saponin biosynthesis in Medicago truncatula. Plant Physiol 170(1):194–210. https://doi.org/10.1104/pp.15.01645

Meyer BN, Heinstein PF, Burnouf-Radosevich M, Delfel NE, McLaughlin JL (1990) Bioactivity-directed isolation and characterization of Quinoside A: one of the toxic/bitter principles of Quinoa seeds (Chenopodium quinoa Willd.). J Agric Food Chem 38 (1):205–208. https://doi.org/10.1021/jf00091a045

Mishustin BAN (1955) Secretion of toxic substances by alfalfa and their effect on cotton and soil microflora. Akademia Nauk USSR Izvestija Ser Biol 6:3–9

Mizui F, Kasai R, OHTANI, K., & Tanaka, O. K (1988) Saponin from brans of Quinoa, Chenopodium quinoa WILLD. I. Chem Pharm Bull 38(4):1415–1418

Mizui F, Kasai R, Othani K, Tanaka O (1990) Saponins from Bran of Quinoa, Chenopodium Quinoa Willd. II. Chem Pharm Bull 38(2):375–377. https://doi.org/10.1248/cpb.38.375

Moses T, Pollier J, Thevelein JM, Goossens A (2013) Bioengineering of plant (tri)terpenoids: from metabolic engineering of plants to synthetic biology in vivo and in vitro. New Phytol 200:27–43

Mroczek A (2015) Phytochemistry and bioactivity of triterpene saponins from Amaranthaceae family. Phytochem Rev 14(4):577–605. https://doi.org/10.1007/s11101-015-9394-4

Muir AD, Paton D, Thompson K, Aubin AA (2008) CA2328946C

Naderi D, Zohrabi Z, Shakib AM, Mahmoudi E, Khasmakhi-Sabet SA, Olfati JA (2012). Optimization of micropropagation and Agrobacterium-mediated gene transformation to spinach (Spinacia oleracea L.). Adv Biosci Biotechnol 03(07):876–880. https://doi.org/10.4236/abb.2012.37109

Nakajima HO (1976) Color Reaction of Some Sapogenins. Planta Med 29:116–122

Ng KG, Price KR, Fenwick GR (1994) A TLC method for the analysis of quinoa (Chenopodium quinoa) saponins. Food Chem 49(3):311–315. https://doi.org/10.1016/0308-8146(94)90177-5

Oda K, Matsuda H, Murakami T, Katayama S, Ohgitani T, Yoshikawa M (2000) Adjuvant and haemolytic activities of 47 Saponins derived from medicinal and food plants. Biol Chem 381(1):67–74. https://doi.org/10.1515/bc.2000.009

Oleszek W, Hamed A (2010). Saponin-based surfactants. Surfactant Renew Resour 239

Oleszek W, Jurzysta M (1987) The allelopathic potential of alfalfa root medicagenic acid glycosides and their fate in soil environments. Plant Soil 98(1):67–80. https://doi.org/10.1007/BF02381728

Oleszek WA (2002) Chromatographic determination of plant saponins. J Chromatogr A 967(1):147–162. https://doi.org/10.1016/S0021-9673(01)01556-4

Pedersen MW (1965) Effect of Alfalfa Saponin on cotton seed germination. Agron J:516–517

Phillips DR, Rasbery JM, Bartel B, Matsuda SP (2006) Biosynthetic diversity in plant triterpene cyclization. Curr Opin Plant Biol 9(3):305–314. https://doi.org/10.1016/j.pbi.2006.03.004

Potter SM, Jimenez-Flores R, Pollack J, Lone TA, Berber-Jimenez MD (1993) Protein-saponin interaction and its influence on blood lipids. J Agric Food Chem 41 (8):1287–1291

Prego I, Maldonado S, Otegui M (1998) Seed structure and localization of reserves in Chenopodium quinoa. Ann Bot 82(4):481–488. https://doi.org/10.1006/anbo.1998.0704

Pulvento C, Riccardi M, Lavini A, Iafelice G, Marconi E, D'Andria R (2012) Yield and quality characteristics of Quinoa grown in open field under different saline and non-saline irrigation regimes. J Agron Crop Sci 198 (4):254–263. https://doi.org/10.1111/j.1439-037X.2012.00509.x

Quiroga Ledezma C (2015) Traditional processes and technological innovations in Quinoa harvesting, processing and industrialization, pp 218–249

Rajput ZI, Hu S-H, Xiao C-W, Arijo AG (2007) Adjuvant effects of saponins on animal immune responses. J Zhejiang Univ Sci B 8(3):153–161. https://doi.org/10.1631/jzus.2007.B0153

Rao AV, Sung MK (1995) Saponins as anticarcinogens. J Nutr 125(suppl_3):717S–724S

Ribeiro BD, Alviano DS, Barreto DW, Coelho MAZ (2013) Functional properties of saponins from sisal (Agave sisalana) and juá (Ziziphus joazeiro): critical micellar concentration, antioxidant and antimicrobial activities. Colloids Surf A 436:736–743. https://doi.org/10.1016/j.colsurfa.2013.08.007

Risi J (1986) Adaptation of the Andean grain crop quinoa (Chenopodium quinoa Willd.) for cultivation in Britain. PhD thesis, University of Cambridge

Roy PK, Munshi JD, Dutta HM (1990) Effect of saponin extracts on morpho-histology and respiratory physiology of an air-breathing fish, Heteropneustes fossilis (Bloch). J Freshwater Biol 2(2):135–145

Ruales J, Nair BM (1993) Saponins, phytic acid, tannins and protease inhibitors in quinoa (Chenopodium quinoa, Willd) seeds. Food Chem 48(2):137–143. https://doi.org/10.1016/0308-8146(93)90048-K

San Martín R, Ndjoko K, Hostettmann K (2008) Novel molluscicide against Pomacea canaliculata based on quinoa (Chenopodium quinoa) saponins. Crop Protect 27(3–5):310–319. https://doi.org/10.1016/j.cropro.2007.03.015

Sen S, Makkar HPS, Becker K (1998) Alfalfa saponins and their implication in animal nutrition. J Agric Food Chem 46(1):131–140. https://doi.org/10.1021/jf970389i

Shany S, Gestetner B, Birk Y, Bondi A (1970) Lucerne saponins III.—effect of lucerne saponins on larval growth and their detoxification by various sterols. Sci Food Agric 21(10):508–510

Sindambiwe JB, Calomme M, Geerts S, Pieters L, Vlietinck AJ, Vanden Berghe DA (1998) Evaluation of biological activities of triterpenoid saponins from Maesa lanceolata. J Nat Prod 61(5):585–590. https://doi.org/10.1021/np9705165

Soleimani M, Kirk W (2012) Enhance resistance to alternaria alternata causing potato brown leaf spot disease by using some plant defense inducers. J Plant

Protect Res 52(1):83–90. https://doi.org/10.2478/v10045-012-0014-7

Solíz JB, Rodriguez DJD, Rodríguez-García R, Angulo-Sánchez JL, Méndez-Padilla G (2002) Quinoa Saponins: concentration and composition analysis. Trends in new crops and new uses, pp 110–114

Sparg SG, Light ME, Van Staden J (2004) Biological activities and distribution of plant saponins. J Ethnopharmacol 94(2–3):219–243. https://doi.org/10.1016/j.jep.2004.05.016

Städler E, Rosenthal GA, Berenbaum MR (1992) Behavioral responses of insects to plant secondary compounds. In: Herbivores: their interactions with secondary plant metabolites: ecological and evolutionary processes, vol 2, pp 45–88

Stuardo M, San Martín R (2008) Antifungal properties of quinoa (Chenopodium quinoa Willd) alkali treated saponins against Botrytis cinerea. Ind Crops Prod 27(3): 296–302. https://doi.org/10.1016/j.indcrop.2007.11.003

Takagi S, Otsuka H, Akiyama T, Sankawa U (1986) Digitonin-cholesterol complex formation: effects of varying the length of the side-chain. Chem Pharm Bull 30(11):3485–3492. https://doi.org/10.1248/cpb.30.3485

Takao T, Watanabe N, Yuhara K, Itoh S, Suda S, Tsuruoka Y, Nakatsugawa K, Konishi Y (2005) Hypocholesterolemic effect of protein isolated from quinoa (Chenopodium quinoa Willd.) seeds. Food Sci Technol Res 11(2):161–167

Teixeira JR, Lapa AJ, Souccar C, Valle JR (1984) Timbós: ichthyotoxic plants used by Brazilian Indians. J Ethnopharmacol 10(3):311–318. https://doi.org/10.1016/0378-8741(84)90018-7

Thimmappa R, Geisler K, Louveau T, O'Maille P, Osbourn A (2014) Triterpene biosynthesis in plants. Annu Rev Plant Biol 65(1):225–257. https://doi.org/10.1146/annurev-arplant-050312-120229

USEPA (2002) Saponins of Chenopodium quinoa (097094) Fact sheet. United States Environmental Protection Agency. https://www3.epa.gov/pesticides/chem_search/reg_actions/registration/fs_PC-097094_30-Jan-02.pdf

Van Atta GR, Guggolz J, Thompson CR (1961) Plant analysis: determination of Saponins in Alfalfa. J Agric Food Chem 9(1):77–79. https://doi.org/10.1021/jf60113a022

van Dyck S, Flammang P, Meriaux C, Bonnel D, Salzet M, Fournier I, Wisztorski M (2010) Localization of secondary metabolites in marine invertebrates: contribution of MALDI MSI for the study of saponins in Cuvierian tubules of H. forskali. PLOS ONE 5(11). https://doi.org/10.1371/journal.pone.0013923

Varriano-Marston E, Defrancisco A (1984) Ultrastructure of quinoa fruit (Chenopodium quinoa Willd). J Food Struct 3(3):165–173

Vincken J-P, Heng L, de Groot A, Gruppen H (2007) Saponins, classification and occurrence in the plant kingdom. Phytochemistry 68(3):275–297. https://doi.org/10.1016/j.phytochem.2006.10.008

Voutquenne L, Lavaud C, Massiot G, Men-Olivier LL (2002) Structure-activity relationships of haemolytic Saponins. Pharm Biol 40(4):253–262. https://doi.org/10.1076/phbi.40.4.253.8470

Wang Y, Zhang Y, Zhu Z, Zhu S, Li Y, Li M, Yu B (2007) Exploration of the correlation between the structure, hemolytic activity, and cytotoxicity of steroid saponins. Bioorg Med Chem 15(7):2528–2532. https://doi.org/10.1016/j.bmc.2007.01.058

Ward SM (2002) A recessive allele inhibiting Saponin synthesis in two lines of Bolivian Quinoa (Chenopodium quinoa Willd.). J Heredity 92(1):83–86. https://doi.org/10.1093/jhered/92.1.81

Weickert MJ, Pagratis M, Curry SR, Blackmore R (1997) Stabilization of apoglobin by low temperature increases yield of soluble recombinant hemoglobin in *Escherichia coli*. Appl Environ Microbiol 63(11):4313–4320

Woldemichael GM, Wink M (2001) Identification and biological activities of triterpenoid saponins from Chenopodium quinoa. J Agric Food Chem 49 (5):2327–2332. https://doi.org/10.1021/jf0013499

Yábar E, Gianoli E, Echegaray ER (2002) Insect pests and natural enemies in two varieties of quinua (Chenopodium quinoa) at Cusco, Peru. J Appl Entomol 126 (6):275–280

Yang A, Akhtar SS, Iqbal S, Qi Z, Alandia G, Saddiq MS, Jacobsen S-E (2018) Saponin seed priming improves salt tolerance in quinoa. J Agron Crop Sci 204(1):31–39. https://doi.org/10.1111/jac.12229

Yang AF, Duan XG, Gu XF, Gao F, Zhang JR (2005) Efficient transformation of beet (Beta vulgaris) and production of plants with improved salt-tolerance. Plant Cell, Tissue Organ Cult 83(3):259–270. https://doi.org/10.1007/s11240-005-6670-9

Yaroshko O, Vasylenko M, Gajdošová A, Morgun B, Khrystan O, Velykozhon L, Kuchuk M (2019) "Floral-dip" transformation of Amaranthus caudatus L. and hybrids A. caudatus × A. paniculatus L. Biologija 64 (4). https://doi.org/10.6001/biologija.v64i4.3904

Zhu N, Sheng S, Sang S, Jhoo JW, Bai N, Karwe MV, Ho CT (2002) Triterpene saponins from debittered quinoa (Chenopodium quinoa) seeds. J Agric Food Chem 50 (4):865–867. https://doi.org/10.1021/jf0110021

Zimmer DE, Pedersen MW, McGuire CF (2010) A bioassay for Alfalfa Saponins using the fungus, Trichoderma viride Pers. ex Fr.1. Crop Sci 7(3):223. https://doi.org/10.2135/cropsci1967.0011183x000700030015x

Zurita-Silva A, Fuentes F, Zamora P, Jacobsen S-E, Schwember AR (2014) Breeding quinoa (Chenopodium quinoa Willd.): potential and perspectives. Mol Breed 34(1):13–30. https://doi.org/10.1007/s11032-014-0023-5

Abiotic Stress Tolerance in Quinoa

9

Luke Grenfell-Shaw and Mark Tester

Abstract

Quinoa (*Chenopodium quinoa* Willd.) is a highly resilient crop, displaying high levels of salinity and drought tolerance, although only low levels of heat tolerance. In this chapter, current knowledge of the response of quinoa to abiotic stresses is discussed, focussing on physiological responses, putative molecular mechanisms and genetic control. Relatively little research has been conducted into quinoa, and even less has been done to elucidate the genetics underpinning quinoa's tolerance to abiotic stresses. The quinoa genes *CqSOS1*, *CqNHX1*, *CqBADH*, *CqHSP70*, *CqHSP20* and *CqABAs* are discussed, and areas requiring further research at the genetic level are highlighted. There are currently extensive gaps in our understanding of the response of quinoa to certain abiotic stresses, with virtually no studies covering the effects of soil acidity, boron toxicity and nutrient deficiency. Whilst there is significant genetic diversity within the quinoa species, significant research is required to understand how this genetic variation can be harnessed. The recent release of a high-quality quinoa reference genome provides a useful tool to greatly facilitate the characterisation of genes involved in abiotic stress tolerance and taking a genetics-led approach should aid the rapid advancement of our understanding of quinoa's abiotic stress tolerance mechanisms and lead to more effective improvement in engineering and breeding strategies. With these improvements, quinoa can fulfil a role urgently needed to provide food security in arid and semi-arid regions.

9.1 Introduction

Declining freshwater resources and increasing levels of soil salinisation are two challenges that affect agriculture in semi-arid and arid areas. These problems will become more widespread and severe because of global climate change, particularly in the Middle East and North Africa region (IPCC 2014). Halophytic crops, such as quinoa, have the potential to sustain agriculture in such areas, and could be at the forefront of a saline irrigation revolution (Koyro and Lieth 2011). Quinoa (*Chenopodium quinoa* Willd.) has been cultivated in the Andes region for more than 7000 years (Dillehay et al. 2007). It is remarkable for its high nutritional value: its seed consists of ~16% protein, with, importantly, an optimal balance of essential amino acids for sustained human health; 4–9% oil; 64% carbohydrates and vitamins (A, B2, E) and minerals

L. Grenfell-Shaw · M. Tester (✉)
Division of Biological and Environmental Sciences and Engineering (BESE), King Abdullah University of Science and Technology (KAUST), Thuwal 23955-6900, Saudi Arabia
e-mail: mark.tester@kaust.edu.sa

S. M. Schmöckel (ed.), *The Quinoa Genome*, Compendium of Plant Genomes,
https://doi.org/10.1007/978-3-030-65237-1_9

(calcium, iron, copper, manganese, zinc) (Repo-Carrasco et al. 2003; Escuredo et al. 2014; Bhargava et al. 2006; Vega-Gálvez et al. 2010).

Additionally, quinoa exhibits high tolerance to certain abiotic stresses, particularly salinity and drought. Some cultivars of quinoa have been shown to grow in saline conditions similar to seawater (40 dS m^{-1}, about 400 mM NaCl) (Jacobsen et al. 2001; Bazile et al. 2015; Adolf et al. 2013). There are a diverse set of ecotypes; the five main ecotypes are variously adapted to valley, *altiplano (*mountainous, up to 4000 m), salt desert, sea level and tropic environments, suggesting excellent genetic diversity, which should assist in the development of superior quinoa strains (Bendevis et al. 2014; Jacobsen et al. 2003; Vega-Gálvez et al. 2010; Bertero et al. 2004). Quinoa's high genetic diversity has been confirmed by (Christensen et al. 2007) and (Fuentes et al. 2009). However, this makes direct comparison of studies which use different quinoa cultivars difficult; for example, the resistance of *altiplano* varieties to salinity often exceeds that of valley varieties (Shabala et al. 2013). This variation could explain some of the discrepancies seen between studies. Therefore, direct comparison between studies with different cultivars must be done with caution, with the potential for confusion regarding resistance to abiotic stress. It would be an important step forward for the research community to select a model cultivar for each ecotype in order to aid comparison between studies. However, this variation between cultivars will prove invaluable in the future as their genetic variation is exploited through identification and selection of superior genes and alleles to develop highly resistant quinoa strains.

Despite its long history of cultivation and its central role in agriculture in parts of South America, quinoa is yet to undergo an intensive domestication programme, making it a less attractive option for commercial agriculture (Morales et al. 2017). It has enormous potential to become a food security crop in marginal soils, providing a more reliable food source. To be successful in these areas, such as Pakistan and Bangladesh, where high temperature can be a major abiotic stress, quinoa's heat tolerance will need to be improved. Understanding and evaluating the wealth of genetic diversity present within the quinoa species will give researchers the opportunity to optimise different quinoa accessions to withstand the specific stresses faced in key environments.

In this chapter, the stresses of salinity, drought, temperature, nutrient deficiency, boron toxicity, soil acidity and metal toxicity are considered. For each, the following themes will be examined: the geographic extent of the stress; the general symptoms exhibited by quinoa; the physiological response of quinoa; the mechanisms of stress tolerance and their molecular basis; our current genetic understanding and future research prospects and opportunities offered by the recent release of the high-quality genome sequence of quinoa (Jarvis et al. 2017).

The majority of research to date has focussed on the response of quinoa to salinity, drought and cold stresses. Many of the physiological mechanisms of tolerance are still poorly understood, and our knowledge at the molecular and genetic level is mostly sparse. Thus, this chapter also serves as a road map to indicate current gaps in our knowledge.

Whilst these stresses are considered in turn, it is important to remember that an abiotic stress rarely occurs in isolation: salinity stress is often accompanied by drought stress, and drought stress often occurs alongside heat stress (Savin and Nicolas 1996). Additionally, boron toxicity is often observed in conjunction with soil salinity and alkalinity (Goldberg 1997). However, the combined effects of abiotic stresses are generally beyond the scope of this chapter.

This review aims to amalgamate key advances yielded by research in the field of quinoa. Whilst relatively little research has been conducted at the genetic level there are a few studies which have investigated the genetics of quinoa to great effect (Schmöckel et al. 2017; Jiang et al. 2016; Morales et al. 2017; Liu et al. 2018a; Maughan et al. 2009). As this more penetrating approach is increasingly employed, it has the potential to rapidly develop our understanding and ability to engineer quinoa to realise its long-predicted potential.

9.2 Soil Salinity

Soil salinity is a major constraint on global agriculture and impacts both food security and political stability. Currently, an estimated 1,128 Mha, including 20% of irrigated land, is affected by soil salinity. The largest single area, 189 Mha, is situated in the Middle East (Wicke et al. 2011). This area is predicted to increase because of climate change and poor irrigation practices (Hayes et al. 2015). The impact of soil salinity has been estimated to cause an annual loss of US$27.3 billion (Qadir et al. 2014). Salinity tolerance allows plants to continue growing and producing photoassimilates under saline conditions; this is governed by a polygenic mechanism (Roy et al. 2014; Munns 2002). Quinoa has been successfully grown in NaCl concentrations equivalent to that of seawater, up to 40 dS m^{-1}, or 400 mM NaCl, making it the most salt-tolerant crop (Jacobsen et al. 2001; Bazile et al. 2015; Adolf et al. 2013; Munns and Tester 2008).

As detailed by Ruiz et al. (2016), there is wide-ranging variability in the tolerance of quinoa ecotypes to salinity stress. For example, seeds of the Peruvian cultivar, Kancolla, could germinate at salinity levels of up to 57 dS m^{-1}. By contrast, Gómez-Pando et al. (2010) studied 182 Peruvian accessions, and only the 15 most tolerant accessions had a germination rate above 60% at salinity levels of 25 dS m^{-1}.

Quinoa exhibits a wide range of tolerance mechanisms, including shoot ion-independent tolerance (osmotic tolerance), ionic tolerance and tissue tolerance (Munns and Tester 2008; Roy et al. 2014), but also other putative mechanisms which are not seen in other crops and are as yet unverified, such as "salt bladders" (Shabala et al. 2014; Kiani-Pouya et al. 2017; Adolf et al. 2013; Shabala 2013). Whilst these mechanisms are at least partially understood, relatively little is known about their genetic control and only a small number of salinity tolerance genes have been identified in quinoa.

Symptoms of salinity stress generally include growth inhibition, accelerated development and senescence and under prolonged exposure, death. Growth inhibition is the predominant factor, from which the other symptoms follow. Abscisic acid (ABA) is synthesised when the plant experiences salt stress, causing closure of stomata which adversely affects the rate of photosynthesis. Cell expansion is limited either directly through osmotic stress, or by ABA (Munns and Tester 2008).

First, morphological traits will be considered, followed by physiological and metabolic parameters. Genetic analysis will be included where possible, and highlighted where it is lacking.

9.2.1 Morphology

9.2.1.1 Seed Structure

NaCl has been shown to markedly affect germination of seeds in halophytes (Debez et al. 2004). A study by (Prado et al. 2000) found that in the presence of 400 mM NaCl, only 14% of seeds had germinated after 14 h, compared with the control, where 87% had germinated. Reduced glucose and fructose levels were found in the embryonic axes in the presence of NaCl.

In the Peruvian cultivar Hualhuas, grown under varying salinity conditions (up to 500 mM NaCl), the distribution of minerals was altered, but was highly regulated. The changes in distribution did not appear to damage seed or viability. It has been suggested that seed structure and compartmentalisation give quinoa seeds added tolerance to salt stress (Koyro and Eisa 2008). The embryo is located peripherally, around the perisperm storage tissue. The seed coat appeared to limit the entry of Na^+ and Cl^- into the interior, as over 90% of the Na^+ and Cl^- was found in the seed coat. This might suggest that the seed coat and perisperm are important protective barriers for excluding Na^+ and Cl^-. It might also help maintain a high K^+/Na^+ ratio, as whilst Na^+ levels still increased, the K^+/Na^+ ratio never went below 1, probably aiding germination. Hariadi et al. (2011) suggested the ability of a seed to exclude Na^+ from the developing embryo was crucial for seed viability.

In an experiment by Ruiz-Carrasco et al. (2011), seeds of four genotypes were grown on agar plates with 0, 150 or 300 mM NaCl. At 150 mM NaCl, seed germination was not significantly affected. However, at 300 mM NaCl, the germination rate for the cultivar BO78 was almost 0 and had dropped by 15–30% in the other cultivars. Whilst at 150 mM NaCl root length was not reduced, and even increased in some cultivars, at 300 mM NaCl root growth was inhibited by 25% in the PRJ cultivar, and by about 70% in BO78. Shoot growth was negatively affected at both 150 and 300 mM NaCl. Thus, in this instance the root/shoot ratio was differentially affected by salt. Interestingly, under 300 mM NaCl conditions, RT-PCR analyses of *CqSOS1* and *CqNHX1* showed that their expression was also differentially expressed in the shoots and roots, as well as between genotypes. However there did not appear to be a correlation between levels of *CqSOS1* and *CqNHX1* and the length of root or shoot. Orsini et al. (2011) reported that in vitro germination of BO78 was delayed by 150 mM NaCl, but ultimately reached the same level as the control. In seedlings, root growth increased under 150 mM NaCl, compared to the control. However, both germination rate and root growth decreased under 300 mM NaCl.

Interestingly, saponin seed priming has been reported to improve salt tolerance in quinoa (Yang et al. 2018). Seeds were primed in seven solutions of different saponin concentrations. Saponin concentrations between 10% and 25% were most effective for alleviating adverse effects of salt stress during seed germination. Improved growth, physiology and yield performance were linked with low ABA concentration, improved water potential, higher photosynthetic rate and stomatal conductance, and lower Na^+ and K^+ in the leaves. A further study would be to investigate which genes are differentially expressed upon saponin treatment. If these genes can be upregulated without saponin application then it could improve quinoa salinity tolerance.

The high tolerance of quinoa seeds might stem from both structural and physiological features; however, little follow-up work has been done to date. The recent sequencing of the high-quality quinoa genome should assist research through aiding mutagenesis, knockout or RNAi studies to investigate the genes involved in the seed coat/Na^+ exclusion hypothesis and the salinity tolerance displayed by quinoa seeds (Jarvis et al. 2017).

9.2.1.2 Stomata

Part of the osmotic (ion-independent) response of plants to salinity is the rapid closure of stomata, leading to a reduction in growth rate (Munns 2002; Flowers 2004). When exposed to long-term salt stress, the density and size of stomata also tend to decrease (Munns and Tester 2008). Stomata also close in response to a decrease in leaf turgor, high atmospheric vapour pressure deficit and by water deficit in the root zone (Schachtman and Goodger 2008; Roelfsema and Hedrich 2005). This is thought to be a response to improve water use efficiency (WUE). A 54% reduction of stomatal density at 750 mM NaCl in the relatively salt-sensitive cultivar, BO78, was seen, and at 300 mM NaCl, there was around a 30% reduction (Orsini et al. 2011). A study of 14 varieties of quinoa reported that, whilst stomatal density was reduced for all varieties under saline conditions (incremental salinity increase up to 400 mM NaCl), the degree to which this occurred varied between genotypes (mostly by 20–30%, but up to around 45%) (Adolf et al. 2013; Shabala et al. 2013).

However, a study conducted by Becker et al. (2017) observed the opposite response. In the cultivar Achachino, they reported the number of stomata increased in response to salt stress. However, the size of the stomata were smaller and the percentage leaf area covered by stomata was lower in salt-treated plants and stomatal conductance was reduced. This was the suggested cause for the reduction in photosynthesis, which in turn was reflected in decreased plant biomass and yield. Under saline conditions, when water availability is reduced, findings have been inconsistent, showing either increased or decreased stomatal densities (Flowers et al. 1977; Hetherington and Woodward 2003; Orsini et al. 2011; Shabala et al. 2012;).

The picture is further complicated by the study of Prado et al. (2017). For two cultivars, Pasankalla and CICA, there was an increase in stomatal area and density under saline conditions; in another cultivar (Wariponcho), stomatal area decreased under saline treatment. There was further variation between cultivars, regarding stomatal density and area on the adaxial and abaxial leaf surfaces. In Wariponcho, abaxial stomata area and stomatal density was reduced by saline treatment, and in Witulla, on the adaxial surface. In Wariponcho, the lower stomatal density was proposed to increase WUE. Therefore, the stomatal area and density does not appear to be directly related to WUE. How CICA and Pasankalla increase their WUE under saline conditions whilst increasing the stomatal area needs further investigation.

The importance of stomatal number, area and density is still unclear. Understanding the genetic control over stomatal size and density will allow the precise investigation of the role of stomata in WUE and salinity tolerance, and likely also drought tolerance.

9.2.2 Stomatal Conductance, Gas Exchange and Photosynthetic Rate

Stomatal conductance decreases under saline conditions, and is regulated by the size of the stomatal pores (Munns and Tester 2008). Ion fluxes are associated with the rapid and reversible opening and closing of the stomata by the guard cells. ABA controls this process. Accordingly, an increase in ABA, along with decreased leaf and soil water potential, is symptomatic of osmotic stress caused by salinity.

A decrease in stomatal conductance not only reduces water loss, but also CO_2 uptake (Iyengar and Reddy 1996). A study by Adolf et al. (2013) compared stomatal conductance and photosynthetic CO_2 assimilation of two quinoa cultivars, Utusaya and Titicaca, under saline conditions. The Utusaya cultivar only showed a 25% reduction in net CO_2 assimilation under 400 mM NaCl conditions, compared to a 67% reduction in Titicaca. However, the results are more nuanced. The stomatal conductance and thus the photosynthetic rate are significantly lower in Utusaya than Titicaca under non-saline conditions. The actual rate of CO_2 assimilation for each cultivar under salinity stress was roughly the same. This suggests that Titicaca is better able to exploit more benign conditions than Utusaya, but might indicate an intra-species baseline of WUE. An ideal cultivar would have maximum growth possible under both favourable and stressful conditions. Investigating quinoa's plasticity and its genetic control over stomatal conductance and establishing lower and upper boundaries could be an important step forward. Concurring with this work, Becker et al. (2017) found that stomatal conductance and the internal CO_2 concentration were reduced under saltwater conditions.

Miranda-Apodaca et al. (2018) investigated stomatal conductance under saline conditions, as well as drought conditions. Under the highest saline conditions (500 mM NaCl), an 83% reduction was seen in stomatal conductance and a 65% reduction in the net photosynthetic rate was observed. The intercellular CO_2 concentration also decreased as the stress became more severe. Their results also indicate that up to 240 mM NaCl, the photosynthetic rate is largely limited by stomatal conductance. Under saline conditions, osmotic adjustment capacity was higher whilst the photosynthetic non-stomatal limitations were lower than under drought stress. Additionally, the shoot/root ratio did not change. Interestingly, a greater non-stomatal effect was seen under drought conditions, as demonstrated by an increase in intercellular CO_2 concentration. Whilst WUE was maintained under salinity stress, it increased under drought conditions.

A study by Killi and Haworth (2017) found that under saline conditions of 300 mM NaCl, there was a reduction in stomatal and mesophyll conductance—similar to that seen under drought-only conditions—but also impaired PSII electron transport, reduced carboxylation of RUBISCO and regeneration of ribulose-1, 5-bisphosphate, alongside increased heat dissipation of non-photochemical energy. This could suggest that photosynthesis was reduced by ion toxicity, potentially through interference with photosynthetic enzymes and degradation of protein-pigment complexes within the thylakoid membrane.

Orsini et al. (2011) found that transpiration and conductance only decreased at salinity levels in excess of 600 mM NaCl, in contrast to the other studies which report affects at much lower concentrations of NaCl. This variation could be due to different experimental techniques, cultivars or developmental stage studied.

Significant research is required to elucidate the genetic control over stomatal conductance and photosynthetic rate in quinoa. Additionally, understanding and then selecting plants that display a high degree of flexibility towards their environment—exploiting benign conditions, but highly tolerant to stress—would provide an important step forward in the development of quinoa. Transient expression analysis could be utilised to analyse expression variation under salinity stress and the physiological effects of RNAi could be studied.

9.2.3 Ionic Tolerance

9.2.3.1 Sodium Loading and Translocation

Na^+ exclusion has been recognised as a key trait for salinity tolerance for some time (Munns and Tester 2008). In quinoa, ion toxicity is mitigated by loading Na^+ into the xylem and into the vacuoles. The genes *SOS1* and *NHX1* are suggested to perform this function, respectively. In *Arabidopsis*, the gene known as *Salt Overly Sensitive 1* (*SOS1*) codes for a Na^+/H^+ exchanger, which transports Na^+ out of the cell. It is situated at the plasma membrane of epidermal root cells and in parenchyma cells at the xylem/symplast boundary of roots, stems and leaves (Blumwald 2000; Qiu et al. 2002; Shi 2002). Maughan et al. (2009) cloned and characterised two *SOS1* homologs in quinoa. These were found to have a high level of similarity to *SOS1* homologs of other species. Interestingly, *CqSOS1A* and *CqSOS1B* were more highly expressed in roots compared to the leaves under non-saline conditions, but under salt stress, expression of both genes was upregulated in the leaves but not the roots. This suggests that *CqSOS1* expression levels in the roots do not respond to saline conditions. Ruiz-Carrasco et al. (2011) confirmed that *CqSOS1* expression differed between roots and shoots under salinity stress, however whether there was an increase or decrease depended on the cultivar; in contrast to Maughan's results there was not a common expression pattern. The most tolerant genotype studied, PRJ, was the only one to have enhanced expression of *CqSOS1* in the roots. This suggests salt tolerance could be associated with increased Na^+ efflux from the plant. Supporting this, a study in *Arabidopsis thaliana* reported an inverse relationship between expression of *AtSOS1* and levels of Na^+ in the plant (Jha et al. 2010).

Highlighting the importance of SOS1 in halophytes, a genetic study into *Eutrema salsugineum* and *Eutrema parvulum* found that knocking out *SOS1* decreases their stress tolerance (Oh et al. 2009; Wu et al. 2012; Oh et al. 2010). An equivalent study in quinoa would be very useful, to improve our understanding of the importance of *SOS1*, *NHX1* and *HKT1;1* in quinoa's resistance to salinity.

NHX1 is a counterpart to SOS1, and transports Na^+ into the vacuole (Shi et al. 2000). In *Arabidopsis NHX1* codes for a tonoplast-localised vacuolar Na^+/H^+ antiporter, which is believed to cause Na^+ compartmentalisation, and potentially also K^+ homeostasis in the vacuole (Bassil et al. 2011). Other transporters may be involved (Volkov 2015). *CqNHX1* expression was reported to be enhanced by 300 mM NaCl in the seedlings of several cultivars. However, in the cultivar most sensitive to salt, BO78, expression was unchanged, contrasting with the most salt-resistant cultivars which reported enhanced *CqNHX1* expression (Ruiz et al. 2016). The increased expression of *CqNHX1* in the roots of the salt-tolerant cultivars indicates that compartmentalisation of Na^+ or K^+ is an important tolerance mechanism in this species. Curiously, in the roots, the two most tolerant cultivars had different expression patterns: in PRJ, an upregulation of *CqNHX1* was seen, whereas in PRP no increase was seen. Thus, it has been suggested, as with *CqSOS1*, that quinoa tolerance may depend on other mechanisms of ion uptake, exclusion, translocation and compartmentalisation (Ruiz et al. 2016). Like *CqSOS1*, the action of *CqNHX1* in different parts of the plant, under differing levels of salinity and between cultivars is yet to be fully understood.

Morales et al. (2011) conducted a comparative analysis of *CqSOS1* and *CqNHX1* under salinity (450 mM NaCl). Stable expression of each was seen in the leaves, suggesting constant transcription of those genes. However, in the roots there was variable expression of *CqSOS1* between ecotypes, the highest levels of expression being seen in the *salares* (salt-flat) ecotype, suggesting cytoplasmic Na^+ was being transported out of the roots. *Salares* is also highly salt resistant, suggesting that *CqSOS1* expression in roots improves salt tolerance. It would be a useful experiment to overexpress *CqSOS1* and measure the salt tolerance. However, expression of *CqNHX1* in the roots did not appear to be affected by salt stress.

Shabala et al. (2013) reported that differences in Na^+ uptake were associated with genotypic differences. The most tolerant cultivars displayed lower xylem Na^+ content, implying they had elevated SOS1 and NHX1 activity.

The recent paper by Zou et al. (2017) also looked into the gene expression of ion transporters in the cultivar Real under saline conditions of 100 mM NaCl. They identified several genes for ion transporters, which are more highly expressed in epidermal bladder cells (EBCs), including the anion transporters *SLAH*, *NRT* and *CIC*, and the cation transporters *NHX1* and *NKT1*. In addition, the plasma membrane and vacuolar H^+-ATPases were upregulated. Interestingly, they reported that the transcript level was raised for 10 sugar transporters in the EBCs which they suggested was to support their energy and nutrient requirements.

Enhanced transporter activity was also seen, with higher levels of sodium-proton exchangers, *HKT*s and *SLAH*s seen.

Schmöckel et al. (2017) aimed to identify salt tolerance genes in quinoa through RNAseq analyses, comparative genomics and topology prediction. They identified 15 genes which could be important for salt tolerance. They focussed on proteins which had a transmembrane domain, as typically exhibited by transporter proteins. One candidate gene identified shared homology with ankyrin repeat proteins and is related to *Shaker*-like K^+ channels, suggesting a role in maintaining high K^+ concentrations under salinity stress, allowing the plant to maintain a preferable Na^+/K^+ ratio (Becerra et al. 2004). K^+ homeostasis is important for maintaining enzyme functionality, necessary for plant growth (Munns and Tester 2008). Other candidate genes postulated to mediate salt tolerance include sulphate transporters, an amino acid permease, a photosystem II D2 protein, a flavonoid 3'-monooxygenase and a receptor protein kinase 1. Further studies are now required to investigate the role and importance of these candidate genes.

An interesting gene that has been linked with salinity tolerance in barley is *MSBP* (membrane steroid binding protein). Three candidate genes, involved in sterol binding or phospholipid synthesis were upregulated. Overexpression of *MSBP* in yeast gave increased salinity tolerance, whilst the knockout of *AtMSBP1* in *Arabidopsis* increased its salt sensitivity. *Atmsbp1* plants displayed a reduced number of lateral roots under salinity. However, in barley, increased levels of MSBP correlated with reduced root length and lateral root formation, plus increased levels of auxin under salinity (Witzel et al. 2018). It would be interesting to investigate if a homolog of *MSBP* existed in quinoa, and if it played a role in salinity tolerance.

Ma et al. (2006) reported that fewer than 25% of the salt-regulated genes that had been identified were known to be salt-stress-specific in *Arabidopsis*. This suggests many abiotic stress tolerance genes have a broad function and interact in complex, interrelated systems. Stress responses are interrelated, so it is likely that many genes which lead to an increase in salt tolerance also increase the tolerance to other stresses.

To gain a deeper understanding of the role of *SOS1*, *NHX1*, *HKT1;1* and the other genes mentioned, further genetic studies are required to investigate how they can be harnessed to increase salinity tolerance.

9.2.4 Tissue Tolerance

9.2.4.1 Osmoprotective Molecules and Other Protective Mechanisms

Quinoa appears to have relatively effective and multifaceted tissue tolerance. This is through a combination of osmoprotective molecules, carbohydrate metabolism and sequestering of toxic ions into the vacuole.

When seedlings, carbohydrate metabolism appears to play an important role in salinity tolerance, allowing osmotic adjustment. One study showed changes in the content of glucose, sucrose and fructose between salt-treated and non-treated seedlings (Prado et al. 2000). There were higher levels of soluble sugars under distilled water conditions, except for in cotyledons, where the levels of sucrose were higher. Another study investigated sucrose-starch partitioning and related enzymes in seedlings of the cultivar Sajama under low temperature (Rosa et al. 2009). In salt-stressed plants, higher levels of sucrose phosphate synthase and soluble acid invertase activity were seen. Concurrently, there was an increase in soluble sugars which help to preserve the optimal osmotic balance under saline conditions. This might suggest that sugars increase salt tolerance.

It has also been suggested that a reduced matric potential in the seed interior may decrease water loss under conditions of high external osmolality (Bazile et al. 2015). Following the observation that seeds from salt-treated quinoa plants have higher levels of protein, Koyro and Eisa (2008) suggested that increased protein levels may contribute to lowering the matric potential.

Dehydrins appear to have osmoprotective properties. They have been found in nearly all vegetative tissues under drought, cold and salinity stress (Battaglia et al. 2008; Rorat et al. 2006). When wheat dehydrin was ectopically expressed in *Arabidopsis thaliana*, increased tolerance to salinity and drought stress as seen. Other studies have indicated dehydrins play a key role in stress tolerance (Saavedra et al. 2006). Dehydrins have been observed in quinoa. Carjuzaa et al. (2008) conducted Western blot analysis of embryo tissues of two quinoa cultivars, one adapted to high altitude and one to sea-level environments. Dehydrin bands common to both cultivars were identified, although the levels of 30 and 32 kDa bands differed. From this, the authors concluded that whilst most of the dehydrins seen were constitutive, some may be associated with giving tolerance to certain environmental conditions.

Building on this, Burrieza et al. (2012) investigated the effect of different salinity levels on dehydrin composition in Hualhuas cultivar embryos exposed to arid and salty conditions. Whilst four dehydrins were immunodetected in both the control and salt-stressed plants, no additional bands were seen under saline conditions. Only one band (the 30-kDa dehydrin) increased under 300 and 500 mM NaCl conditions. Therefore, it is unclear whether dehydrins play a specific role in giving salinity tolerance or give general resistance.

Further investigations into dehydrins are required—particularly if the 30 kDa dehydrin provides increased salinity tolerance. Therefore, identifying the 30 kDa gene and overexpressing it could prove to be a valuable way to improve our understanding of the role of dehydrins and if increased levels would lead to more resistant quinoa cultivars.

Osmoprotective molecules have also been suggested to be important for maintaining cell turgor and enabling cell expansion, even under high salinity. However, their role as "compatible solutes" is currently unclear (Flowers 2004; Shabala and Mackay 2011). Inorganic ions have been suggested to play a more important role than organic osmolytes in maintaining the correct osmotic balance. It has been suggested that accumulating organic osmolytes to maintain turgor is metabolically expensive and inefficient and that inorganic ions, particularly K^+, would be more efficient (Hasegawa et al. 2000; Shabala et al. 2012; Flowers 2004). K^+ is likely to play a more significant role than organic osmolytes in regulating osmotic adjustment. Studying the BO78 cultivar, under the highest saline conditions, of 600 mM NaCl, levels of K^+ were three times higher than in the controls and plants under more moderate saline conditions (less than 450 mM of NaCl), where internal K^+ was unaffected. There was also a modest dose-dependent proline accumulation observed. Whilst the authors note the importance of inorganic ions, it is important to highlight no increase in K^+ concentration was seen until NaCl concentrations in excess 450 mM (Orsini et al. 2011). Quinoa is affected by salt concentrations in excess of 150 mM NaCl, so this would indicate that whilst K^+ may play a role at very high salinity conditions, this may be a "last resort" by the plant to maintain turgor and cellular functions; however, at more commonly encountered saline conditions in the field, other mechanisms might play a more important role.

Whilst the role of organic osmolytes in maintaining the optimal osmolality is unclear, it seems more certain that they play an important role in mitigating the effects of reactive oxygen species (ROS). ROS can cause protein denaturation, lipid peroxidation in cellular membranes, carbohydrate oxidation, pigment breakdown, DNA damage and interference with enzyme activity (Noctor and Foyer 1998). Under salinity stress, there is impaired photosynthetic activity, owing to reduced stomatal conductance and the toxic effects of Na^+ accumulation (Munns and Tester 2008; Bazile et al. 2015). This can cause levels of ROS to increase.

The four major classes of organic osmolyte are found in quinoa—amino acids, sugars, polyols and quaternary amines and their putative importance is of much interest (Aguilar et al. 2003; Ruffino et al. 2010; Orsini et al. 2011; Ruiz-Carrasco et al. 2011). However, again, there is still much uncertainty over the role and importance of different osmolytes.

Some studies indicate proline plays an important role in salt tolerance, whilst others suggest that it is unimportant. According to Szabados and Savouré (2010), proline appears to play important roles as a compatible solute, osmoprotectant and signalling molecule. In the BO78 cultivar, proline concentrations in the stem and leaf increased tenfold under the highest salinity conditions (600 and 750 mM NaCl) compared to control conditions (Orsini et al. 2011). Ruiz-Carrasco et al. (2011) noted a 2- to fivefold increase in proline levels under 300 mM NaCl, with the authors noting that greater increases seemed to be associated with cultivars from more stress-prone habitats. Interestingly, it has been observed that the absolute levels of proline, and other osmolytes, remain very low, despite a large relative increase. (Ruffino et al. 2010) and (Hariadi et al. 2011) suggest this indicates their osmoprotective role instead of an osmotic one.

However, other studies have thrown doubt on the importance of proline. Morales et al. (2011) observed that whilst there were high levels of betaine, trehalose and trigonelline in the cultivars Chipaya and Ollague, there were negligible levels of proline, sorbitol and inositol. Under salt stress, it was levels of trigonelline, betaine and trehalose which increased. Another study suggested that other osmoprotective molecules play a more important role than proline. Ismail et al. (2016) studied two quinoa genotypes and found that proline accumulation appeared to have little role in either tissue tolerance or osmotic adjustment. They also suggested that rutin plays a key role as an antioxidant under salt stress, with a 25-fold increase seen in quinoa leaves upon salt stress. Exogenous application to the glycophyte broad bean was reported to improve tissue tolerance.

The conflicting results of these studies show that more research is required to understand the function and importance of proline. Genetic knockout and gene expression studies provide a powerful tool for doing so. This has become more straightforward since the release of the high-quality quinoa genome.

9.2.4.2 Enzymes

Glycine betaine appears to have a protective function, preserving photosystem II activity against oxidative stress. Exogenous application of glycine betaine significantly ameliorated the negative effects of UV-induced oxidative stress on photosynthetic efficiency (Shabala et al. 2012). Betaine aldehyde dehydrogenase synthesises glycine betaine, which can protect plants from the detrimental effects of temperature change, salinity and dissolved oxygen (Stephens-Camacho et al. 2015).

Jiang et al. (2016) were the first to characterise the BADH gene in quinoa (*CqBADH1*). They observed that under NaCl stress (100 and 250 mM), there were increased levels of *CqBADH1* mRNA and glycine betaine. Interestingly, according to expression profile analyses, *CqBADH1* was predominantly expressed in the root and displayed time-dependent expression profiles under NaCl stress. Whilst expression levels were over 5-times higher in the root, levels of glycine betaine were highest in the leaves, suggesting glycine betaine is transported throughout the plant, and provides an important protective function in both roots and leaves. Overexpression of the BADH gene has led to increased abiotic stress tolerance in sweet potato and rice (Fan et al. 2012; Tang et al. 2014). Conducting the equivalent experiment in quinoa would be very informative, as would identifying cultivars which have naturally high levels of BADH.

Several other enzymes have also been reported to increase tissue tolerance. Notably, superoxide dismutase (SOD), catalase (CAT), ascorbate peroxidase (APX) and peroxidase activity has been reported to increase by a factor of 2.3–5.6 times under salinity stress (Amjad et al. 2015). Miranda-Apodaca et al. (2018) reported that the activity of ascorbate peroxidase (APX) was only affected under conditions of 500 mM NaCl, 58% higher than the control values. CAT activity was increased by 81%. Under drought conditions, APX increased by 43%, however CAT activity was unaffected.

Boas et al. (2016) studied early seedling tissue under conditions of NaCl up to 200 mM. At higher concentrations of NaCl, higher levels of SOD were seen in both the roots and shoots. However, high SOD activity only resulted in increased H_2O_2 production in the shoots. The group hypothesised that H_2O_2 produced by SOD was removed by CAT, which increased in a similar manner to SOD. This metabolic adjustment had no impact on seedling dry matter accumulation. The role of SOD in salinity tolerance was reported by Ismail et al. (2016), who said that SOD activity was correlated with tolerance to salinity.

9.2.5 The Role of ABA

The importance of ABA for abiotic stress tolerance in quinoa has emerged during recent years. It has been long recognised as an important stress signal and several studies have indicated its importance in quinoa (Razzaghi et al. 2011). The recent paper on the quinoa genome by Zou et al. (2017) suggested that stress tolerance in quinoa is associated with the expansion of genes involved with ABA homeostasis and signalling, along with enhanced basal-level ABA responses. It also suggested that ABA synthesis, transport and signalling may be essential for quinoa's tolerance as a halophyte. Higher expression of genes associated with ABA biosynthesis and energy import was seen in the epidermal bladder cells (EBCs). They found that the pathway of ABA de novo synthesis was upregulated in the bladder cells. Additionally, transcript levels of two *NCED* genes, which are predicted to encode the rate-limiting enzyme for ABA biosynthesis were 4–sixfold higher in the EBCs than the leaves. In the EBCs, there was also elevated expression of ABA transporters, including ABCG40 and ABCG25, as well as most of the PYL family of the ABA receptor proteins (Zou et al. 2017). They note that quinoa contains a high number of ABCGs, a group of ABC transporters which are involved in ABA transportation. They suggest that the enhanced regulation of ABA homeostasis and signalling may contribute to abiotic stress tolerance in quinoa.

In contrast, Morales et al. (2017) reported ABA-independent expression patterns of quinoa, under transcriptional analysis. Studying the R49 genotype, they determined that *CqABA1*, *CqABA2* and *CqABA4* were repressed. A possible explanation is that in quinoa a corresponding gene family might be present as a tetraploid species, and the designed primers did not differentiate between members of the family. Another study also concluded that ABA seems to play a minor role under drought conditions; quinoa might synthesis other anti-transpirant compounds in the xylem sap, rather than ABA (Jacobsen et al. 2009).

Under salt stress, Morales et al. (2017) reported the upregulation of many heat shock proteins (HSPs). The highest upregulation observed was of CqHSP20 (200-fold). In tomatoes this is induced under a variety of abiotic stresses, including heat, salt and drought suggesting the importance of ABA-independent responses (Yu et al. 2016). They also reported the *CqLEA* gene was upregulated 140-fold in quinoa. Interestingly, expression of *CqLEA* has been reported to be mediated by ABA. *CqAP2/ERF* was also upregulated, which plays a central role in regulating plant adaptation to stresses through ABA-dependent and -independent pathways (Golldack et al. 2014).

9.2.6 Protein Composition Under Salinity Stress

Research by (Aloisi et al. 2016) investigated the composition of seeds from quinoa plants under drought stress. A *salares* cultivar (R49) was compared with two lowland cultivars (VI-1 and VR). Protein expression changed most in VR; whereas several free amino acids were unchanged or increased by salinity in R49 compared to VR and VI-1. Total polyphenol content increased most in R49, and enhanced radical scavenging capacity was seen in R49 and VR. Salinity has a powerful effect in changing the seed amino acid and proteome profiles and generally increased the concentration of bioactive molecules and amino acids from protein extracts. This was genotype-dependent, and concurring with other studies it appeared that the *salares* ecotype, R49, was most tolerant to salinity stress.

9.3 Drought

9.3.1 Role of ABA in Drought

Drought can be defined as a lack of precipitation over an extended period of time, affecting specific sectors and activities. Drought can be considered from the following perspectives: meteorological, hydrological, agricultural and socioeconomic. This section will focus on agricultural drought, defined as a level of soil moisture which does not meet the needs of certain crops at a given growth stage. Climate change has increased the occurrence of drought, which could have highly detrimental consequences for agriculture, economic stability and food security. For example, in 2009, a severe drought in Kenya caused a 45% reduction in the wheat yield compared to the following year. The impact is felt even in highly developed countries; the 2011 drought that affected large parts of the U.S. agricultural belt increased food prices across the globe and increased food prices in the USA by around 4% (FAO 2013).

Quinoa is naturally tolerant to drought as it has low water requirements. Some of the key physiological factors are a low osmotic potential, growth plasticity and tissue elasticity, low turgid weight/dry weight ratio, low elasticity and maintenance of positive turgor at low water potentials (Vacher 1998; Jensen et al. 2000). Quinoa's high WUE is thought to balance its decreased leaf stomatal conductance and optimise carbon gain. The results indicated that photosynthesis was maintained even after stomatal closure and ABA was produced in response to mild water deficit (Jacobsen et al. 2009).

Some symptoms of drought stress are similar to those of salinity stress, such as stomatal closure—which limits shoot growth and accelerates leaf senescence—and the accumulation of compatible solutes to adjust the cytoplasmic osmotic potential, some of which might also play a protective against ROS. Additionally, deeper root systems and leaf expansion are also seen.

Vacher (1998) noted the superior responses of quinoa to drought conditions. On the Bolivian altiplano, there is a severe water deficit during the dry season. Drought caused a large drop in the leaf water potential, stomatal conductance, transpiration and photosynthesis rate in quinoa. Vacher also observed there was rapid and pervasive stomatal closure, associated with a decrease in transpiration and photosynthesis rate along with a large reduction in minimal water potential at the beginning of the drought. However, as the drought continued, the stomatal conductance, transpiration and photosynthesis rate remained stable. Under drought conditions, the WUE increased and was then maintained. A high dry matter yield was observed. This is in contrast to the glycophyte bitter potato; stomatal conductance, transpiration and photosynthesis rate greatly decreased after 4 weeks of drought and yield decreased, displaying poor tolerance to drought.

Killi and Haworth (2017) observed that whilst there were stomatal and mesophyll limitations to CO_2 transport under drought conditions, photosynthetic capacity and photosystem II performance were unaffected. This is in contrast to Vacher (1998), who reported a reduction in the rate of photosynthesis. This highlights the need for genetic studies to clarify the importance of various physiological components. However, a study by Fghire et al. (2015) inn which quinoa drought tolerance in a Puno cultivar was investigated noted a significant decrease in the number of "Photosystem II reaction centres per exited cross section", when the plants were exposed to the highest level of drought stress (33% of crop evapotranspiration). The discrepancy between the studies could be down to the varying levels of drought to which the plants were exposed or the different cultivars of quinoa used. The Fghire study used a Puno cultivar from a highland region of Peru, which tends to display high stress tolerance, whilst the Killi study used the Redhead cultivar, which is not a highly tolerant cultivar.

Al-Naggar et al. (2017) studied five quinoa genotypes under drought stress to identify the most drought-tolerant, according to a drought tolerance index (DTI), which was CICA-17. The DTI was strongly correlated with physiological traits such as seed yield, WUE, branching, chlorophyll concentration index and leaf area, in addition to inflorescence traits. Whilst such studies are useful to identify tolerant cultivars, to create a step-change in quinoa development, the underlying genes and superior alleles need to be identified.

Yang et al. (2016) investigated the response of quinoa to drought and different watering strategies. Under drought conditions, height and shoot dry weight decreased. However, under the alternate root zone drying irrigation strategy (ARD)—where they watered alternate sides of the roots, whilst using the same volume of water as for deficit irrigation—plants showed significantly higher plant height and shoot dry weight compared to deficit irrigation. This difference was greater at higher temperatures and was associated with higher WUE, ABA concentration and nutrient contents. The authors concluded that under "high temperature" (25 °C/20 °C) an ARD is an effective irrigation strategy. These findings merit further investigation. Is ARD a consistently superior irrigation method? Is it effective at all developmental stages of quinoa and between cultivars? Is it still effective under varying environmental conditions?

Interestingly, González et al. (2009a, b) report that for the cultivar Sajama under drought conditions, plant and dry root weight decreased. Under drought conditions, levels of proline increased, but there was no difference in the levels of starch, sucrose or fructose. This could suggest that the proline fulfils an osmoprotective role, given the levels of other osmolytes remained unchanged, so the osmotic potential would be little changed. Additionally, they also observed that root weight ratio, specific leaf area, number of leaves, chlorophyll content and height of plants were unaffected by drought conditions compared to the control conditions. Whilst the plant weight and dry root weight decreased, it is interesting that other many other physiological aspects remained unchanged.

A pioneering study by Maliro et al. (2017) looked at the performance of 11 quinoa cultivars under irrigated and rainfed conditions in two locations in central Malawi. Under irrigated conditions, the highest-yielding genotype was Titicaca (3019 kg/ha; Denmark, bred by Jacobsen), whereas under rainfed conditions, a black-seeded variety (2050 kg/ha; Colorado, USA) was the highest yielding. Generally, yields were reduced under rainfed conditions. The significant variation seen between genotypes under different watering regimes and different locations, highlights the importance of conducting extensive breeding programmes to ascertain the most suitable genotype for a given location. It also emphasises the need for genetic studies, given the difficulty in resolving such differences. Owing to the recent high-quality genome sequence, we are now at the point and have the ability to leverage genomics to gain key insights (Jarvis et al. 2017).

9.3.2 Transcriptomic Studies

Raney et al. (2014) did a transcriptomic analysis of drought-induced stress in quinoa, under four different drought conditions and with two cultivars of quinoa, representative of valley and altiplano salares ecotypes. Altiplano salares ecotypes generally display greater tolerance than the valley ecotypes to drought-like conditions. Gene expression analysis identified 462 putative gene products that showed differential expression depending on treatment. 27 putative gene products were expressed differentially with respect to variety and treatment, with the root tissues displaying significant expression differences in response to increasing water stress. Two of these genes have a regulatory function—one with a high amino acid sequence homology to Naringenin, 2-oxogluturate 3-dioxygenase; and the other with a high amino acid sequence homology with a receptor-like cytosolic serine threonine-protein kinase rbk1-like protein (Raney et al. 2014). The function of the former is the catalysis of 3-beta-hydroxylation of 2S-flavanones to 2R, 3R-dihydroflavenols. These are intermediates in the flavonoid biosynthesis pathway. Flavonoids have previously been shown to increase drought tolerance through increasing antioxidant capabilities (Hernandez et al. 2004). The function of the latter could be similar to that of SnRK2.6/OST1 (OPEN STOMATA 1), which is a serine threonine-protein kinase in *Arabidopsis thaliana*, which has been implicated in drought tolerance (Yunta et al. 2011). The altiplano salares cultivar, Ollague, overexpressed a gene which codes for a protein which shares high homology with a putative chaperone-1-like protein in *Zea mays*, which responds to high light intensity and heat. Other contigs identified show shared homology with secologanin synthase, heat shock proteins, proteins which have ROS-protective roles and a pathogenesis-related gene-protein (López-Frías et al. 2011; Yamamoto et al. 2000; Martindale and Holbrook 2002; Kitajima and Sato 1999). BLAST searches and gene ontology analysis show an overlap between drought tolerance stress and other abiotic, and biotic, stress mechanisms, thus suggesting that these genes have far-reaching and broad areas of action (Raney et al. 2014).

Liu et al. (2018b) conducted a pioneering study into HSPs in quinoa. The *Hsp70s* genes code for a set of chaperone proteins that are central to cellular networks as folding catalysts and molecular chaperones which are widely recognised to be important for plant stress tolerance (Tripp et al. 2009; Mayer and Bukau 2005, 70). Several *Hsp70s* have already been identified as important to drought tolerance in other plants (Lin et al. 2001; Sarkar et al. 2013; Tang et al. 2016). The group identified 16 *Hsp70* genes in the quinoa genome, made up of eight paralogous pairs. Different CqHSP70s were predicted to localise to the ER, plastid and mitochondria, variously, suggesting differing functionality. According to qPCR, there were significant variations in the levels of *Cqhsp70s* in response to drought stress. About half of the *Cqhsp70s* displayed a "drop-climb-drop" expression pattern—similar to their *Arabidopsis* homologs. The other genes had varying responses, some highly induced by drought, others being induced at a later time point. This suggests various roles for *Cqhsp70* genes in response to drought. Further investigation is required to measure the importance of these genes through knockout studies or through overexpression studies. Genetic knockout and overexpression experiments are required to aid our understanding of the importance of HSP70s to drought and other abiotic stresses, given their non-specific protective effects (Ma et al. 2006).

A study by Morales et al. (2017) examined the transcriptional responses of three Chilean genotypes of quinoa under drought conditions. They identified Genotype R49 (a *salares* cultivar) as the most drought-tolerant, according to its relative water content, electrolyte leakage and the maximum efficiency of photosystem II. Notably, the sHSP families were highly upregulated. *CqHSP20* expression increased 200-fold, suggesting it is part of quinoa's response to drought stress. However, the importance of HSP20 is unclear. A study in tomato showed that *Hsp20* was highly expressed under various abiotic and biotic stress, including under drought, salt and heat treatments (Yu et al. 2016). A recent study in potatoes indicated that *Hsp20* was similarly expressed in response to a wide range of abiotic stresses, suggesting a high degree of conservation between species (Zhao et al. 2018). Given its wide-ranging putative involvement in the response to abiotic stress, *Hsp20* merits significant further investigation, through knockout and overexpression studies.

The expression pattern for ABA biosynthesis and other genes involved in archetypal drought responses were also assessed by qPCR (Morales et al. 2017). The group observed that *CqABA1*, *CqABA2* and *CqABA4* from the ABA biosynthesis pathway were repressed at two points. The reason is unclear. The authors suggest there could be a family of corresponding genes which are present as a tetraploid species, and the primers did not differentiate between the members of the family, possibly leading to inhibitory effects. An alternate suggestion is that ABA plays only a minor role under drought conditions in quinoa, and antitranspirants are produced other than ABA in the xylem sap (Jacobsen et al. 2009). Other hormonal stress signals may also be important, suggesting studies should be conducted into cytokinin and ethylene reactions.

Morales et al. (2017) also reported a 140-fold upregulation of the *CqLEA* gene, a gene which has been shown to be responsive to ABA in rice (Battaglia et al. 2008). This suggests, despite the repression of *ABA* genes at certain points, enough ABA was produced to trigger the upregulation of *CqLEA*. *CqAP2/ERF* was also upregulated, which generally plays a key role in regulating plant adaptation to stress via both ABA-dependent and -independent pathways, suggesting that it might be required for coordinating regulatory networks which underpin abiotic stress tolerance (Golldack et al. 2014; Yamaguchi-Shinozaki and Shinozaki 2006). A study in berries showed the concurrent upregulation of HSP chaperones alongside ERF subfamily transcription factors and increased ABA levels, possibly indicating an interconnected response (Carbonell-Bejerano et al. 2013).

Morales et al. (2017) also categorised the type of process that the over-represented genes were associated with: "cellular response to stimulus" (18 genes); "single organism signalling" (18 genes); "development process involved in reproduction" (13 genes) and "single organism reproductive process" (13 genes). They state that, in general, this concurs with the results found for wheat, rice, cassava, chickpea and *Arabidopsis* (Wong et al. 2005; Mochida et al. 2006; Gorantla et al. 2007; Lokko et al. 2007; Huang et al. 2008; Varshney et al. 2009), indicating that most responses to drought stress are conserved in quinoa. However, considering quinoa has a higher tolerance to drought than the aforementioned species, the authors postulate that as yet unidentified genes could be responsible for the difference in tolerance.

Plasma membrane-related genes have been suggested to be linked with cell turgor maintenance or membrane stability (Razzaghi et al. 2015; Morales et al. 2017) noted an upregulation of *ABC, ERD6-like, MATE and SWEET-like*, which have transporter activity, whilst cell wall-modifying genes were downregulated. The authors suggest these changes are linked with the remobilisation of assimilates and ROS detoxification.

The research of Schmöckel et al. (2017) further reinforces the importance of membrane-related genes in their study of transmembrane proteins for tolerance to abiotic stress. They identified 15 genes specifically for salinity tolerance, however their study could provide a valuable starting point for further studies into drought tolerance (Schmöckel et al. 2017).

Zurita-Silva et al. (2014) report that *HSP20* (hsp20-putative chaperones superfamily protein), *LEA* (late embryogenesis abundant protein family), *AP2/ERF* (integrase-type DNA-binding superfamily protein), *HSP83* (chaperone protein, HTPG family protein), *PP2C* (protein phosphatase 2C family protein) and *P5CS* (delta-1-pyroroline-5-carboxylate synthase 2) are all drought-responsive genes. Their importance and function in quinoa's drought tolerance are yet to be elucidated.

9.4 Temperature

Temperature stress has a strong effect on the growth and development of quinoa. Temperature can be broadly split into the following categories: low temperature, associated with cold stress responses; optimum temperature range for plant growth and development and high-temperature, associated with heat stress responses.

9.4.1 Low Temperature

9.4.1.1 Physiological Studies

Quinoa has been grown in the mountainous Andes for thousands of years, at an altitude in excess of 4000 m. In these areas, frosts are common and quinoa in these regions has been noted for its relatively high level of tolerance to frost and cold stress, although this is dependent on the developmental stage (Jacobsen et al. 2003; Rea et al. 1979). Quinoa can grow at −5 °C and it can survive in a vegetative state at −16 °C (Vacher 1998; Catacora and Canahua 1991). However, at the anthesis stage, even a mild frost (−2 °C), can cause serious damage.

Frost resistance consists of avoidance and tolerance. Avoidance refers to the ability of a plant to prevent ice formation in its tissues; quinoa does this through supercooling. Tolerance refers to the ability of a plant to survive extracellular ice formation and cell dehydration without irreversible damage (Jacobsen et al. 2007).

Quinoa appeared able to acclimate to the cold, with the LT50 declining linearly according to days of cold exposure. The osmotic potential increased with acclimation, the major changes coming from changes to sugars. Interestingly, when supercooling was prevented using an artificial nucleator, the LT50 values were 3 °C higher than nucleation temperatures, indicating that quinoa could not tolerate freezing. The ice nucleation temperature was around −8 °C, depending on variety and days of acclimation. The average lethal temperature for the altiplano cultivar, Witulla, was −5.1 °C. Jacobsen et al. conclude that supercooling is the main mechanism of freezing tolerance in quinoa.

The relatively high soluble sugar content in quinoa likely depresses the freezing point. There is often a strong correlation between sugar content and freezing tolerance (Levitt 1980). After 10 days of cold acclimation, a tenfold increase in sugar content was observed in quinoa (Jacobsen et al. 2007). The concentration of soluble sugars may provide an indicator of frost tolerance. However, the effect of proline—associated with salinity and drought stress tolerance—was less pronounced, showing a weak correlation with freezing temperature and a non-significant correlation with LT50 (Jacobsen et al. 2007). Genetic studies are needed to further explore and deepen our understanding into the mechanisms, and importance of supercooling, and freezing tolerance in quinoa. Gene expression analysis would provide a valuable way forward.

Jacobsen et al. (2005) investigated the tolerance of quinoa to cold stress at various developmental stages. At the two-leaf stage, cultivars from the Peruvian altiplano (Wariponcho, Witulla, Ayara and Lp-4B) could tolerate −8 °C for 4 h, whereas a cultivar from the Peruvian valleys (Quillahuaman) could only tolerate the same temperature for 2 h. At a high relative humidity at −4 °C (causing white frost), the death rate was 25%, rising to 56% at a low relative humidity (causing black frost, leading to necrosis). In general, the Quillahuaman cultivar displayed twice the level of negative effects compared to the altiplano cultivars. After frost treatment of −4 °C for 4 h at the two-leaf stage, the seed yield decreased by 9%. However, at the 12-leaf and flowering stages, seed yield was much reduced, by 51 and 66%, respectively. Under conditions of −2 °C for 6 h, there was very little damage at the 12-leaf stage to either cultivar, but at anthesis some damage was seen in Quillahuaman after 4 h. At −4 °C there was damage to both Witulla and Quillahuaman, but significantly more to the valley ecotype.

This highlights the varying tolerance of quinoa to cold stress depends on the cultivar, developmental stage and the degree of cold stress. It is also an important consideration for the cultivation of quinoa in an increasingly erratic climate—late frosts have the potential for much more damage than earlier ones. Understanding how and why quinoa loses its tolerance to cold stress at later developmental stages is likely to be important for safeguarding quinoa agriculture in the face of climate change.

Concurring with the previous study, both cultivars accumulated soluble sugars, though this was measured on dry matter basis; differences in concentration in living tissue may be very slight. The altiplano cultivar had a higher water content. High sugar content is associated with hardiness. Interestingly, the valley cultivar had higher accumulation of proline. The importance of proline in cold tolerance is unclear, requiring further investigation.

Bois et al. (2006) found that at 2 °C, germination was delayed but not prevented. Interestingly, thermal sensitivity at the germination stage did not appear to be related to the geographic origin of the cultivars. In contrast to Jacobsen, for the two cultivars investigated, Surumi and Chucapaca, at the vegetative stage (15 leaves visible in Surumi and 18 in Chucapaca) all plants died after 4 h at −6 °C, but no serious effect was seen down to −3 °C. Ice nucleation was noted between −5 and −6 °C. According to Bois, low leaf water delayed the freezing process—which is surprising given the more cold tolerant altiplano cultivar in Jacobsen's study had a higher water content.

Rosa et al. (2009) conducted a study into the effects of low temperature on the enzyme activities involved in sucrose-starch partitioning in salt-stressed and salt-acclimated cotyledons. They found that low temperature (5 °C) affected sucrose synthase activity in salt-treated cotyledons. Levels of fructose and starch were higher under low-temperature stress. The results demonstrate differing responses to sucrose-starch partitioning in salt-stressed and salt-acclimated cotyledons.

Whilst it has not been specifically studied in quinoa, plants undergoing cold acclimation generally see changes in membrane composition, and it would be expected to see similar changes in the quinoa (Nishida and Murata 1996). Additionally, the accumulation of protective compounds is often seen, such as sugars, ABA, free amino acids and polyamines. Whilst an increase in sugar and proline levels has already been noted under cold stress, it would be worth investigating the importance and role of these other compounds. As shown by Jacobsen et al. (2007), proline, sucrose and glycine betaine concentration increase with stress duration.

9.4.1.2 Genetic Studies

Morales et al. (2017) briefly note that CqCAP160 (cold acclimation protein) increases under drought conditions, suggesting that CqCAP160 responds to stress in a more general manner. No investigation has been done into CqCAP160 related to its role in temperature stress tolerance.

Looking at model plants such as *Arabidopsis* for genes which help regulate temperature tolerance presents an important starting point for understanding the genetics of temperature tolerance in quinoa. Through BLAST and other genomic tools, homologs can be identified, which can then be tested in the lab and field. For example, looking for a homolog to *Arabidopsis'* DREB1A gene reveals a sequence in quinoa with 72% identity, called LOC110697411, which could be important for cold resistance. However, genetic studies are required to identify its importance in quinoa.

Despite physiological and mechanistic research into quinoa's resistance to cold temperature, very little is known at the genetic level, for example, over the sugar conversion enzymes, genetic control over supercooling and cell wall enzymes. The recent high-quality quinoa genome should assist with this, making comparative analyses between quinoa and, for example, Arabidopsis possible, helping to identify loci of interest (Jarvis et al. 2017).

9.4.2 Effects of Temperature on Germination

An early study by Jacobsen and Bach (1998) investigated the effect of temperature on seed germination rate. They studied a temperature range of 8–35 °C, they estimated that the base temperature, at which germination rate is zero, was 3 °C. Since this time, other quinoa cultivars have been shown to have a lower germination temperature. They state that whilst there was variation within the population studied, the optimal temperature for 80% of the population for germination was between 30 and 35 °C. They estimated that the maximum temperature for germination would be around 50 °C.

In contrast, a study by Tan and Emre (2017) investigated how temperature affected the germination rate of various cultivars. Studying the effects of temperatures between 5 and 25 °C, they observed that generally the optimum temperature was 20 °C, with germination rates tending to be slightly decreased at 25 °C. The Q-52 cultivar had the highest rates of germination at 10 °C and above.

Chilo et al. (2009) reported a similar effect; they studied the effect of temperature and salinity on germination and seedling growth. Under temperature conditions of between 5 and 20 °C (and control saline conditions), they found that as temperature decreased, they observed a reduction in the germination percentage and rate of seedling growth.

Yang et al. (2016) studied the growth and physiological responses of quinoa to drought and temperature stress. Supporting the results of Tan and Akcay, they found that quinoa displayed optimal growth at higher temperatures (25/20 °C vs. 18/8 °C).

Temperature appears to be the main determinant of the rate of leaf appearance in plants (Kiniry et al. 1991; Bertero 2001). The photoperiod, rate of change of the photoperiod and radiation are secondary factors. (Bertero et al. 1999) noted the importance of temperature and photoperiod during quinoa development. Specifically, the duration of the phase emergence-to-visible flower buds was studied. He noted that in the Narino cultivar, at 10 °C under a 16-hour photoperiod, this stage took 97 days. In the Sajama cultivar, at 21 °C, but a shorter photoperiod of 10.25 h, this stage took 17 days. Despite the difference in cultivar, this indicates the importance of heat for plant development. Investigating the genetic control to induce a similar response even under colder conditions could prove a valuable step forward.

A study by Becker et al. (2017) investigated the combined effects of soil salinity and relatively high temperature in the cultivar Achachino found that under high-temperature seeds germinated earlier, developed faster and matured sooner than at low temperatures. High temperature was also found to increase the stomata size on the abaxial leaf surface. The group also reported that the final developmental stages, after anthesis, might have been prolonged under high temperature, perhaps suggesting the conditions were above the optimal temperature for that developmental stage in the Achachino cv. However, the high-temperature treatment was not too extreme: 23 °C from days 0 to 20; 25.5 °C until day 40; 28 °C until day 80 and then 25.5 °C until harvesting. Whilst this study demonstrates that quinoa can effectively grow at temperatures in the mid to high 20 s °C, it does not establish an upper limit of quinoa temperature tolerance.

9.4.3 High Temperature

Hot climates and deserts can pose very significant challenges to the cultivation of quinoa, causing abortion of flowers and death of pollen (Jacobsen et al. 2003). Relatively little research has been conducted specifically into the effects of high-temperature stress on quinoa. This might be because quinoa's traditional area of cultivation in South America is rarely exposed to high-temperature stress, however hot environments are often very dry and saline and if the heat tolerance of quinoa can be increased then quinoa could be very well-suited to providing food security in such regions, where high temperatures are common.

González et al. (2017) investigated the cardinal temperatures for 10 quinoa cultivars. They calculated the optimum temperature for seed germination and divided the cultivars into two groups: Kamiri, Robura, Kancolla and Sajama, which had an optimal temperature of less than 33 °C; and CICA, Amilda, Sayana, Ratuqui and Smaranti which had an optimal temperature greater than 33 °C. This work provides a useful starting point into categorising heat-resistant cultivars. A similar study was carried out by Mamedi et al. (2017). They found that for the Sajama and Santamaria cvs, the highest germinate percentage was between 15 and 35 °C. For Titicaca the highest germination percentage was between 5 and 35 °C. However, the effect of high temperature needs to be investigated at all developmental stages. Crucially, the genes responsible for this increased tolerance need to be investigated. This wide and high-temperature range suggests other growth stages need to be investigated in order to identify the most temperature-sensitive points of quinoa development.

Bunce (2017) conducted an inventive study, looking at the possible effects of climate change on quinoa, through studying elevated CO_2 and high temperature during anthesis. At high temperatures (35 °C/29 °C) under ambient CO_2 levels, seed yield was significantly altered compared to the control temperature of 20 °C/14 °C, but the response greatly depended on the genotype. The yield of the Cherry Vanilla cultivar was 0.3× the control yield, whilst Salcedo yield was 1.7× the control yield. The Redhead genotype was not significantly affected. Interestingly, at elevated CO_2 levels (600 μmol mol-1 compared to 400 μmol mol-1 ambient level), under high-temperature treatment, seed yield increased by 12–19% in all cultivars. This suggests that for the cultivars investigated, 35 °C is a heat stress that can be tolerated, or even beneficial, under high CO_2 conditions. This is a surprising result given by Becker et al. (2017) suggested that even temperatures of 28 °C negatively affected plant development. However, not enough research has been done on the effect of high temperature on the later stages of plant development. Redhead and Cherry Vanilla are cultivars developed in the USA and are generally accepted to have reduced yields under hot climates, so it is surprising a decrease in yield was not observed at high temperatures. Salcedo is from the highland region of Puno, from which cultivars tend to be highly resilient to abiotic stress and, indeed, it appeared Salcedo thrived at high temperatures. Identifying differential gene expression between these cultivars could provide an important insight into genes which are important for heat tolerance. A broader investigation into identifying the quinoa cultivars which exhibit the highest heat tolerance would also assist future breeding programmes.

Lesjak and Calderini (2017) investigated the effect of increased night temperatures on quinoa yield, at two phases: flowering and grain filling. Temperature was increased 4 °C above the ambient temperature (which varied between 14 and 18 °C) between 8 pm and 9am. However, they prevented the daytime temperature from reaching 30 °C. They found that the rise in temperature reduced grain yield by 31% at the point of flowering and by 12% at the grain filling stage in the Regalona cultivar (southern Chile variety), and by 23% and 26% in the BO5 No191 accession (a sea-level cultivar). Grain number was significantly affected and associated with yield. This study could be particularly prescient given the changing climatic conditions across the world. As to why there was such a large impact on yield requires further research. Identifying quinoa cultivars which display a greater tolerance variable temperature, as might be expected with increasing climate change, and the genes underpinning this, would be a valuable step forward.

For comparison, Prasad et al. (2008) studied sorghum under high temperatures. The first 29 days after sowing the plants were grown at 32 °C/22 °C (day/night). After that, they were grown either at 40 or 30 °C. When grown constantly at 40 °C, delayed panicle emergence and decreased plant height, and reduced seed set, seed numbers, seed yield, seed size and harvest indices were observed. However, leaf photosynthesis was not influenced. Even short exposures to 40 °C 10 days prior to flowering or at flowering caused maximum decreases in seed set and seed yield. 40 °C stress during post flowering stages decreased seed yield, larger reductions occurring at the early stages of seed development. If similar patterns were seen in quinoa, it could change agriculture practices, if it were shown even short exposure to high temperatures was damaging. Similar results were seen with chickpea. A study by (Wang et al. 2006), found that at high-temperature stress (35 °C) at the early flower stage, pod production was decreased by 34% and 22% for Myles and Xena, respectively. During pod development, high temperature caused a reduction in seed yield by 59% and 53% in Myles and Xena, respectively.

9.5 Nutrient Deficiency Stress (Nitrogen and Phosphorus)

Crop yield is closely linked with nutrition, and relies on a sufficient amount of micro- and macronutrients to ensure full plant development and production of grain. Whilst fertiliser is a major cost for farmers, it is often overused and it has been estimated that for cereal crops, only 40–60% of applied nitrogen fertiliser is taken up (Raun and Johnson 1999). Excess nitrogen can leach from the soil, or be lost through volatilisation (Sylvester-Bradley and Kindred 2009). This excess N is not only a wasteful economic loss, but also an environmental pollutant which leads to the release of greenhouse gases and the eutrophication of aquatic environments (Zhang et al. 2015).

A lack of nitrogen and phosphorus can have the biggest impact on crop yield (Hawkesford et al. 2012). This is because radiation capture is limited, preventing proper growth and development. A higher supply of nitrogen is needed as it is required in nearly all major plant macromolecules. Plants which get insufficient nitrogen may yellow, particularly the older leaves, owing to chlorophyll loss. The leaves of phosphorus-deficient crops tend to become dark-green, due to the accumulation of sugars and other compounds. Root growth can also be reduced, decreasing the uptake of water and nutrients. Akula and Ravishankar (2011) also note that deficiencies in nitrogen and phosphate directly influence the accumulation of phenylpropanoids, which are thought to play various important roles for plant development and growth (Solecka, 1997).

Improving nutrient use efficiency (NUE)—the crop yield relative to the level of nutrients available—is an important strategy to improve yield for a given crop. NUE can be increased through changing agronomic practices, such as crop rotations and timing of fertiliser application (Fageria and Baligar 2005). However, the factors affecting NUE greatly depend on the agricultural environment—a high rainfall, high input area will need to managed very differently to a low input, dry area (Cossani et al. 2010). To increase phosphorus use efficiency in phosphorus-fixing soils is requires a different approach to that for phosphorus-available soils.

There are several mechanisms that would have ubiquitous application, regardless of environment and management and can be applied to quinoa breeding. Firstly, root architecture can be altered to improve uptake efficiency, either through deep rooting systems, or through shallow systems, depending on where the fertiliser is concentrated. Whilst transporters of nitrogen and phosphorus have been described for barley, no research has been conducted in quinoa. Through selecting superior alleles in rice, researchers have improved the efficiency of these transporters (Fan et al. 2016; Hu et al. 2015)—this must surely be a goal when breeding quinoa.

Secondly, the utilisation efficiency of nutrients can be improved. The biochemical pathways for the conversion of inorganic nutrients into organic building blocks for protein synthesis have been well studied (Xu et al. 2012). In barley, glutamine synthetase's (GS) role in nitrogen assimilation has been well studied, and *GS* is seen as a potential target gene for improving NUE. However, little is known about whether quinoa has a similar gene, or if the effects would be similar. Perhaps more promisingly, the over-expression of the gene encoding alanine aminotransferase has improved the NUE in both canola and rice (Good et al. 2007; Shrawat et al. 2008), giving the possibility that if a homolog was identified in quinoa a similar effect might be seen.

Thirdly, the feedback signalling between the uptake and utilisation systems needs to be considered. The status of one system affects the capacity of the other through signalling. Remobilisation of nutrients, such as nitrogen, proteins and other molecules also determines NUE. In barley genetic loci have been identified which regulate leaf senescence and N reallocation (Jukanti et al. 2007; Heidlebaugh et al. 2008; Cai et al. 2013; Bezant et al. 1997). Similar loci need to be identified in quinoa and the causal gene(s) elucidated. The recent release of the high-quality quinoa genome will assist in this search.

9.6 Boron Toxicity

Boron is a nutrient essential for plants, however it becomes toxic at soil water concentrations only slightly higher than the optimum for plant growth. These toxic concentrations tend to be reached only in arid areas, and commonly are associated with salinity (Dhankhar and Dahiya 1980). Irrigation is the primary source of increasing boron levels, when water comes from wells with high levels of boron. Additionally, igneous and sedimentary rocks can contain high levels of boron. Boron less readily leaches from soils than Na^+ and Cl^-, making it a more intractable problem. Boron symptoms typically include chlorosis and necrosis at the margins of the leaves, and necrotic spots in the central leaf. Symptoms are not universal, and visible leaf symptoms are not tantamount to yield loss; conversely, there can be no observed leaf symptoms but a decline in yield (Maas 1986). Soil analysis only gives an approximate risk assessment for boron toxicity, and cannot be used to precisely predict plant growth.

Unfortunately, almost no research has been conducted into the boron tolerance of quinoa. Even the boron tolerance limits have not been investigated. Quinoa is part of the Amaranthaceae family, the same as sugar beet, which has a boron yield threshold of 4–6 g m^{-3}, classed as good tolerance (Hall 2018). However, without studies it is only possible to speculate. Barley, generally considered to be the most salt-tolerant of the cereals, exhibits only a moderate tolerance to boron toxicity, in comparison to sorghum, which is highly tolerant. Therefore, boron toxicity does not appear to closely correlate with salinity tolerance (Munns and Tester 2008; Hall 2018).

The boron tolerance mechanism exhibited by most studied plant species is based on reducing the uptake of boron in the roots and shoots, and exists at a cellular, organ and organism level. Between species boron concentration levels in shoots and leaves tend not to be closely related to boron tolerance. However, between closely related species—or within a species—genotypes susceptible to boron toxicity tend to exhibit higher levels of boron in the shoots and leaves. It is possible that quinoa, as a halophyte, exhibits additional internal tolerance. According to Rozema et al. (1992), solutes such as sorbitol in certain halophytes might inactivate excess boron.

Beyond exclusion of boron, other accumulation restriction mechanism suggested are: the chelation of boron by polysaccharides and organic acids; the inactivation of boron in cell wall or cytoplasm of root cells; the ability of $B(OH)_3$ to cross the lipid bilayer; differences in transpiration rates and transport of boron in the xylem. Whether any or all of these mechanisms occur in quinoa needs to be investigated, as does the genetic control of such mechanisms.

Several major additive genes which appear to control boron levels have been identified, and in some species the chromosomal locations have been identified (Nable et al. 1997). Significant intra-specific variation of boron tolerance has been noted, for example in wheat (Chatterjee and Das 1980; Paull et al. 1988; Jamjod 1996). The genes identified in wheat and barley, for example, provide an useful starting point for uncovering the genetic control of boron toxicity in quinoa. Whilst the genes and chromosomal location is likely to be different in quinoa, looking at other species could still provide important clues for quinoa.

9.7 Soil Acidity

All the world's regions are affected by soil acidity to some extent. In particular, agricultural management in developing countries in ineffective at limiting the decrease in pH (FAO 2015). Aluminium can reach toxic levels when soil has a pH less than 5.5, as it is solubilised to Al^{3+}, which inhibits root elongation and affects the absorption of several plant nutrients such as calcium, manganese and phosphorus (Foy et al. 1978). Aluminium toxicity is the biggest single factor limiting crop productivity in acidic soils, which comprised around 40% of the world's arable land in 1995 (Kochian 1995). Exploiting variation between cultivars through identifying aluminium-resistant cultivars has partly provided a solution.

The initial symptom of aluminium toxicity is the inhibition of root elongation, and the root apex is most affected by aluminium toxicity (Kochian 1995). Two types of mechanism are responsible for aluminium resistance and tolerance: aluminium exclusion from the symplasm in the root apex; and the detoxification of aluminium after it enters the symplasm (Blum 1988). Exclusion of aluminium might involve the release of dicarboxylic acids in the roots, which chelate Al^{3+}. It has been noted that Al-tolerant lines release considerably more organic acid than more susceptible lines (Kochian 1995).

Virtually no research has been conducted into quinoa's tolerance to aluminium and thus much research is required to uncover the mechanisms of quinoa resistance to aluminium, as well as identifying key genes involved in the process in order to focus breeding programmes. The recent release of the high-quality quinoa genome should assist with this.

9.8 UV-B

Huarancca Reyes et al. (2018) investigated the physiological effects of short acute UV-B treatments in quinoa. They found no severe alterations of photosynthetic pigments and flavonoids, but significant increase in antioxidant capacity. UV-B dramatically decreased stomatal conductance, which was possibly associated with ROS production. Inhibition of photosynthetic electron transport was also observed, suggested as a possible response to reduce ROS. Irreversible damage to the photosynthetic apparatus was found after 60 min of exposure to UV-B, probably due to severe ROS overproduction. Rubisco activity and photosynthetic electron transport were also affected.

González et al. (2009b) investigated the effects of strongly reduced solar UV-B in quinoa. The Cristalina cultivar had reduced leaf number but increased cotyledon area and seedling height, whilst Chucapaca appeared to be unaffected by the same level of UV-B. Interestingly, the highest chlorophyll content in Cristalina was seen under normal UV-B conditions, whereas in Chucapaca it was under reduced UV-B conditions. The authors conclude there is significant variation between the cultivars and the metabolic pathways show a degree of plasticity. These studies highlight the impact that UV-B can have on quinoa growth, but the genetics underpinning quinoa tolerance to UV-B require research.

9.9 Copper Toxicity

Buss et al. (2012) investigated the effect of copper toxicity on quinoa plants, and the ameliorating effects of biochar. Young quinoa plants were grown in sandy soil with 0, 50 or 200 μg/g of Cu^{2+}. The quinoa plants displayed reduced growth and severe stress symptoms shortly after the application of 50 μg/g of Cu. At 200 μg/g Cu, they died. However, the effect of biochar was striking. When biochar was added to make up 4% of the dry soil weight, even under 200 μg/g of Cu^{2+}, the quinoa plants reached almost the same biomass as under control conditions. The effect of biochar is not to enhance quinoa tolerance, but probably acts to attract positively charged Cu^{2+} to its negatively charged surfaces; less Cu^{2+} entered the plant tissues, which had reduced Cu^{2+} concentrations. This indicates a method to increase the productivity, and quality, of plants grown on soil which have heavy metal poisoning.

9.10 Other Stresses

Other stresses affect quinoa, such as waterlogging and manganese and cadmium toxicity, pollutants and physical stresses like high wind. Little or no research has been conducted into their effects on quinoa specifically. However, it is likely quinoa is affected by these stresses in a manner analogous to other crops.

González et al. (2009b) studied quinoa responses to waterlogging. They found the plant and root dry weights were lower under waterlogged conditions—and lower than under drought conditions. Leaf area was also reduced by waterlogging. Chlorophyll a and chlorophyll b levels were adversely affected by waterlogging. Soluble protein content was increased in waterlogged conditions. It has been suggested that this is to ensure a sufficient carbohydrate supply to assist in plant survival at low oxygen concentrations (Zeng et al. 1999; Barta 1988; Huang 1995). However, a decrease in sucrose cleavage due to reduced enzyme activity could also be responsible. At low oxygen levels, sucrose synthase and root invertase activities have been reported to be inhibited (Zeng et al. 1999; Drew 1997). This suggests that the high levels of soluble sugars and starch under waterlogged conditions could be related to a reduction in the carbohydrate sink strength, necessary because of lower root respiration. Additionally, protein degradation was observed and increased stomatal closure. Investigating changing genetic expression under waterlogged conditions will help elucidate the genetic control and genes of importance which should help develop highly resistant quinoa.

9.11 Conclusions

The responses of plants to abiotic stresses are complex and interrelated. Quinoa is particularly interesting for its high tolerance to salinity, as well as robust resilience to drought stress. However, currently our genetic understanding of how this tolerance is achieved is limited. Even our understanding of the physiological mechanisms that quinoa utilises is poorly understood in many areas. Certain genes have been identified and their function explored, such as *CqSOS1*, *CqNHX1*, *BADH*, *HSP20* and *HSP70*, however even here our knowledge is limited. As the importance of quinoa in contributing to global food security in the face of climate change and a rising population is increasingly recognised, research efforts need to match quinoa's rising profile, with studies that will allow key tolerance genes to be enhanced, so specific, effective and fast breeding programmes can be enacted.

In this chapter, the physiological and genetic components of the quinoa stress response have been covered to the extent of our current knowledge. The recent release of the high-quality quinoa reference genome should greatly aid research efforts to identify and locate QTLs, genes and alleles which play a role in quinoa's tolerance to abiotic stress. Utilising this knowledge could make quinoa a crop which plays an integral role in resolving future global food security challenges.

Acknowledgements Support from King Abdullah University of Science and Technology (KAUST) is gratefully acknowledged.

References

Adolf VI, Jacobsen SE, Shabala S (2013) Salt tolerance mechanisms in quinoa (*Chenopodium quinoa* Willd.). Environ Exp Bot 92:43–54

Aguilar PC, Cutipa Z, Machaca E, López M, Jacobsen SE (2003) Variation of Proline Content of Quinoa (*Chenopodium quinoa* Willd.) in High Beds (Waru Waru). Food Rev Int 19(1–2):121–127

Akula R, Ravishankar GA (2011) Influence of abiotic stress signals on secondary metabolites in plants. Plant Signal Behav 6(11):1720–1731

Al-Naggar A, El-Salam R, Badran A, El-Moghazi M (2017) Genotype and drought effects on morphological, physiological and yield traits of Quinoa (*Chenopodium quinoa* Willd.). Asian J Adv Agric Res 3 (1):1–15

Aloisi I, Parrotta L, Ruiz KB, Landi C, Bini L, Cai G, Biondi S, Del Duca S (2016) New insight into Quinoa seed quality under salinity: changes in proteomic and amino acid profiles, phenolic content, and antioxidant activity of protein extracts. Front Plant Sci 7:656

Amjad M, Akhtar SS, Yang A, Akhtar J, Jacobsen SE (2015) Antioxidative response of Quinoa exposed to iso-osmotic, ionic and non-ionic salt stress. J Agron Crop Sci 201(6):452–460

Barta AL (1988) Response of Field Grown Alfalfa to Root Waterlogging and Shoot Removal. I. Plant Injury and Carbohydrate and Mineral Content of Roots. Agron J 80(6):889

Bassil E, Tajima H, Liang YC, Ohto M, Ushijima K, Nakano R, Esumi T, Coku A, Belmonte M, Blumwald E (2011) The *Arabidopsis* Na^+/H^+ Antiporters NHX1 and NHX2 control vacuolar pH and K^+ homeostasis to regulate growth, flower development, and reproduction. Plant Cell 23 (9):3482–3497

Battaglia M, Olvera-Carrillo Y, Garciarrubio A, Campos F, Covarrubias AA (2008) The enigmatic LEA proteins and other hydrophilins. Plant Physiol 148 (1):6–24

Bazile D, Bertero HD, Nieto C (2015) Tolerance to saline conditions. In: State of the Art report on Quinoa around the World in 2013. Oficina Regional de la FAO para America Latina y el Caribe, p 149

Becerra C, Jahrmann T, Puigdomènech P, Vicient CM (2004) Ankyrin repeat-containing proteins in *Arabidopsis*: characterization of a novel and abundant group of genes coding ankyrin-transmembrane proteins. Gene 340(1):111–121

Becker VI, Goessling JW, Duarte B, Cacador I, Liu F, Rosenqvist E, Jacobsen SE (2017) Combined effects of soil salinity and high temperature on photosynthesis and growth of quinoa plants (*Chenopodium quinoa*). Funct Plant Biol 44(7):665

Bendevis MA, Sun Y, Shabala S, Rosenqvist E, Liu F, Jacobsen SE (2014) Differentiation of photoperiod-induced ABA and soluble sugar responses of two Quinoa (*Chenopodium quinoa* Willd.) cultivars. J Plant Growth Regul 33(3):562–570

Bertero H (2001) Effects of photoperiod, temperature and radiation on the rate of leaf appearance in Quinoa (*Chenopodium quinoa* Willd.) under field conditions. Ann Bot 87(4): 495–502

Bertero H., King R., Hall A. (1999) Modelling photoperiod and temperature responses of flowering in quinoa (*Chenopodium quinoa* Willd.). Field Crops Res 63 (1):19–34

Bertero H., Vega A. de la, Correa G, Jacobsen S., Mujica A (2004) Genotype and genotype-by-environment interaction effects for grain yield and grain size of quinoa (*Chenopodium quinoa* Willd.) as revealed by pattern analysis of international multi-environment trials. Field Crops Res 89(2–3):299–318

Bezant H, Laurie DA, Pratchett N, Chojecki J, Kearsey MJ (1997) Mapping of QTL controlling NIR predicted hot water extract and grain nitrogen content in a spring barley cross using marker-regression. Plant Breed 116(2):141–145

Bhargava A, Shukla S, Ohri D (2006) *Chenopodium quinoa*—An Indian perspective. Ind Crops Prod 23 (1):73–87

Blum A (1988) Plant breeding for stress environments. CRC Press, Boca Raton, Florida, USA, pp 133–162

Blumwald E (2000) Sodium transport and salt tolerance in plants. Curr Opin Cell Biol 12(4):431–434

Boas LVV, Brandão IR, Silva DM, Santos MO, Souza KRD, Alves JD (2016) Antioxidant metabolism of *Chenopodium quinoa* Willd. under salt stress. Rev Bras Ciênc Agrár—Braz J Agric Sci 11(4):281–288

Bois JF, Winkel T, Lhomme JP, Raffaillac JP, Rocheteau A (2006) Response of some Andean cultivars of quinoa (*Chenopodium quinoa* Willd.) to temperature: Effects on germination, phenology, growth and freezing. Eur J Agron 25(4):299–308

Bunce J (2017) Variation in yield responses to elevated CO2 and a brief high temperature treatment in Quinoa. Plants 6(4):26

Burrieza HP, Koyro H-W, Tosar LM, Kobayashi K, Maldonado S (2012) High salinity induces dehydrin accumulation in *Chenopodium quinoa* Willd. cv. Hualhuas embryos. Plant Soil 354(1–2):69–79

Buss W, Kammann C, Koyro H-W (2012) Biochar reduces copper toxicity in Willd. in a sandy soil. J Environ Qual 41(4):1157

Cai S, Yu G, Chen X, Huang Y, Jiang X, Zhang G, Jin X (2013) Grain protein content variation and its association analysis in barley. BMC Plant Biol 13(1):35

Carbonell-Bejerano P, Santa María E, Torres-Pérez R, Royo C, Lijavetzky D, Bravo G, Aguirreolea J, Sánchez-Díaz M, Antolín MC, Martínez-Zapater JM (2013) Thermotolerance responses in ripening berries of *Vitis vinifera* L. cv Muscat Hamburg. Plant Cell Physiol 54(7):1200–1216

Carjuzaa P, Castellión M, Distéfano AJ, Vas M del, Maldonado S (2008) Detection and subcellular localization of dehydrin-like proteins in quinoa (*Chenopodium quinoa* Willd.) embryos. Protoplasma 233(1–2):149–156

Catacora C, Canahua A (1991) Selección de genotipos dequinua (Chenopodium quinoa Willd.) resistentes a heladas, yperspectivas de producción en camellones. In: VII Congreso Internacional de Cultivos Andinos. La Paz, Bolivia, pp 53–56

Chatterjee B, Das N (1980) Note on the differences in the response of wheat varieties to boron. Indian J Agric Sci 50:796

Chilo G, Molina MV, Carabajal R, Ochoa M (2009) Temperature and salinity effects on germination and seedling growth on two varieties of *Chenopodium quinoa*. AgriScientia 26(1/6):15–22

Christensen SA, Pratt DB, Pratt C, Nelson PT, Stevens MR, Jellen EN, Coleman CE, Fairbanks DJ, Bonifacio A, Maughan PJ (2007) Assessment of genetic diversity in the USDA and CIP-FAO international nursery collections of quinoa (*Chenopodium quinoa* Willd.) using microsatellite markers. Plant Genet Resour Charact Util 5(02):82–95

Cossani CM, Slafer GA, Savin R (2010) Co-limitation of nitrogen and water, and yield and resource-use efficiencies of wheat and barley. Crop Pasture Sci 61 (10):844

Debez A, Ben Hamed K, Grignon C, Abdelly C (2004) Salinity effects on germination, growth, and seed production of the halophyte *Cakile maritima*. Plant Soil 262(1/2):179–189

Dhankhar O, Dahiya S (1980) Effect of different levels of boron and soil salinity on the yield of dry matter and its mineral composition in ber (*Zizyphus rotundifolia*). Int Symp Salt Affect Soils

Dillehay TD, Rossen J, Andres TC, Williams DE (2007) Preceramic adoption of peanut, squash, and cotton in Northern Peru. Science 316(5833):1890–1893

Drew MC (1997) Oxygen deficiency and root metabolism: injury and Acclimation Under Hypoxia and Anoxia. Annu Rev Plant Physiol Plant Mol Biol 48 (1):223–250

Escuredo O, González Martín MI, Wells Moncada G, Fischer S, Hernández Hierro JM (2014) Amino acid profile of the quinoa (*Chenopodium quinoa* Willd.) using near infrared spectroscopy and chemometric techniques. J Cereal Sci 60(1):67–74

Fageria NK, Baligar VC (2005) Enhancing nitrogen use efficiency in crop plants. Adv Agron 88:97–185

Fan W, Zhang M, Zhang H, Zhang P (2012) Improved tolerance to various abiotic stresses in transgenic sweet potato (*Ipomoea batatas*) expressing spinach betaine aldehyde dehydrogenase. PLoS ONE 7(5):e37344

Fan X, Tang Z, Tan Y, Zhang Y, Luo B, Yang M, Lian X, Shen Q, Miller AJ, Xu G (2016) Overexpression of a pH-sensitive nitrate transporter in rice increases crop yields. Proc Natl Acad Sci 113(26):7118–7123

Prado FE, Boero C, Gallardo M, Gonzalez JA (2000) Effect of NaCl on germination, growth, and soluble sugar content in *Chenopodium quinoa* Willd. seeds. Bot Bull Acad Sin 41:27–34

Prado FE, Hilal MB, Albornoz PL, Gallardo MG, Ruiz VE (2017) Anatomical and physiological responses of four Quinoa cultivars to salinity at seedling stage. Indian J Sci Technol 10(8):1–12

Fghire R, Anaya F, Ali OI, Benlhabib O, Ragab R, Wahbi S (2015) Physiological and photosynthetic response of quinoa to drought stress. Chil J Agric Res 75(2):174–183

Flowers TJ (2004) Improving crop salt tolerance. J Exp Bot 55(396):307–319

Flowers TJ, Troke PF, Yeo AR (1977) The mechanism of salt tolerance in halophytes. Annu Rev Plant Physiol 28(1):89–121

Foy CD, Chaney RL, White MC (1978) The Physiology of Metal Toxicity in Plants. Annu Rev Plant Physiol 29(1):511–566

Fuentes FF, Martinez EA, Hinrichsen PV, Jellen EN, Maughan PJ (2009) Assessment of genetic diversity patterns in Chilean quinoa (*Chenopodium quinoa* Willd.) germplasm using multiplex fluorescent microsatellite markers. Conserv Genet 10(2):369–377

Goldberg S (1997) Reactions of boron with soils. Plant Soil 193(1–2):35–48

Golldack D, Li C, Mohan H, Probst N (2014) Tolerance to drought and salt stress in plants: unraveling the signaling networks. Front Plant Sci 5:151

Gómez-Pando LR, Álvarez-Castro R, Eguiluz-de la Barra A (2010) Effect of salt stress on Peruvian Germplasm of *Chenopodium quinoa* Willd.: a promising crop. J Agron Crop Sci 196(5):391–396

González JA, Buedo SE, Bruno M, Prado FE (2017) Quantifying cardinal temperatures in Quinoa (*Chenopodium quinoa*) Cultivars. Lilloa 54(2):179–194

González JA, Gallardo M, Hilal M, Prado FE (2009a) Physiological responses of quinoa (*Chenopodium quinoa* Willd.) to drought and waterlogging stresses: dry matter partitioning. Bot Stud 50(1):35–42

González JA, Rosa M, Parrado MF, Hilal M, Prado FE (2009b) Morphological and physiological responses of two varieties of a highland species (*Chenopodium quinoa* Willd.) growing under near-ambient and strongly reduced solar UV–B in a lowland location. J Photochem Photobiol B 96(2):144–151

Good AG, Johnson SJ, De Pauw M, Carroll RT, Savidov N, Vidmar J, Lu Z, Taylor G, Stroeher V (2007) Engineering nitrogen use efficiency with alanine aminotransferase. Can J Bot 85(3):252–262

Gorantla M, Babu P, Reddy Lachagari V, Reddy A, Wusirika R, Bennetzen JL, Reddy AR (2007) Identification of stress-responsive genes in an indica rice (*Oryza sativa* L.) using ESTs generated from drought-stressed seedlings. J Exp Bot 58(2):253–265

Hall AE (2018) Crop responses to flooding, salinity, and other limiting soil conditions. Crop responses to environment: adapting to global climate change. CRC Press, Taylor & Francis Group, Boca Raton, FL, pp 197–199

Hariadi Y, Marandon K, Tian Y, Jacobsen S-E, Shabala S (2011) Ionic and osmotic relations in quinoa (*Chenopodium quinoa* Willd.) plants grown at various salinity levels. J Exp Bot 62(1):185–193

Hasegawa PM, Bressan RA, Zhu J-K, Bohnert HJ (2000) Plant cellular and molecular responses to high salinity. Annu Rev Plant Physiol Plant Mol Biol 51(1):463–499

Hawkesford M, Horst W, Kichey T, Lambers H, Schjoerring J, Møller IS, White P (2012) Functions of Macronutrients. In: Marschner's mineral nutrition of higher plants. Elsevier, pp 135–189

Hayes JE, Pallotta M, Garcia M, Öz MT, Rongala J, Sutton T (2015) Diversity in boron toxicity tolerance of Australian barley (*Hordeum vulgare* L.) genotypes. BMC Plant Biol 15(1):231

Heidlebaugh NM, Trethewey BR, Jukanti AK, Parrott DL, Martin JM, Fischer AM (2008) Effects of a barley (*Hordeum vulgare*) chromosome 6 grain protein content locus on whole-plant nitrogen reallocation under two different fertilisation regimes. Funct Plant Biol 35(7):619

Hernandez I, Alegre L, Munne-Bosch S (2004) Drought-induced changes in flavonoids and other low molecular weight antioxidants in *Cistus clusii* grown under Mediterranean field conditions. Tree Physiol 24 (11):1303–1311

Hetherington AM, Woodward FI (2003) The role of stomata in sensing and driving environmental change. Nature 424(6951):901–908

Hu B, Wang W, Ou S, Tang J, Li H, Che R, Zhang Z, Chai X, Wang H, Wang Y, Liang C, Liu L et al. (2015) Variation in NRT1.1B contributes to nitrate-use divergence between rice subspecies. Nat Genet 47 (7):834–838

Huang B (1995) Root respiration and carbohydrate status of two wheat genotypes in response to Hypoxia. Ann Bot 75(4):427–432

Huang D, Wu W, Abrams SR, Cutler AJ (2008) The relationship of drought-related gene expression in *Arabidopsis thaliana* to hormonal and environmental factors. J Exp Bot 59(11):2991–3007

Huarancca Reyes T, Scartazza A, Castagna A, Cosio EG, Ranieri A, Guglielminetti L (2018) Physiological effects of short acute UVB treatments in *Chenopodium quinoa* Willd. Sci Rep 8(1):371

IPCC (2014) Climate Change 2014: impacts, adaptation and vulnerability. IPCC, Geneva, Switzerland

Ismail H, Maksimovic JD, Maksimovic V, Shabala L, Zivanovic BD, Tian Y, Jacobsen SE, Shabala S (2016) Rutin, a flavonoid with antioxidant activity, improves plant salinity tolerance by regulating K + retention and Na + exclusion from leaf mesophyll in quinoa and broad beans. Funct Plant Biol 43(1):75–86

Iyengar ERR, Reddy MP (1996) Photosynthesis in highly salt tolerant plants. In: Pessaraki M (ed) Handbook of photosynthesis. Marcel Dekker, Boca Raton, USA, p 909

Jacobsen SE, Bach AP (1998) The influence of temperature on seed germination rate in quinoa (*Chenopodium quinoa* Willd.). Seed Sci Technol Switz 26(2):515–523

Jacobsen SE, Liu F, Jensen CR (2009) Does root-sourced ABA play a role for regulation of stomata under drought in quinoa (*Chenopodium quinoa* Willd.). Sci Hortic 122(2):281–287

Jacobsen SE, Monteros C, Christiansen JL, Bravo LA, Corcuera LJ, Mujica A (2005) Plant responses of quinoa (*Chenopodium quinoa* Willd.) to frost at various phenological stages. Eur J Agron 22(2):131–139

Jacobsen SE, Monteros C, Corcuera LJ, Bravo LA, Christiansen JL, Mujica A (2007) Frost resistance mechanisms in quinoa (*Chenopodium quinoa* Willd.). Eur J Agron 26(4):471–475

Jacobsen SE, Mujica A, Jensen CR (2003) The resistance of quinoa (*Chenopodium quinoa* Willd.) to adverse abiotic factors. Food Rev Int 19(1–2):99–109

Jamjod S (1996) Genetics of boron tolerance in durum wheat. PhD thesis Univ Adel

Jarvis DE, Ho YS, Lightfoot DJ, Schmöckel SM, Li B, Borm TJA, Ohyanagi H, Mineta K, Michell CT, Saber N, Kharbatia NM, Rupper RR et al. (2017) The genome of *Chenopodium quinoa*. Nature 542(7641):307–312

Jensen CR, Jacobsen S-E, Andersen MN, Nuñez, N, Andersen SD, Rasmussen L, Mogensen VO (2000) Leaf gas exchange and water relation characteristics of fieldquinoa (Chenopodium quinoa Willd.) during soil drying. Eur J Agronomy 13:11–25

Jha D, Shirley N, Tester M, Roy SJ (2010) Variation in salinity tolerance and shoot sodium accumulation in *Arabidopsis* ecotypes linked to differences in the natural expression levels of transporters involved in sodium transport. Plant, Cell Environ 33(5):793–804

Jiang Y, Zhu S, Yuan J, Chen G, Lu G (2016) A betaine aldehyde dehydrogenase gene in quinoa (*Chenopodium quinoa*): structure, phylogeny, and expression pattern. Genes Genomics 38(11):1013–1020

Jukanti AK, Heidlebaugh NM, Parrott DL, Fischer IA, McInnerney K, Fischer AM (2007) Comparative transcriptome profiling of near-isogenic barley (*Hordeum vulgare*) lines differing in the allelic state of a major grain protein content locus identifies genes with possible roles in leaf senescence and nitrogen reallocation. New Phytol 177(2):333–349

Kiani-Pouya A, Roessner U, Jayasinghe NS, Lutz A, Rupasinghe T, Bazihizina N, Bohm J, Alharbi S, Hedrich R, Shabala S (2017) Epidermal bladder cells confer salinity stress tolerance in the halophyte quinoa and *Atriplex* species: EBC in salinity stress responses. Plant, Cell Environ 40(9):1900–1915

Killi D, Haworth M (2017) Diffusive and metabolic constraints to photosynthesis in Quinoa during drought and salt stress. Plants 6(4):49

Kiniry J, Rosenthal W, Jackson B, Hoogenboom G (1991) Predicting leaf development of crop plants. In: Hodges T (ed) Predicting crop phenology. CRC Press, Boca Raton, USA, pp 29–39

Kitajima S, Sato F (1999) Plant pathogenesis-related proteins: molecular mechanisms of gene expression and protein function. J Biochem (Tokyo) 125(1):1–8

Kochian LV (1995) Cellular mechanisms of aluminum toxicity and resistance in plants. Annu Rev Plant Physiol Plant Mol Biol 46(1):237–260

Koyro H, Lieth M (2011) Halophytic crops: A resource for the future to reduce the water crisis? Emir J Food Agric 23(1):1–16

Koyro HW, Eisa SS (2008) Effect of salinity on composition, viability and germination of seeds of Chenopodium quinoa Willd. Plant Soil 302(1–2):79–90

Lesjak J, Calderini DF (2017) Increased night temperature negatively affects grain yield, biomass and grain number in Chilean Quinoa. Front Plant Sci 8:352

Levitt J (1980) Freezing Temperatures. Responses of plants to environmental stresses: chilling, freezing and high temperature stresses. Academic Press, New York, pp 67–344

Lin B, Wang J, Liu H, Chen R, Meyer Y, Barakat A, Delseny M (2001) Genomic analysis of the Hsp70 superfamily in *Arabidopsis thaliana*. Cell Stress Chaperones 6(3):201–208

Liu J, Wang R, Liu W, Zhang H, Guo Y, Wen R (2018a) Genome-wide characterization of heat-shock protein 70s from *Chenopodium quinoa* and expression analyses of Cqhsp70s in response to drought stress. Genes 9(2):35

Liu Z, Zhu J, Yang X, Wu H, Wei Q, Wei H, Zhang H (2018b) Growth performance, organ-level ionic relations and organic osmoregulation of *Elaeagnus angustifolia* in response to salt stress. PLoS ONE 13 (1):e0191552

Lokko Y, Anderson JV, Rudd S, Raji A, Horvath D, Mikel MA, Kim R, Liu L, Hernandez A, Dixon AGO, Ingelbrecht IL (2007) Characterization of an 18,166 EST dataset for cassava (*Manihot esculenta* Crantz) enriched for drought-responsive genes. Plant Cell Rep 26(9):1605–1618

López-Frías G, Martínez LM, Ponce G, Cassab GI, Nieto-Sotelo J (2011) Role of HSP101 in the stimulation of nodal root development from the coleoptilar node by light and temperature in maize (*Zea mays* L.) seedlings. J Exp Bot 62(13):4661–4673

Ma S, Gong Q, Bohnert HJ (2006) Dissecting salt stress pathways. J Exp Bot 57(5):1097–1107

Maas EV (1986) Salt tolerance of plants. Appl Agric Res 1:12–26

Maliro MFA, Guwela VF, Nyaika J, Murphy KM (2017) Preliminary studies of the performance of Quinoa (*Chenopodium quinoa* Willd.) genotypes under irrigated and rainfed conditions of Central Malawi. Front Plant Sci 8:227

Mamedi A, Afshari RT, Oveisi M (2017) Cardinal temperatures for seed germination of three Quinoa (*Chenopodium quinoa* Willd.) cultivars. Iranian J Field Crop Sci 47(1):89–100

Martindale JL, Holbrook NJ (2002) Cellular response to oxidative stress: signaling for suicide and survival. J Cell Physiol 192(1):1–15

Maughan PJ, Turner TB, Coleman CE, Elzinga DB, Jellen EN, Morales JA, Udall JA, Fairbanks DJ, Bonifacio A (2009) Characterization of *Salt Overly Sensitive 1* (*SOS1*) gene homoeologs in quinoa (*Chenopodium quinoa* Willd.). Genome 52(7):647–657

Mayer MP, Bukau B (2005) Hsp70 chaperones: Cellular functions and molecular mechanism. Cell Mol Life Sci 62(6):670–684

Miranda-Apodaca J, Yoldi-Achalandabaso A, Aguirresarobe A, del Canto A, Pérez-López U (2018) Similarities and differences between the responses to osmotic and ionic stress in quinoa from a water use perspective. Agric Water Manag 203:344–352

Mochida K, Kawaura K, Shimosaka E, Kawakami N, Shin-I T, Kohara Y, Yamazaki Y, Ogihara Y (2006) Tissue expression map of a large number of expressed sequence tags and its application to in silico screening of stress response genes in common wheat. Mol Genet Genomics 276(3):304–312

Morales AJ, Bajgain P, Garver Z, Maughan PJ (2011) Physiological responses of *Chenopodium quinoa* to salt stress. Int J Plant Physiol Biochem 3(13):219–232

Morales A, Zurita-Silva A, Maldonado J, Silva H (2017) Transcriptional responses of Chilean Quinoa (*Chenopodium quinoa* Willd.) under water deficit conditions uncovers ABA-independent expression patterns. Front Plant Sci 8:216

Munns R (2002) Comparative physiology of salt and water stress. Plant, Cell Environ 25(2):239–250

Munns R, Tester M (2008) Mechanisms of Salinity Tolerance. Annu Rev Plant Biol 59(1):651–681

Nable R, Banuelos G, Paull J (1997) Boron Toxicity. Plant Soil 193(1–2):181–198

Nishida I, Murata N (1996) Chilling sensitivity in plants and cyanobacteria: the crucial contribution of membrane lipids. Annu Rev Plant Physiol Plant Mol Biol 47(1):541–568

Noctor G, Foyer CH (1998) Ascorbate and glutathione: keeping active oxygen under control. Annu Rev Plant Physiol Plant Mol Biol 49(1):249–279

Oh D-H, Dassanayake M, Haas JS, Kropornika A, Wright C, Urzo MP d', Hong H, Ali S, Hernandez A, Lambert GM, Inan G, Galbraith DW et al. (2010) Genome structures and halophyte-specific gene expression of the extremophile *Thellungiella parvula* in Comparison with *Thellungiella salsuginea* (*Thellungiella halophila*) and *Arabidopsis*. Plant Physiol 154 (3):1040–1052

Oh D-H, Leidi E, Zhang Q, Hwang S-M, Li Y, Quintero FJ, Jiang X, D'Urzo MP, Lee SY, Zhao Y, Bahk JD, Bressan RA et al. (2009) Loss of halophytism by interference with SOS1 expression. Plant Physiol 151(1):210–222

Orsini F, Accorsi M, Gianquinto G, Dinelli G, Antognoni F, Carrasco KBR, Martinez EA, Alnayef M, Marotti I, Bosi S, Biondi S (2011) Beyond the ionic and osmotic response to salinity in *Chenopodium quinoa*: functional elements of successful halophytism. Funct Plant Biol 38(10):818

Paull J, Rathjen A, Cartwright B (1988) Genetic control of tolerance to high concentrations of soil boron in wheat. In: Proceedings of the 7th international wheat genetics symposium, pp 871–877

Prasad PVV, Pisipati SR, Mutava RN, Tuinstra MR (2008) Sensitivity of grain sorghum to high temperature stress during reproductive development. Crop Sci 48(5):1911

Qadir M, Quillérou E, Nangia V, Murtaza G, Singh M, Thomas RJ, Drechsel P, Noble AD (2014) Economics of salt-induced land degradation and restoration. Nat Resour Forum 38(4):282–295

Qiu Q-S, Guo Y, Dietrich MA, Schumaker KS, Zhu J-K (2002) Regulation of SOS1, a plasma membrane Na +/H + exchanger in *Arabidopsis thaliana*, by SOS2 and SOS3. Proc Natl Acad Sci 99(12):8436–8441

Raney JA, Reynolds DJ, Elzinga DB, Page J, A. Udall J, Jellen EN, Bonfacio A, Fairbanks DJ, Maughan PJ (2014) Transcriptome analysis of drought induced stress in *Chenopodium quinoa*. Am J Plant Sci 05 (03):338–357

Raun WR, Johnson GV (1999) Improving nitrogen use efficiency for cereal production. Agron J 91(3):357

Razzaghi F, Ahmadi SH, Adolf VI, Jensen CR, Jacobsen SE, Andersen MN (2011) Water relations and transpiration of Quinoa (*Chenopodium quinoa* Willd.) under salinity and soil drying: water relations and transpiration of Quinoa. J Agron Crop Sci 197 (5):348–360

Razzaghi F, Jacobsen SE, Jensen CR, Andersen MN (2015) Ionic and photosynthetic homeostasis in quinoa challenged by salinity and drought – mechanisms of tolerance. Funct Plant Biol 42(2):136

Rea J, Tapia M, Mujica A (1979) Practicas agronomicas. In: Tapia M, Gandarillas H, Alandia S, Cardozo A, Mujica A (eds) Quinoa y Kaniwa, Cultivos Andinos. FAO, Rome, Italy, pp 83–120

Repo-Carrasco R, Espinoza C, Jacobsen S-E (2003) Nutritional value and use of the Andean Crops Quinoa (*Chenopodium quinoa*) and Kañiwa (*Chenopodium pallidicaule*). Food Rev Int 19(1–2):179–189

Roelfsema MRG, Hedrich R (2005) In the light of stomatal opening: new insights into 'the Watergate': Tansley review. New Phytol 167(3):665–691

Rorat T, Szabala BM, Grygorowicz WJ, Wojtowicz B, Yin Z, Rey P (2006) Expression of SK3-type dehydrin in transporting organs is associated with cold acclimation in Solanum species. Planta 224(1):205–221

Rosa M, Hilal M, González JA, Prado FE (2009) Low-temperature effect on enzyme activities involved in sucrose–starch partitioning in salt-stressed and salt-acclimated cotyledons of quinoa (*Chenopodium quinoa* Willd.) seedlings. Plant Physiol Biochem 47 (4):300–307

Roy SJ, Negrão S, Tester M (2014) Salt resistant crop plants. Curr Opin Biotechnol 26:115–124

Rozema J, Bruin J, Broekman RA (1992) Effect of boron on the growth and mineral economy of some halophytes and non-halophytes. New Phytol 121 (2):249–256

Ruffino AMC, Rosa M, Hilal M, González JA, Prado FE (2010) The role of cotyledon metabolism in the establishment of quinoa (*Chenopodium quinoa*) seedlings growing under salinity. Plant Soil 326(1–2):213–224

Ruiz KB, Biondi S, Martínez EA, Orsini F, Antognoni F, Jacobsen SE (2016) Quinoa—a model crop for understanding salt-tolerance mechanisms in halophytes. Plant Biosyst: Int J Deal Asp Plant Biol 150 (2):357–371

Ruiz-Carrasco K, Antognoni F, Coulibaly AK, Lizardi S, Covarrubias A, Martínez EA, Molina-Montenegro MA, Biondi S, Zurita-Silva A (2011) Variation in salinity tolerance of four lowland genotypes of quinoa (*Chenopodium quinoa* Willd.) as assessed by growth, physiological traits, and sodium transporter gene expression. Plant Physiol Biochem 49(11):1333–1341

Jacobsen SE, Quispe H, Mujica A (2001) Quinoa: an alternative crop for saline soils in the Andes. In: Scientist and farmer: partners in research for the 21st century. Program Report 1999–2000, pp 403–408

Saavedra L, Svensson J, Carballo V, Izmendi D, Welin B, Vidal S (2006) A dehydrin gene in *Physcomitrella patens* is required for salt and osmotic stress tolerance. Plant J 45(2):237–249

Sarkar NK, Kundnani P, Grover A (2013) Functional analysis of Hsp70 superfamily proteins of rice (*Oryza sativa*). Cell Stress Chaperones 18(4):427–437

Savin R, Nicolas M (1996) Effects of short periods of drought and high temperature on grain growth and starch accumulation of two malting barley cultivars. Aust J Plant Physiol 23(2):201

Schachtman DP, Goodger JQD (2008) Chemical root to shoot signaling under drought. Trends Plant Sci 13 (6):281–287

Schmöckel SM, Lightfoot DJ, Razali R, Tester M, Jarvis DE (2017) Identification of Putative Transmembrane Proteins Involved in Salinity Tolerance in *Chenopodium quinoa* by Integrating Physiological Data, RNAseq, and SNP Analyses. Front Plant Sci 8:1023

Shabala L, Mackay A, Tian Y, Jacobsen SE, Zhou D, Shabala S (2012) Oxidative stress protection and stomatal patterning as components of salinity tolerance mechanism in quinoa (*Chenopodium quinoa*). Physiol Plant 146(1):26–38

Shabala S (2013) Learning from halophytes: physiological basis and strategies to improve abiotic stress tolerance in crops. Ann Bot 112(7):1209–1221

Shabala S, Bose J, Hedrich R (2014) Salt bladders: do they matter? Trends Plant Sci 19(11):687–691

Shabala S, Hariadi Y, Jacobsen SE (2013) Genotypic difference in salinity tolerance in quinoa is determined by differential control of xylem Na + loading and stomatal density. J Plant Physiol 170(10):906–914

Shabala S, Mackay A (2011) Ion Transport in halophytes. In: Advances in botanical research. Elsevier, pp 151–199

Shi H (2002) The putative plasma membrane Na +/H + antiporter SOS1 controls long-distance

Na + transport in plants. Plant Cell Online 14(2):465–477
Shi H, Ishitani M, Kim C, Zhu J-K (2000) The *Arabidopsis thaliana* salt tolerance gene SOS1 encodes a putative Na +/H + antiporter. Proc Natl Acad Sci 97(12):6896–6901
Shrawat AK, Carroll RT, DePauw M, Taylor GJ, Good AG (2008) Genetic engineering of improved nitrogen use efficiency in rice by the tissue-specific expression of *alanine aminotransferase*. Plant Biotechnol J 6(7):722–732
Solecka D (1997) Role of phenylpropanoid compounds in plant responses to different stress factors. Acta Physiol Plant 19(3):257–268
Stephens-Camacho N, Muhlia-Almazan A, Sanchez-Paz A, Rosas-Rodriguez J (2015) Surviving environmental stress: the role of betaine aldehyde dehydrogenase in marine crustaceans. J Invertebr Pathol 12:66–74
Sylvester-Bradley R, Kindred DR (2009) Analysing nitrogen responses of cereals to prioritize routes to the improvement of nitrogen use efficiency. J Exp Bot 60(7):1939–1951
Szabados L, Savouré A (2010) Proline: a multifunctional amino acid. Trends Plant Sci 15(2):89–97
Tan M, Emre A Temperature and salinity effects on germination of some Quinoa (*Chenopodium Quinoa* Willd.) cultivars. In: Proceedings of the 89th IRES international conference
Tang T, Yu A, Li P, Yang H, Liu G, Liu L (2016) Sequence analysis of the Hsp70 family in moss and evaluation of their functions in abiotic stress responses. Sci Rep 6(1):33650
Tang W, Sun J, Liu J, Liu F, Yan J, Gou X, Lu B-R, Liu Y (2014) RNAi-directed downregulation of betaine aldehyde dehydrogenase 1 (OsBADH1) results in decreased stress tolerance and increased oxidative markers without affecting glycine betaine biosynthesis in rice (*Oryza sativa*). Plant Mol Biol 86(4–5):443–454
Tripp J, Mishra SK, Scharf KD (2009) Functional dissection of the cytosolic chaperone network in tomato mesophyll protoplasts. Plant, Cell Environ 32 (2):123–133
Vacher J (1998) Responses of two main Andean crops, quinoa (*Chenopodium quinoa* Willd) and papa amarga (*Solanum juzepczukii* Buk.) to drought on the Bolivian Altiplano: significance of local adaptation. Agric Ecosyst Environ 68(1–2):99–108
Varshney RK, Hiremath PJ, Lekha P, Kashiwagi J, Balaji J, Deokar AA, Vadez V, Xiao Y, Srinivasan R, Gaur PM, Siddique KH, Town CD, et al. (2009) A comprehensive resource of drought- and salinity-responsive ESTs for gene discovery and marker development in chickpea (*Cicer arietinum* L.). BMC Genom 10(1):523
Vega-Gálvez A, Miranda M, Vergara J, Uribe E, Puente L, Martínez EA (2010) Nutrition facts and functional potential of quinoa (*Chenopodium quinoa* willd.), an ancient Andean grain: a review. J Sci Food Agric 90(15):2541–2547
Volkov V (2015) Quantitative description of ion transport via plasma membrane of yeast and small cells. Front Plant Sci 6:425
Wang J, Gan YT, Clarke F, McDonald CL (2006) Response of Chickpea yield to high temperature stress during reproductive development. Crop Sci 46 (5):2171
Wicke B, Smeets E, Dornburg V, Vashev B, Gaiser T, Turkenburg W, Faaij A (2011) The global technical and economic potential of bioenergy from salt-affected soils. Energy Environ Sci 4(8):2669
Witzel K, Matros A, Møller ALB, Ramireddy E, Finnie C, Peukert M, Rutten T, Herzog A, Kunze G, Melzer M, Kaspar-Schoenefeld S, Schmülling T et al. (2018) Plasma membrane proteome analysis identifies a role of barley membrane steroid binding protein in root architecture response to salinity: Barley plasma membrane proteins under salinity. Plant, Cell Environ 41(6):1311–1330
Wong CE, Li Y, Whitty BR, Díaz-Camino C, Akhter SR, Brandle JE, Golding GB, Weretilnyk EA, Moffatt BA, Griffith M (2005) Expressed sequence tags from the Yukon ecotype of *Thellungiella* reveal that gene expression in response to cold, drought and salinity shows little overlap. Plant Mol Biol 58(4):561–574
Wu H-J, Zhang Z, Wang J-Y, Oh D-H, Dassanayake M, Liu B, Huang Q, Sun H-X, Xia R, Wu Y, Wang Y-N, Yang Z et al. (2012) Insights into salt tolerance from the genome of *Thellungiella salsuginea*. Proc Natl Acad Sci 109(30):12219–12224
Xu G, Fan X, Miller AJ (2012) Plant nitrogen assimilation and use efficiency. Annu Rev Plant Biol 63(1):153–182
Yamaguchi-Shinozaki K, Shinozaki K (2006) Transcriptional regulatory networks in cellular responses and tolerance to dehydration and cold stresses. Annu Rev Plant Biol 57(1):781–803
Yamamoto H, Katano N, Ooi A, Inoue K (2000) Secologanin synthase which catalyzes the oxidative cleavage of loganin into secologanin is a cytochrome P450. Phytochemistry 53(1):7–12
Yang A, Akhtar SS, Amjad M, Iqbal S, Jacobsen SE (2016) Growth and physiological responses of Quinoa to drought and temperature stress. J Agron Crop Sci 202(6):445–453
Yang A, Akhtar SS, Iqbal S, Qi Z, Alandia G, Saddiq MS, Jacobsen SE (2018) Saponin seed priming improves salt tolerance in quinoa. J Agron Crop Sci 204(1):31–39
Yu J, Cheng Y, Feng K, Ruan M, Ye Q, Wang R, Li Z, Zhou G, Yao Z, Yang Y, Wan H (2016) Genome-wide identification and expression profiling of tomato Hsp20 gene family in response to biotic and abiotic stresses. Front Plant Sci 7:1215
Yunta C, Martinez-Ripoll M, Albert A (2011) SnRK2.6/OST1 from *Arabidopsis thaliana* : cloning, expression, purification, crystallization and preliminary X-ray analysis of K50N and D160A mutants. Acta Crystallogr Sect F Struct Biol Cryst Commun 67 (3):364–368

Zeng Y, Wu Y, Avigne WT, Koch KE (1999) Rapid repression of maize invertases by low oxygen. Invertase/sucrose synthase balance, sugar signaling potential, and seedling survival. Plant Physiol 121 (2):599–608

Zhang X, Davidson EA, Mauzerall DL, Searchinger TD, Dumas P, Shen Y (2015) Managing nitrogen for sustainable development. Nature 528:51–59

Zhao P, Wang D, Wang R, Kong N, Zhang C, Yang C, Wu W, Ma H, Chen Q (2018) Genome-wide analysis of the potato Hsp20 gene family: identification, genomic organization and expression profiles in response to heat stress. BMC Genom 19(1):61

Zou C, Chen A, Xiao L, Muller HM, Ache P, Haberer G, Zhang M, Jia W, Deng P, Huang R, Lang D, Li F et al. (2017) A high-quality genome assembly of quinoa provides insights into the molecular basis of salt bladder-based salinity tolerance and the exceptional nutritional value. Cell Res 27(11):1327–1340

Zurita-Silva A, Fuentes F, Zamora P, Jacobsen SE, Schwember AR (2014) Breeding quinoa (*Chenopodium quinoa* Willd.): potential and perspectives. Mol Breed 34(1):13–30

Flowering in *Chenopodium* and Related Amaranths

10

Helena Štorchová

Abstract

The transition from the vegetative to the reproductive phase is a crucial event in plant development. The floral induction is tightly controlled at multiple levels. While physiological and anatomical studies of flowering have begun in the nineteenth century, the genetic basis of the floral induction remained concealed until the end of the twentieth century. The molecular regulatory pathways mediating the responses to environmental and endogenous cues were first revealed in the model plant *Arabidopsis thaliana*, later in *Oryza sativa* (rice), and other crops. Knowledge on flowering in wild species proceeded in much slower pace. Little research was devoted to the family Amaranthaceae, except for the agriculturally important sugar beet. Nowadays, this picture starts to change owing to the availability of genomic and transcriptomic resources in non-model organisms. This review outlines basic characteristics of the regulation of flowering in model plants and summarizes current knowledge on this topic in the genus *Chenopodium* and its relatives.

H. Štorchová (✉)
Institute of Experimental Botany, Czech Academy of Sciences, Rozvojová 263, 165 02 Prague 6, Lysolaje, Czech Republic
e-mail: storchova@ueb.cas.cz

10.1 Introduction

The appropriate timing of developmental processes is the fundamental prerequisite of plant survival and crop yield. Leaf abscission, seed germination, tuber formation, fruit ripening, and flowering should occur at the most appropriate time of year. The onset of the respective developmental programs is triggered by a combination of endogenous and exogenous factors, such as age, assimilate concentration and distribution, ambient temperature, light quality and intensity, water and nutrient availability, and day length (photoperiod) (Bernier et al. 1993; Blazquez et al. 2003). The change in photoperiod is an important environmental cue in temperate climate with seasonal latitude-dependent fluctuations. It also affects the most prominent event in plant life—the transition from vegetative growth to a flowering state. Angiosperm species differ in their response to day length. Short-day plants remain vegetative until they perceive a photoperiod shorter than a threshold (Kojima et al. 2002), whereas long-day plants accelerate flowering when a day is sufficiently long (Koornneef et al. 1991). The third category, comprised of photoperiod-neutral plants, does not exhibit a specific requirement for day length (Lifschitz et al. 2006, 2014).

Classical studies of flowering revealed that the inductive photoperiod was perceived by leaves, which resulted in the floral bud formation in

S. M. Schmöckel (ed.), *The Quinoa Genome*, Compendium of Plant Genomes,
https://doi.org/10.1007/978-3-030-65237-1_10

apical meristems (Chailakhyan 1936; King 1972; Zeevart 1976). The factor "florigen" was proposed to transmit the signal from leaves to the meristematic apex. It was identified as the product of the gene *FLOWERING LOCUS T* (*FT*) in *Arabidopsis thaliana* (Koornneef et al. 1991; Corbesier et al. 2007), which became the first model plant for the study of the molecular basis of floral induction.

10.2 Flowering in *Arabidopsis thaliana*

The FT protein is produced in *A. thaliana* leaves under long days. It moves through the companion cells of the phloem (Liu et al. 2012) to the apex, where it interacts with the transcription factor FLOWERING LOCUS D (FD) (Abe et al. 2005) to activate *APETALA 1* (*AP1*) and other genes participating in floral development (Benlloch et al. 2007). Information about photoperiod is mediated by the protein CONSTANS (CO), which regulates *FT* transcription (An et al. 2004). The abundance of the CO protein, which is degraded under darkness (Valverde et al. 2004), is controlled by the endogenous clock (Suarez-Lopez et al. 2001) and day length (Izawa et al. 2002). In addition to the photoperiodic floral induction pathway, the *FT* gene integrates signals from the vernalization, autonomous, and gibberellin-dependent pathways (Mutasa-Göttgens and Hedden 2009; Amasino 2010; Andres and Coupland 2012), illustrating its central role in the process of floral induction (Fig. 10.1a).

Vernalization is the promotion of flowering by a prolonged period of low temperature. It represents an adaptation of plants to seasonal changes in temperature in temperate regions with harsh winters. The key integrator of the vernalization pathway in *A. thaliana* is the gene *FLOWERING LOCUS C* (*FLC*) (Johanson et al. 2000; Sheldon et al. 2000). *FLC* codes for a MADS-box transcription factor that inhibits *FT* transcription. During the cold period, *FLC* expression is suppressed by epigenetic chromatin remodeling, which makes *FT* activation possible under appropriate light and temperature conditions (Bastow et al. 2004; Michaels 2009). *FLC* also helps to maintain the juvenile phase (Deng et al. 2011), which is an early developmental period during which plants cannot be induced to flowering. The juvenile phase may last from a few days in short-living annuals to more than 20 years in trees (Matsoukas et al. 2012).

A. thaliana is a well-understood and powerful plant model. However, the immense diversity of life history traits among higher plants demands investigation of other species differing in photoperiod, temperature, and vernalization requirements, in the duration of juvenile phase, and in the resistance to abiotic and biotic stresses. Stress-response signaling pathways are interconnected with floral regulatory pathways and stressful conditions may both accelerate or delay flowering (Samad et al. 2017; Cho et al. 2017).

10.3 Short-Day Plant Models

The first short-day models adopted for the investigation of flowering at the molecular level were *Pharbitis nil* (Liu et al. 2001) and rice (Yano et al. 2001). Two *FT* orthologs in *P. nil* were expressed under inductive short days, peaking at the end of nights (Hayama et al. 2007). Only one of them, *PnFTL2*, was upregulated by poor nutrition stress (Yamada and Takeno 2014), which suggested functional diversification between both *FTL* genes. The *CO/FT* regulatory module, the backbone of the flowering regulatory network in *A. thaliana*, is not conserved in *P. nil*, because *PnCO* transcription does not correlate with the expression of *PnFTL1* and *PnFTL2* (Hayama et al. 2007). This example documents, that not all the regulatory mechanisms adopted by *A. thaliana* are used by other species.

Two floral activators orthologous to *FT* were also confirmed in rice, which was studied in great detail owing to its agricultural importance (Sun et al. 2014). *Heading date 3a* (*Hd3a*) accelerates flowering transition under inductive short days, whereas *RICE FLOWERING LOCUS T 1* (*RFT1*) is responsible for flowering under non-inductive long days (Komiya et al. 2008, 2009). The two

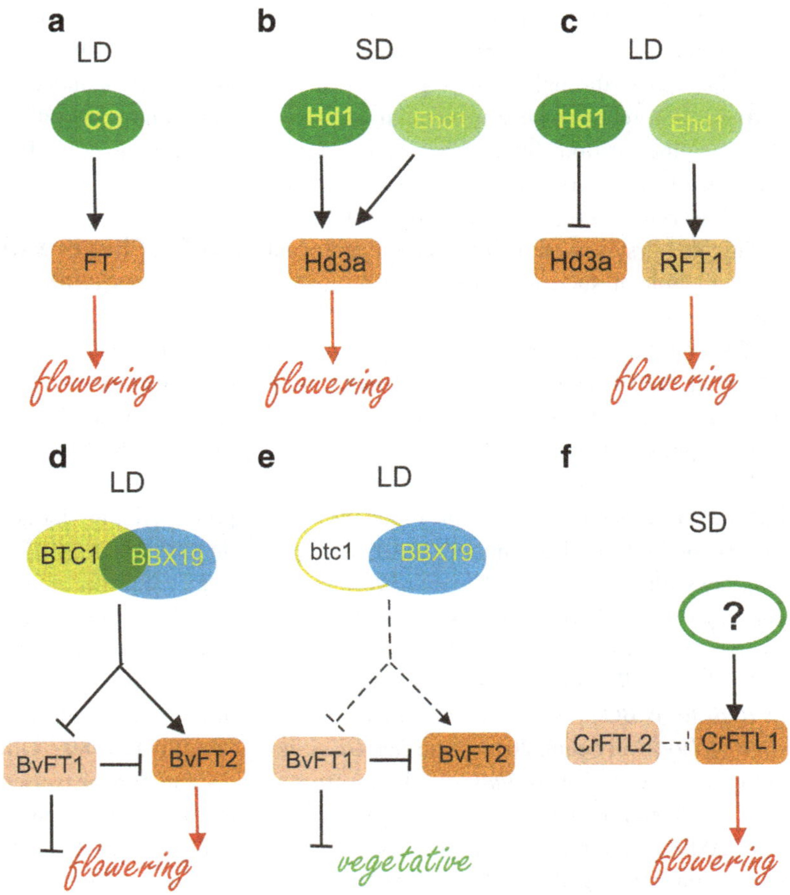

Fig. 10.1 **a** The key regulatory module regulating flowering in angiosperms. *A. thaliana* under long day **b** Rice under short day. The Hd1 protein activates the *HD3a* expression. **c** Rice under long day. The Hd1 protein inhibits the *HD3a* expression. **d** Annual sugar beet under long day. BTC1 and BBX19 proteins activate *BvFT2* and suppress *BvFT1*. **e** Biennial sugar beet under long day. Mutation in BTC1 impairs its function, *BvFT2* is not transcribed, plants do not flower. **f** *C. rubrum* under short day. *CrFTL2* does not inhibit floral induction

flowering integrators are controlled by distinct upstream regulators. The *CO* homolog *Heading date 1* (*Hd1*) upregulates *Hd3a* under short day, but inhibits it under long day. In contrast, the B-type response regulator *Early heading date 1* (*Ehd 1*) activates *RFT1* under long day via the pathway unique to rice (Doi et al. 2004) (Fig. 10.1b).

Genetic studies in just a small handful of model species revealed considerable variation in the composition of signaling pathways participating in floral induction, but they also identified a conserved essential element integrating the outputs of various pathways—the *FT* gene. The subsequent investigation of the molecular basis of flowering in crops such as maize (Meng et al. 2011), wheat (Li and Dubcovsky 2008), sunflower (Blackman et al. 2010), onion (Lee et al. 2013), legumes (Weller and Ortega 2015), or ornamental plants (Li et al. 2014; Randoux et al. 2014) confirmed this conclusion (reviewed by Wickland and Hanzawa 2015). Flowering research in wild species has lagged behind in economically important crops. Large taxonomic groups remain unexplored. An example is the order Caryophyllales comprising also carnivorous and succulent plants. Within this clade, only the family Amaranthaceae attracted substantial research interest owing to the agriculturally important sugar beet.

10.4 Flowering in Sugar Beet

Sugar beet (*Beta vulgaris*) represents the sole source of sugar grown in temperate climates. It was domesticated from sea beet (*Beta vulgaris* subsp. *maritima*) about 200 years ago. Sea beet is a long-day plant which may or may not require vernalization (Lexander 1980). Cultivated sugar beet has gone through a strong selection for vernalization requirement (or biennial life habit) during domestication. Sugar beet is grown as a biennial plant, where bolting (shoot elongation) and flowering is prevented (by not vernalizing) during the first season to ensures an abundant accumulation of sucrose in taproots and a high yield.

Two *FT* homologs were found in sugar beet (Pin et al. 2010). The *BvFT2* gene encodes floral inducer, whereas the second paralog *BvFT1* acts in the opposite manner—like a floral inhibitor. *BvFT1* is transcribed during the first season in biennial accessions, which prevents *BvFTL2* expression and floral promotion under inductive long days (Pin et al. 2010). The two *FT* genes are controlled by two upstream elements—the pseudo response regulator *BOLTING TIME CONTROL 1* (*BTC1*) (Pin and Nilsson 2012), and *DOUBLE B-BOX TYPE ZINC FINGER 19* (*BvBBX19*) (Dally et al. 2014). The *BvCOL* genes homologous to *CO* in *A. thaliana* do not participate in the floral induction in sugar beet (Chia et al. 2008). An intriguing hypothesis suggests an interaction between the BTC1 and BBX19 proteins, which forms a complex functioning as the CO protein in *A. thaliana* (Dally et al. 2014). BTC1 and BBX19 activate the floral inducer *BvFT2* and repress the floral inhibitor *BvFT1* in annual sea beet (Fig. 10.1c). Biennial cultivars of sugar beet possess non-functional alleles of *BTC1*, which allow the high expression of the floral suppressor *BvFTL1* and inhibit flowering until vernalization occurs (Fig. 10.1d). The *FLC-LIKE 1* (*BvFL1*) gene orthologous to *FLC* does not play a major role in the control of vernalization (Vogt et al. 2014), the vernalization signaling pathway has not yet been described in sugar beet.

The availability of genetic resources and genomic sequences has focused research on a few model species, while overlooking many interesting plants. Classical physiological experiments on flowering have been performed in some of these species since the middle of the twentieth century. *Chenopodium* from the family Amaranthaceae was the favorite object of plant physiologists since as early as 1940 (Allard and Garner 1940). Recent advances in next-generation sequencing, genomics, and transcriptomics have paved the way for genetic studies in the previously neglected amaranths.

10.5 Flowerering in *Chenopodium rubrum*

The short-day plant *Chenopodium rubrum* (synonymum *Oxybasis rubra*, Fuentes-Bazan 2012) was introduced as a model to study floral induction by Cumming (1959). The ecotype 374 originating from Alaska is highly sensitive to photoperiodic induction—a single period of darkness (12–14 h) induces flowering in seedlings shortly after opening the cotyledons (Cumming et al. 1965, Cumming 1967). The peak of inducibility coincides with the highest growth of cotyledons and is followed by decreased sensitivity in 6–12 day-old plantlets (Seidlová 1980; Ullmann et al. 1985). This decline may be caused by cotyledon aging before the first pair of leaves is fully developed. Owing to the sensitivity of this species to photoperiod in seedling stage, experiments with flower induction in *C. rubrum*, as well as detailed anatomical studies, are highly reproducible and fast.

Two *FT* homologs have been found in *C. rubrum* (Cháb et al. 2008). The *CrFTL1* gene was activated by a dark period, which also induced flowering. It complemented *ft* mutation and accelerated flowering in *A. thaliana* (Drabešová et al. 2014). *CrFTL1* is a floral activator orthologous to *BvFT2* in sugar beet. The *CrFTL2* gene orthologous to the sugar beet inhibitor *BvFT1* exhibited invariable expression under various photoperiodic regimes, which was inconsistent with any function in flowering in *C. rubrum* (Cháb et al. 2008; Drabešová et al. 2014). *CrFTL2* underwent an unusual structural rearrangement accompanied by the acquisition of a novel exon, which may have contributed to a possible functional shift (Drabešová et al. 2016). The loss of floral this inhibitor may be associated with its annual growth habit and the absence of a juvenile phase in *C. rubrum* (Fig. 10.1d).

Expression patterns of the *CrCOL1* and *CrCOL2* genes, the closest *CO* homologs in *C. rubrum*, did not support their role in floral induction (Drabešová et al. 2014). The upstream regulator of *CrFTL1* is currently unknown. It will be interesting to determine whether the homologs of *BTC1* and *BBX19* activate *CrFTL1* and fulfill similar functions in *C. rubrum* as they do in sugar beet (Dally et al. 2014). Although a complete genomic sequence of *C. rubrum* is not yet available, its comprehensive transcriptome represents a useful source for gene identification (Drabešová et al. 2016).

C. rubrum also served as a model to study the role of phytohormones in flowering. Exogenous indole-3-acetic acid (IAA) inhibited flowering when applied before, during, or several hours after the inductive dark period (Pavlová and Krekule 1990). A similar effect was observed after the application of synthetic cytokinin 6-benzyladenine (BAP) at the beginning of the permissive dark period (Blažková et al. 2001). Levels of endogenous cytokinins showed rhythms with minima at dark and maxima at light in shoots and roots of *C. rubrum*. However, cytokinin content in apices rose substantially at the end of the dark period (Macháčková et al. 1993). This increase in cytokinin concentration in apical parts indicates a role for cytokinins (increased cell division) in apex evocation.

Melatonin (N-acetyl-5-methoxytryptamine) is an animal hormone, involved in regulation of an array of physiological processes in vertebrates. It scavenges free radicals and alleviates oxidative stress in plants (Reiter et al. 2015). It is rhythmically produced in plants, but the diurnal changes vary among plant species (Arnao and Hernandez-Ruiz 2015). Melatonin concentration oscillates with peaks at night in *C. rubrum* (Wolf et al. 2001). When applied during the first half of inductive dark period, melatonin inhibited flowering in *C. rubrum* (Kolář et al. 2003). Melatonin may therefore be expected to play a role in floral induction.

Numerous experiments and observations confirmed the importance of *C. rubrum* as a short-day model plant to investigate flowering. Because it is closely related to quinoa (*Chenopodium quinoa*), it may also help to clarify the control of flowering in this valuable crop.

10.6 Flowering in *Chenopodium quinoa* (Quinoa)

Quinoa is a high protein food crop native to South America (Wilson 1990). It is capable of growth at high altitudes and in harsh environments (Risi and Galway 1984), conditions which preclude the cultivation of traditional grain crops such as wheat or maize. Therefore, it has been considered as a crop with tremendous potential outside of its original cultivation area (Jacobsen et al. 2012). Understanding flowering control in *C. quinoa* has a high practical value.

Fuller (1949) described *C. quinoa* as a facultatively short-day species. *C. quinoa* is distributed over a large latitudinal range (50°), from Ecuador to southern Chile, and over a huge altitudinal range spanning from more than 4000 m nearly to sea level in southern Chile (Wilson 1990). Bertero et al. (1999a) compared two *C. quinoa* cultivars—Kanckolla, a cultivar from the Andean plateau in Southern Peru (latitude 15.20 °S), and Blanca de Junín, originating from the tropical valleys of central Peru (latitude 12.02 °S). Photoperiod had a high impact on seed size in both cultivars. Plants grown under short days produced seeds four-fold larger in diameter. In contrast, the duration of the vegetative phase (from emergence to floral initiation) was not affected by photoperiod in Kanckolla, and only slightly shorter in Blanca de Junín under short days. The same authors (Bertero et al. 1999b) carried out a comparative study of quinoa cultivars from almost the entire latitudinal range of cultivation. The duration of their vegetative phase was negatively correlated with the latitude of origin of the accession. It was the shortest in plants coming from sea level site in southern Chile and the longest in cultivars from Ecuador. Recent studies confirmed the effect of photoperiod on seed development and size in short day and neutral cultivars of quinoa (Christiansen et al. 2010; Bendevis et al. 2014).

No gene expression or functional studies of flowering-related genes have been performed in quinoa. However, the recent publication of the complete genomic sequence of quinoa (Jarvis et al. 2017) makes it possible to identify essential genes according to their sequence similarity, which will spur the understanding of flowering regulation in this important crop. Two copies of the floral inducer *FTL1* originating from two diploid ancestors of tetraploid quinoa were recognized in its genomic sequence. The tandem duplication of the floral inhibitor *FTL2* had occurred after the divergence of the *Chenopodium* ancestor from the sugar beet lineage (Storchova, personal communication). The quinoa genome therefore contains four *FTL2* copies, each pair derived from the respective ancestral diploid. It will be interesting to analyze how the particular *FTL2* copies participate in floral regulation and whether their functions have diversified.

Numerous *Chenopodium* species are widely distributed around the world, many of them are weedy plants. They became adapted to changing day length associated with different seasons of the year in various latitudes. For example, Nakatani et al. (2009) observed a clinal pattern for the number of days from sowing to flower bud initiation (vegetative phase) in nine *C. album* accessions collected in Japan. As in *C. quinoa*, the shortest vegetative phase was found in plants originating from the highest latitudes, which were also the most sensitive to photoperiod (Nakatani et al. 2009).

Studies of the genetic basis of flowering requirement adaptations to specific day length and latitude will be the most interesting in the diploid species, which are closely related to the ancestors of quinoa, e.g. *Chenopodium ficifolium* and *Chenopodium suecicum* (Walsh et al. 2015; Štorchová et al. 2015). These will help to clarify the function of the genes responsible for the control of flowering in various quinoa cultivars.

10.7 Future Perspectives

The family Amaranthaceae includes important crops—e.g. sugar beet, quinoa, spinach. It also includes weedy species which harm farmers' yield, such as the noxious weed *C. album*. To facilitate crop cultivation while controlling weeds, understanding the regulation of flowering is necessary. Analyses of the molecular background of flowering in amaranths had been delayed due to the lack of DNA sequence information. Recent advances in genomic tools have created novel genomic and transcriptomic resources which accelerate the amaranth research, including studies on flowering.

Acknowledgements The author is grateful to James D. Stone for critically reading the manuscript and for linguistic correction. This study was supported by the Czech Science Foundation (13-02290S).

References

Abe M, Kobayashi Y, Yamamoto S, Daimon Y, Yamaguchi A, Ikeda Y, Ichinoki H, Notaguchi M, Goto K, Araki T (2005) FD, a bZIP protein mediating signals from the floral pathway integrator FT at the shoot apex. Science 309:1052–1056

Allard HA, Garner WW (1940) Further observations on the response of various species of plants to length of day. US Department of Agriculture Technical Bulletin, p 727

Amasino R (2010) Seasonal and developmental timing of flowering. Plant J 61:1001–1013

An H, Roussot C, Suárez-López P et al (2004) CONSTANS acts in the phloem to regulate a systemic signal that induces photoperiodic flowering of *Arabidopsis*. Development 131:3615–3626

Andres F, Coupland G (2012) The genetic basis of flowering responses to seasonal cues. Nat Rev Genet 13:627–639

Arnao MB, Hernandez-Ruiz J (2015) Functions of melatonin in plants: a review. J Pineal Res 59:133–150

Bastow R, Mylne JS, Lister C, Lippman Z, Martienssen RA, Dean C (2004) Vernalization requires epigenetic silencing of *FLC* by histone methylation. Nature 427:164–167

Bendevis MA, Sun YJ, Shabala S, Rosenqvist E, Liu FL, Jacobsen SE (2014) Differentiation of photoperiod-induced ABA and soluble sugar responses of two Quinoa (*Chenopodium quinoa* Willd.) Cultivars. J Plant Growth Regul 33:562–570

Benlloch R, Berbel A, Serrano-Mislata A, Madueno F (2007) Floral initiation and inflorescence architecture: a comparative view. Ann Bot 100:659–676

Bertero HD, King RW, Hall AJ (1999a) Photoperiod-sensitive development phases in quinoa (*Chenopodium quinoa* Willd.). Field Crop Res 60:231–243

Bernier G, Havelange A, Houssa C, Petitjean A, Lejeune P (1993) Physiological signals that induce flowering. Plant Cell 5:1147–1155

Bertero HD, King RW, Hall AJ (1999b) Modelling photoperiod and temperature responses of flowering in quinoa (*Chenopodium quinoa* Willd.). Field Crop Res 63:19–34

Blackman BK, Strasburg JL, Raduski AR, Michaels SD, Rieseberg LH (2010) The role of recently derived FT paralogs in sunflower domestication. Curr Biol 20:629–635

Blažková A, Macháčková I, Eder J, Krekule J (2001) Benzyladenine-induced inhibition of flowering in *Chenopodium rubrum* in vitro is not related to the levels of isoprenoid cytokinins. Plant Growth Regul 34:159–166

Blazquez MA, Ahn JH, Weigel D (2003) A thermosensory pathway controlling flowering time in *Arabidopsis thaliana*. Nat Genet 33:168–171

Cháb D, Kolář J, Olson MS, Štorchová H (2008) Two *Flowering Locus T* (*FT*) homologs in *Chenopodium rubrum* differ in expression patterns. Planta 228:929–940

Chailakhyan MK (1936) New facts in support of the hormonal theory of plant development C R Acad Sci URSS 13:79–83

Chia TYP, Muller A, Jung C, Mutasa-Gottgens ES (2008) Sugar beet contains a large *CONSTANS-LIKE* gene family including a *CO* homologue that is independent of the early-bolting (*B*) gene locus. J Exp Bot 59:2735–2748

Cho LH, Yoon J, An G (2017) The control of flowering time by environmental factors. Plant J 90:708–719

Christiansen JL, Jacobsen SE, Jorgensen ST (2010) Photoperiodic effect on flowering and seed development in quinoa (Chenopodium quinoa Willd.). Acta Agric Scand Sect B-Soil Plant Sci 60:539–544

Corbesier L, Vincent C, Jang SH, Fornara F, Fan QZ et al (2007) FT protein movement contributes to long-distance signaling in floral induction of *Arabidopsis*. Science 316:1030–1033

Cumming BG (1959) Extreme sensitivity of germination and photoperiodic reaction in the genus *Chenopodium* (Tourn) L Nature 184:1044–1045

Cumming BG, Hendricks SB, Borthwick HA (1965) Rhythmic flowering responses and phytochrome changes in a selection of *Chenopodium rubrum*. Can J Bot 43:825–853

Cumming BG (1967) Early-flowering plants. In: Wilt FH, Wessels NK (eds) Methods in developmental biology. Crowell Co, New York, pp 277–299

Dally N, Xiao K, Holtgraewe D, Jung C (2014) The *B2* flowering time locus of beet encodes a zinc finger transcription factor. Proc Natl Acad Sci USA 111:10365–10370

Deng WW, Ying H, Helliwell CA, Taylor JM, Peacock WJ, Dennis ES (2011) *FLOWERING LOCUS C* (*FLC*) regulates development pathways throughout the life cycle of *Arabidopsis*. Proc Natl Acad Sci USA 108:6680–6685

Doi K, Izawa T, Fuse T, Yamanouchi U, Kubo T, Shimatani Z, Yano M, Yoshimura A (2004) *Ehd1*, a B-type response regulator in rice, confers short-day promotion of flowering and controls *FT*-like gene expression independently of *Hd1*. Genes Dev 18:926–936

Drabešová J, Cháb D, Kolář J, Haškovcová K, Štorchová H (2014) A darklight transition triggers expression of the floral promoter *CrFTL1* and downregulates *CONSTANS*-like genes in a short-day plant *Chenopodium rubrum*. J Exp Bot 65:2137–2146

Drabešová J, Černá L, Mašterová H, Koloušková P, Potocký M, Štorchová H (2016) The evolution of the FT/TFL1 genes in Amaranthaceae and their expression patterns in the course of vegetative growth and flowering in *Chenopodium rubrum*. G3-genes. Genom Genet 6:3066–3076

Fuentes-Bazan, S, G Mansion, and T Borsch, 2012 Towards a species level tree of the globally diverse genus Chenopodium (Chenopodiaceae) Mol Phylogenet Evol 62: 359–374

Fuller HJ (1949) Photoperiodic responses of *Chenopodium quinoa* Willd. and *Amaranthus caudatus* L. Am J Bot 36:175–180

Hayama R, Agashe B, Luley E, King R, Coupland G (2007) A circadian rhythm set by dusk determines the expression of *FT* homologs and the short-day photoperiodic flowering response in *Pharbitis*. Plant Cell 19:2988–3000

Izawa T, Oikawa T, Sugiyama N, Tanisaka T, Yano M, Shimamoto K (2002) Phytochrome mediates the external light signal to repress FT orthologs in photoperiodic flowering of rice. Genes Dev 16:2006–2020

Jacobsen SE, Jensen CR, Liu F (2012) Improving crop production in the arid Mediterranean climate. Field Crop Res 128:34–47

Jarvis DE, Ho YS, Lightfoot DJ, Schmockel SM, Li B, Borm TJA et al (2017) The genome of *Chenopodium quinoa*. Nature 542:307–312

Johanson U, West J, Lister C, Michaels S, Amasino R, Dean C (2000) Molecular analysis of *FRIGIDA*, a major determinant of natural variation in *Arabidopsis* flowering time. Science 290:344–347

King RW (1972) Timing in *Chenopodium rubrum* of export of the floral stimulus from the cotyledons and its action at the shoot apex. Can J Bot 50:697–702

Kolář J, Johnson CH, Macháčková I (2003) Exogenously applied melatonin (N-acetyl-5-methoxytryptamine) affects flowering of the short-day plant *Chenopodium rubrum*. Physiol Plant 118:605–612

Komiya R, Ikegami A, Tamaki S, Yokoi S, Shimamoto K (2008) *Hd3a* and *RFT1* are essential for flowering in rice. Development 135:767–774

Kojima S, Takahashi Y, Kobayashi Y, Monna L, Sasaki T, Araki T, Yano M (2002) Hd3a, a rice ortholog of the Arabidopsis FT gene, promotes transition to flowering downstream of Hd1 under short day conditions. Plant Cell Physiol 43:1096–1105

Komiya R, Yokoi S, Shimamoto K (2009) A gene network for long-day flowering activates *RFT1* encoding a mobile flowering signal in rice. Development 136:3443–3450

Koornneef M, Hanhart CJ, Vanderveen JH (1991) A genetic and physiological analysis of late flowering mutants in Arabidopsis thaliana. Mol Gen Genet 229:57–66

Lee R, Baldwin S, Kenel F, McCallum J, Macknight R (2013) *FLOWERING LOCUS T* genes control onion bulb formation and flowering. Nat Commun 4:2884

Lexander K (1980) Present knowledge of sugar beet bolting mechanisms. In: Proceedings of the 43rd winter congress of the International Institute of Sugar Beet Research, pp 245–258

Li C, Dubcovsky J (2008) Wheat FT protein regulates *VRN1* transcription through interactions with FDL2. Plant J 55:543–554

Li DM, Lu FB, Zhu GF, Sun YB, Liu HL, Liu JW, Wang Z (2014) Molecular characterization and functional analysis of a *FLOWERING LOCUS T* homolog gene from a *Phalaenopsis* orchid. Genet Mol Res 13:5982–5994

Lifschitz E, Eviatar T, Rozman A, Shalit A, Goldshmidt A, Amsellem Z, Alvarez JP, Eshed Y (2006) The tomato FT ortholog triggers systemic signals that regulate growth and flowering and substitute for diverse environmental stimuli. Proc Natl Acad Sci USA 103:6398–6403

Lifschitz E, Ayre BG, Eshed Y (2014) Florigen and anti-florigen - a systemic mechanism for coordinating growth and termination in flowering plants. Front Plant Sci 5:465

Liu J, Yu J, McIntosh L, Kende H, Zeevaart JA (2001) Isolation of a CONSTANS ortholog from *Pharbitis nil* and its role in flowering. Plant Physiol 125:1821–1830

Liu L, Liu C, Hou XL, Xi WY, Shen LS, Tao Z, Wang Y, Yu H (2012) FTIP1 is an essential regulator required for florigen transport. PLoS Biol 10:e1001313

Macháčková I, Krekule J, Eder J, Seidlová F, Strnad M (1993) Cytokinins in photoperiodic induction of flowering in *Chenopodium* species. Physiol Plant 87:160–166

Matsoukas IG, Massiah AJ, Thomas B (2012) Florigenic and antiflorigenic signaling in plants. Plant Cell Physiol 53:1827–1842

Meng X, Muszynski MG, Danilevskaya ON (2011) The *FT*-like *ZCN8* gene functions as a floral activator and is involved in photoperiod sensitivity in maize. Plant Cell 23:942–960

Michaels SD (2009) Flowering time regulation produces much fruit. Curr Opin Plant Biol 12:75–80

Mutasa-Goettgens E, Hedden P (2009) Gibberellin as a factor in floral regulatory networks. J Exp Bot 60:1979–1989

Nakatani K, Takayanagi S, Noguchi K (2009) Characterization of photoperiodic sensitivity in the Japanese population of *Chenopodium album*. Weed Biol Manag 9:79–82

Pavlová L, Krekule J (1990) The effect of IAA application on endogenous rhythm of flowering in *Chenopodium rubrum* L. Biol Plant 32:277–287

Pin PA, Benlloch R, Bonnet D, Wremerth-Weich E, Kraft T, Gielen JJL, Nilsson O (2010) An antagonistic pair of *FT* homologs mediates the control of flowering time in sugar beet. Science 330:1397–1400

Pin PA, Nilsson O (2012) The multifaceted roles of *FLOWERING LOCUS T* in plant development. Plant, Cell Environ 35:1742–1755

Randoux M, Davie`re, Jeauffre J, JM et al (2014) RoKSN, a floral repressor, forms protein complexes with RoFD and RoFT to regulate vegetative and reproductive development in rose. New Phytol 202:161–173

Risi JC, Galway NW (1984) The *Chenopodium* grains of the Andes: Inca crops for modern agriculture. Advance Appl Biol 10:145–216

Reiter RJ, Tan DX, Zhou Z, Cruz MHC, Fuentes-Broto L, Galano A (2015) Phytomelatonin: assisting plants to survive and thrive. Molecules 20:7396–7437

Samad AFA, Sajad M, Nazaruddin N, Fauzi IA, Murad AM, Zainal Z, Ismail I (2017) MicroRNA and transcription factor: key players in plant regulatory network. Front Plant Sci 8:565

Seidlová F (1980) Sequential steps of transition to flowering in *Chenopodium rubrum* L. Physiol Veg 18:477–487

Sheldon CC, Rouse DT, Finnegan EJ, Peacock WJ, Dennis ES (2000) The molecular basis of vernalization: the central role of *FLOWERING LOCUS C* (*FLC*). Proc Natl Acad Sci USA 97:3753–3758

Štorchová H, Drabešová J, Cháb D, Kolář J, Jellen EN (2015) The introns in *Flowering Locus T-Like (FTL)* genes are useful markers for tracking paternity in tetraploid *Chenopodium quinoa* Willd. Genet Resour Crop Evol 62:913–925

Suarez-Lopez P, Wheatley K, Robson F, Onouchi H, Valverde F, Coupland G (2001) CONSTANS mediates between the circadian clock and the control of flowering in Arabidopsis. Nature 410:1116–1120

Sun CH, Chen D, Fang J, Wang PR, Deng XJ, Chu CC (2014) Understanding the genetic and epigenetic architecture in complex network of rice flowering pathways. Protein Cell 5:889–898

Ullmann J, Seidlová F, Krekule J, Pavlová L (1985) *Chenopodium rubrum* as a model plant for testing the flowering effects of PGRs. Biol Plant 27:367–372

Valverde F, Mouradov A, SoppeW Ravenscroft D, Samach A, Coupland G (2004) Photoreceptor regulation of CONSTANS protein in photoperiodic flowering. Science 303:1003–1006

Vogt SH, Weyens G, Lefebvre M, Bork B, Schechert A, Mueller AE (2014) The *FLC*-like gene BvFL1 is not a major regulator of vernalization response in biennial beets. Front Plant Sci 5:146

Walsh BM, Adhikary D, Maughan PJ, Emshwiller E, Jellen EN (2015) *Chenopodium* polyploidy inferences from *Salt Overly Sensitive 1* (*SOS1*) data. Am J Bot 102:533–543

Wolf K, Kolář J, Witters E, van Dongen W, van Onckelen H, Macháčková I (2001) Daily profile of melatonin levels in Chenopodium rubrum L. depends on photoperiod. J Plant Physiol 158:1491–1493

Weller JL, Ortega R (2015) Genetic control of flowering time in legumes. Front Plant Sci 6:207

Wickland DP, Hanzawa Y (2015) The *Flowering Locus T/Terminal Flower 1* gene family: functional evolution and molecular mechanisms Mol Plant 8:983–997

Wilson HD (1990) Quinoa and relatives (*Chenopodium* sect. *Chenopodium* subsect Cellulata). Econ Bot 44:92–110

Yamada M, Takeno K (2014) Stress and salicylic acid induce the expression of *PnFT2* in the regulation of the stress-induced flowering of *Pharbitis nil*. J Plant Physiol 171:205–212

Yano M, Kojima S, Takahashi Y, Lin H, Sasaki T (2001) Genetic control of flowering time in rice, a short day plant. Plant Physiol 127:1425–1429

Zeevart JAD (1976) Physiology of flower formation. Annu Rev Plant Physiol 27:321–348

Quinoa—Evolution and Future Perspectives

11

Gabriela Alandia, Arnesta Odone,
Juan Pablo Rodriguez, Didier Bazile,
and Bruno Condori

Abstract

Quinoa (*Chenopodium quinoa*, Willd.) is a high-quality protein grain originating in the Andean region. Once a staple of the Incas, from being unknown in the rest of the world, this grain has recently become a global commodity. This is largely due to its nutritional qualities and adaptation to a wide range of environments. While the majority of quinoa is produced in South America, especially in Peru, Bolivia and Ecuador, production is increasingly spreading across the globe. The production in the area of origin of this grain is becoming increasingly intensive and replacing traditional smallholder production. The International Year of Quinoa catalysed the growth of quinoa in 2013 and led to an increased demand, production and research of quinoa worldwide. Nutrition qualities that made quinoa popular are the high-quality protein, a range of functionalities related to the nutrients contained in this grain, besides being a gluten free food. Quinoa is well adapted to different latitudes and production under marginal conditions such as drought or salinity. These qualities are used by breeders to develop high yielding cultivars for their regions. New food products containing quinoa are in continuous development. Additionally, this high-quality protein grain with low glycemic index is promoted as a healthy food for celiac and diabetic patients and in the recent vegan, vegetarian or flexitarian diets. Future perspectives for quinoa point to the expected increase of its production around the world. Environmentally, this can bring positive benefits. It represents a nutritious crop for areas affected by climate change. Quinoa also constitutes an alternative to meat that reduces greenhouse gas emissions, furthermore, using this grain increases the use of biodiversity. There are a number of challenges to be addressed, in particular with regards to research into abiotic and biotic stresses, development of new cultivars, saponin reduction and ensuring recognition and fair sharing of genetic materials.

G. Alandia (✉) · A. Odone
University of Copenhagen, Faculty of Science, Department of Plant and Environmental Sciences, Højbakkegaard Alle 13, DK-2630, Taastrup, Denmark
e-mail: gar@plen.ku.dk

J. P. Rodriguez
Crop Diversification and Genetics Program, International Center for Biosaline Agriculture, ICBA, Dubai 14660 United Arab Emirates

D. Bazile
UPR GREEN, CIRAD, Univ Montpellier, Montpellier, France

B. Condori
Inter-American Institute for Cooperation on Agriculture, Bolivian Representation – IICA, La Paz, Bolivia

S. M. Schmöckel (ed.), *The Quinoa Genome*, Compendium of Plant Genomes,
https://doi.org/10.1007/978-3-030-65237-1_11

11.1 Introduction

Quinoa (*Chenopodium quinoa*, Willd.) is a grain originating in the Andean region that has recently being spread globally due to the increasing attention to its nutritional qualities and adaptation to a wide range of environments. These characteristics have attracted the interest of scientists who started to study and develop the production systems to grow this crop in other areas of the world, and at the same time, market demand has increased globally. The following chapter briefly synthesises quinoa development as a crop, in science and in the food markets. It includes some examples from the region of origin and the new areas of production to show different perspectives for this crop. With an overview and analysis of the past and the current challenges, it is possible to draw quinoa′s future perspectives.

11.2 Evolution and Perspectives of Quinoa as a Crop

11.2.1 Production Areas

Quinoa originated in the southern Andes and domestication began before 3000 BC, independently in the highlands and coastal regions as suggested by relevant studies (Bruno 2006; Jarvis et al. 2017). Quinoa was an important part of the Andean diet, at the arrival of the Spanish in the XV century the crop reduced in importance due to the cultivation of the new species brought from the old world (Gandarillas 1979). However, it has experienced a revival in the last century. According to the latest statistical available data from 2018, Bolivia, Peru and Ecuador are now the major quinoa producers, covering 172,000 ha with this grain and producing around 70–80% of global quinoa volumes (Alandia et al. 2020).

In the 1930s, quinoa experiments began in Kenya, and then in the 1970s and 1980s in North America and Europe, moved by the demand for healthy foods and interest in new crops for diversification (Galwey 1993). By the end of the 1980s quinoa was either under research or produced in 11 countries outside the Andes. By the year 2000, the USA and Canada were producing 10% of global quinoa (Alandia et al. 2011). After the International Year of Quinoa (IYQ) in 2013, 75 countries were testing quinoa, with 26 registered cultivars in existence. Quinoa now is in research or production in 123 countries (Fig. 11.1). By 2018, the United States, Canada, the Netherlands and Spain participated in the 15–20% of global exports in the world. The remaining is still produced by Peru and Bolivia (ITC 2019; Alandia et al. 2020).

11.2.2 Production Systems

In the Andean highlands, quinoa is generally produced on small-medium farms, under traditional production systems dependent on rain and using low input technology (Alandia et al. 2020). On small-scale farming, sowing, management and harvesting are manual or semi-mechanised. Production can be highly impacted by extreme weather events, such as El Niño and la Niña, which can drastically damage seed yields. Crops are often affected by water shortages, poor soils, salinity and frost up to 200 nights of the year (Jacobsen 2003). There is also extensive quinoa production for export, in the southern highlands of Bolivia, which continues to be produced under traditional and extreme environmental conditions, while in the Peruvian coastal region quinoa for export is produced more intensively. In the Andean region, quinoa yields remained below 1 t/ha until 2014. After, they surpassed 1 t/ha with 1.6 t/ha from Peru, 1.2 t/ha from Ecuador, while it remained at only 0.52 t/ha in Bolivia. Around 30–40% of quinoa produced in Peru, Bolivia and Ecuador is organic. The majority of organic quinoa is exported from Bolivia, while Peru exports most conventionally produced quinoa (IBCE 2015; Oficina Internacional del Trabajo 2015; Alandia et al. 2020).

Outside of the Andean region, production is increasing. In North America, the main area of quinoa cultivation is in the prairie areas of Canada, since the 1990s, and in the state of Colorado in the US. It is likely to have most success in high altitude areas in the American Rockies and the Canadian Prairies, as well as in high altitude areas. It also

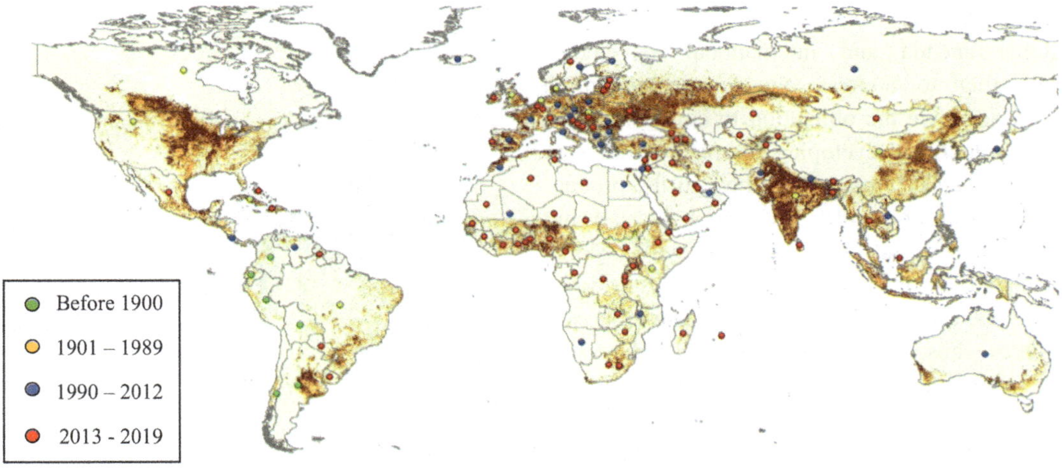

Fig. 11.1 Quinoa presence in the World (adapted with data from Alandia et al. 2020)

showed potential for winter cropping in California (Peterson and Murphy 2013).

In Europe, commercial production of quinoa already exists in Spain, France, the Netherlands, Italy, Denmark to mention some, and this grain is in research in many other countries as it is shown in Fig. 11.1. In many of these countries, production has rapidly increased in the last few years under technified production systems of organic and conventional agriculture. Across Europe, the area under cultivation increased from 0 to 5000 ha and by 2018, countries such as France and Spain were already considered medium producers with areas between 500–5000 ha (Alandia et al. 2020). For example In Denmark, quinoa production area has increased from 6 ha in 2015 to 159 ha in 2018, 26 times more in only three years (Landbrygsstyrelsen 2018).

In 2016, new production was mainly limited to small areas of intensive and experimental agriculture, with predominant mechanised systems. New trials were reported in 95 countries; many of these being countries such as Algeria, Iraq, Mauritania, Yemen, Ethiopia, Senegal, Kyrgyzstan, Sri Lanka and Bhutan, to mention some (Bazile et al. 2016a).

11.2.3 Perspectives

Global quinoa production is expected to increase in the future. This ancient grain started to be produced outside its region of origin in the early 1980s and from being produced mainly in six countries, it is now present across 123 countries of the world. The dispersal of quinoa has been fast compared with other species, mainly due to the high demand and rapid development of the market, its high level of adaptation and tolerance to abiotic stresses. This crop also has the quality to adapt to diverse farming systems as successful trials in different regions of the world have shown. From 2013 to 2018, quinoa was experimentally implemented in 72 countries. Half of the countries with the biggest surface cultivated with quinoa are now located outside the Andean region, i.e. Spain, France, China and Canada (Alandia et al. 2020). Perspectives are that production areas will increase in many of the countries where quinoa is now present only experimentally.

It is well known that meat production is one of the major factors negatively affecting the environment with GHG emissions, land, water and energy use (FAO 2006; Song et al. 2016; Godfray et al. 2018). Projections of global population growth estimate 9.6 billion people by 2050 and 10.6 by 2100 with higher concentrations in Africa and Asia (United Nations 2013). Meat demand was projected to increase by 76% by 2050 with a resulting increase of 80% of GHG (UNEP 2012). This increasing demand can be satisfied with alternative sources of protein to reduce damages to the environment. With the global agenda and international agreements beginning to address climate change, more countries have the obligation to come with efficient solutions. Developing alternative sources to meat is one of the answers through the production of plant protein food. Research is catalysing this process and more funding will become available for this target. With high nutritional qualities, quinoa is already one of the crop candidates chosen in existing projects addressing this objective to increase plant protein in Europe (e.g. PROTEIN2FOOD, Smart Protein). Moreover, quinoa is already well established in the global market and it diversifies global diets away from dependence on few commodities; just three crops make up over 50% of plant calories globally (Bioversity International 2017).

Due to its good adaptation to extreme environments, quinoa is a crop that enhances resilience of farming systems; its implementation in fragile areas might be targeted to increase production. In the future, conditions such as salinity and drought are likely to become problems that farmers have to address more frequently due to climate change. Tolerance to different abiotic stresses is therefore relevant for climate change scenarios, and crops such as quinoa, which produces well under marginal conditions, will be necessary to ensure food security.

Fig. 11.2 Genaro Aroni in light blue, addressing to quinoa producers in the Bolivian southern highlands during a field day (Photo credit: G. Alandia)

Genaro Aroni is a quinoa specialist from the southern highlands of Bolivia. He has 40 years' experience in agricultural production and has worked with quinoa for over 30 years. This interview took place in August 2019.

What changes have you seen in production in the time that you have worked with quinoa? I have seen dramatic changes. Between the 1940s–1960s, the area of the southern highlands was entirely addressed to cattle, and quinoa was produced only on the hillsides. Since the introduction of tractors in the 1970s, to plough the plains, quinoa production has become more widespread and intensive. Farmers moved from camelid livestock production to agriculture. By the 1990s, farmers wanted access to the market, and began to export both conventional and organic quinoa. With the boom of quinoa, production in Bolivia increased to over 90,000 t, mostly of organic. When prices increased, many members of the community returned to claim their land, which created problems. At the same time, there were serious debates about socio-economic problems, migration and population decline in many areas (Fig. 11.2).

Is quinoa production in the southern highlands still for export? Mainly organic quinoa is for export, while conventional quinoa is destined for internal consumption and export to Peru.

How do you evaluate quinoa yields in this region? Quinoa yields are low, and production does not correspond to agricultural areas. The extreme weather and environment of the southern highlands mean that yields have remained low, at around 0.5 t/ha until now. Some studies have shown that organic fertiliser can improve yields, but producers are not willing to pay for it. ***How are production costs***? In general, the cost of production is around €600–780/ha. The main cost is labour. During the quinoa boom, people were hired from outside the region, with salaries of €20/day, as well as accommodation and food, with up to eight or nine workers per plot. Since the price of quinoa fell, the salaries have fallen to €13–16/day, and you often see only one worker per plot.

How do the prices vary? In 2015, quinoa prices reached €5.70/kg, but by 2016 this had fallen to only €0.94/kg. The price of organic quinoa is now €2–2.20/kg, and €1.60/kg for conventional quinoa. Prices have improved slightly together with expectations.

What are the production problems/challenges? In the southern highlands, it is the loss of soil fertility aggravated by the recurring drought; 15 years ago, it rained from mid-December until the first days of March. Now the first rains occur on the first days of February, last only a month and are only half the quantity of water. Climate change is a problem and organic production under these conditions implies higher costs. Pests have increased, and now need treatment four times, when they only needed two treatments before.

What recommendations can you give for improvement? We have to include fallow systems again, add organic matter from camelid livestock, to reduce erosion and benefit reforestation processes, use barriers of native legume species where there has been deforestation and add grasses and biannual legumes that are tolerant to frost. We have tested Andean lupin, which can generate 14 t/ha of green manure. Other native species can produce 6 t of green manure per hectare, and there have been interesting results with fast-growing Andean grasses. We need much more research on wild legumes; we have more than 80 species that could be included. Another recommendation is to improve the sowing technology with the optimal sowing time. Finally, economic support is needed from the government to repopulate the area with llamas. With international production now, there will be competition in cost and quality, therefore Bolivia needs to remain competitive.

Do producers still consume quinoa? ***Do you***? At the national level, there is more consumption now than before, the latest data I have on annual consumption is around 2 kg/person. What I see in the southern highlands is that with the boom of quinoa, and the economic flow that occurred, people diversified their diets and began to access to other products. However, the price is still high, compared to rice and pasta. Yes, I do consume quinoa twice a month.

Anders Nørgaard is an agricultural technician at the University of Copenhagen's experimental farms in Denmark. He has worked with quinoa for over 25 years. This interview took place in August 2019. His answers are complemented with extra information from: Inger Bertelsen (IB), senior consultant in Ecology and Innovation at SEGES and Christian Høegh (CH) Business Developer for Food and Agriculture at Business Lolland-Falster.

How has quinoa production changed along the time you work with this crop? In Denmark, quinoa production started 10–12 years ago after many years of research. Production increase has been slow and concentrated in some islands. Quinoa production system is similar to spring barley. We sow these crops at around the same time, and harvest the quinoa around three weeks after barley. Quinoa used to be slow

Fig. 11.3 Anders Nørgaard after a harvest day at the research station of Copenhagen University (Photo credit: A. Odone)

maturing but now the varieties that we have mature early enough. We now harvest in early September, when we used to harvest in late September or October, which is too late for the Danish climate (Fig. 11.3).

Is production addressed for the local market? Yes, to my knowledge. (IB:) There are local companies buying quinoa produced in Denmark.

What are the yields of quinoa? Average yields here are around 1.2 t/ha. The highest yields were 2.9 t/ha, although sometimes yields can be very low.

How are production costs? (CH:) Production costs including cleaning would reach between €1,100–1,300/ha.

How do prices vary? (CH:) Farmers are hoping to get €3.3–3.4/kg for organic quinoa and €2.1–2.2/kg for conventional; however, not much is sold.

What are the main production problems? Farmers producing quinoa should be careful of three things: sowing depth, weeds and aphids. Harvesting is easy. We tried sowing quinoa at different row spacings, at 12 and 50 cm. This gave the same yields but at 50 cm it is much better for row cleaning.
Weeding is the main problem. The main weeds are 'hvidmelet gåsefod' *Chenopodium album* (hard to distinguish from quinoa) and 'snerlepileurt' *Fallopia convolvulus* (hard to separate). Aphids are a problem; they are controlled when quinoa is flowering.

What inputs are needed for production? Quinoa can be produced organically, for example with clover grown before, although it grows best after oats. (IB:) Also more cultivars are needed, so far only a few are used. (CH:) Approved control methods are needed, both for conventional and organic control of biotic problems.

Do you think there will be more production in Denmark in the future? It depends on the price a farmer can get, but due to the cost of cleaning and processing, higher prices would be needed to make it worthy for the farmer. (IB:) The access to the local market is not easy; consumers still prefer to buy quinoa coming from South America. (CH:) Quinoa is already growing in Denmark and as there is more interest in protein crops, perspectives for it would be that the production area keeps increasing.

Which challenges do you see for the production of quinoa? The challenges are not production related, it is just the problem of selling the quinoa. There is a lack of facilities for processing quinoa in Denmark. If farmers could sell their seed to companies such as DLF, that would be better.

Do you eat quinoa yourself? Not often - I eat it once or twice a year.

11.3 Evolution and Perspectives of Quinoa in Research

11.3.1 Research in the Andean Region

Contemporary research into quinoa in the Andean region began in the 1940s, and established the basis for further, global, research in quinoa. In the 1960s, researchers across the Andean region began collection, characterisation and conservation of local germplasm, plant breeding and research into nutrition, uses and processing of quinoa (Tapia 2014).

A relevant activity in the development of research into quinoa was the first international quinoa workshop, which took place in 2001 in Lima, Peru. This led to a succession of International Quinoa Conferences, and a dramatic increase in the production of quinoa globally (Tapia 2014).

Further growth in quinoa production outside of the Andean region was generated by the United Nations International Year of Quinoa (IYQ), in 2013. The aim of this, according to the FAO, was the 'recognition of ancestral practices of the Andean people' (FAO 2019). A large number of activities were held, from festivals, cooking contests and tasting events to exhibitions, seminars and scientific conferences, alongside the global dissemination of recipes, articles and information about quinoa.

Recent research has included socio-economic studies of the quinoa production chain, conservation of germplasm, molecular analysis, development of products and determination of nutritional value. Breeding has focused on adaptation to climate change (later and shorter rains), resistance to mildew and drought, precocity, as well as grain colour and size, leading to the release of new cultivars (Gandarillas et al. 2014; Tapia 2014).

11.3.2 Research in the Rest of the World

There has been growing interest in quinoa globally over the last few decades, due largely to the high nutritional properties, in particular the quality of protein and lack of gluten. It is also attractive due to its high genetic diversity and adaptability to various climate and soil types.

In Europe and the United States, demand for quinoa has rapidly increased over recent decades. The research focus has been on adapting varieties to longer days and short summers (Jacobsen 2003), reducing saponin content in the seed coat and increasing quinoa's abiotic tolerance mechanisms (Bazile et al. 2016b).

In China, there have also been significant investments into the development of quinoa, particularly in the north of the country (Xiu-shi et al. 2019). On the Indian subcontinent, quinoa is potentially part of a solution to increasing agrobiodiversity and spreading farmers′ risks, as well as a source of protein, although it is still in research stages there (Bhargava and Ohri 2014).

The first quinoa experiments outside of the Andes were in 1935 in Kenya (Bazile et al. 2016b). Across Africa, there is increasing research into quinoa as a crop with potential in drought- and salinity-prone areas (Bazile et al. 2016a) . Quinoa is seen as a crop with high potential in areas of food insecurity and malnutrition, due to the high-quality nutrition that quinoa provides. The main issues for further research have been identified as: acceptability by consumers, best management practices for small-scale farmers and selection of suitable cultivars for different conditions (Maliro and Guwela 2015).

In the Middle East, particularly in Morocco and the UAE, research has focused on quinoa's drought and salinity resistance, for production on marginal land. While yields remain low, the production area is increasing steadily, and beginning to spread further in the Middle East and Central Asia (Choukr-allah et al. 2016; Rodriguez et al. 2020).

11.3.3 Perspectives

Several studies targeted abiotic stress in quinoa, this plant represents a good model to understand the processes and pathways of stress responses. Innovation and research will keep developing this line to keep giving solutions for climate change and provide information towards breeding strategies. Considering the trends of dissemination of quinoa around the world, efforts in research will concentrate in developing new genetic materials for the regions with high potential to grow quinoa. Different technologies might be used for that, such as mixed breeding techniques as it is done in Germany (Kiel University 2018) or novel genomic tools starting to develop in Denmark (Palmgren 2019). Moleculars tools keep in use to characterise the diversity of quinoa collections. Now with the recent quinoa genome description, the identification of genes will be targeted for sources of tolerance for abiotic and biotic stress with the use of traditional and modern phenotyping tools (Jarvis et al. 2017; Tester 2019). From the nutritional side, several studies have described the qualities of quinoa from different points of view such as the recent descriptions of quinoa as a functional food. More investment in the food area will be addressed to develop quinoa as an ingredient for different food products.

Wild species in quinoa production systems, Bolivia

The Bolivian research Foundation PROINPA has recently undertaken research into quinoa interactions with native wild legume and grass species. These provide protection from erosion, fix nitrogen and provide other nutrients for quinoa, wild legume species can also be used as a forage source and in traditional medicine. PROINPA aims to reintroduce these wild species to quinoa production systems, making use of local ancestral knowledge, as well as preserving these species and replicating their interactions with quinoa plants in the wild.

Wild lupin, *Q'ila-q'ila,* is a native legume, which can tolerate frost and drought, and grows well in sandy soil. PROINPA is investigating this in a rotation with quinoa, to provide biomass to the soil and fix nitrogen.

Other wild species can act as natural barriers. These include shrubs such as *Lampaya* and Thola thola (*Parastrephia lepidophylla)*, and grasses such as *Nassella*, *Festuca* and *Stipa* spp. These living barriers help to reduce the wind speed and the loss of soil, thus protecting the quinoa plants. However, their numbers have been reduced due to the increase of the agricultural land area. PROINPA therefore is promoting their use (Fig. 11.4).

Other species are being promoted for their use as fodder for camelids, traditionally raised in the quinoa-camelid production system in southern Bolivia. Ideally, two or three llamas per hectare would produce manure to benefit quinoa production (Bonifacio et al. 2013).

El-Samad et al. (2018) investigated in Egypt the potential use and the nutritional value of quinoa as a leaf crop comparable to spinach. They found that while spinach had higher contents of fibre and carbohydrates, as well as nitrite and nitrate, quinoa leaves had higher content of fats, proteins, antioxidants and higher levels of nitrogen, potassium, calcium and manganese. However, the nutritional value of quinoa leaves could be different depending on cultivars. Despite this, the authors recommend that quinoa leaves could be a good source of proteins, minerals and vitamins since they can grow with minimal input and could be beneficial to communities living on marginal soils and saline water.

Quinoa as a leaf vegetable, Egypt (Photo credit: J.P. Rodriguez)

While quinoa grains are commonly incorporated into different kinds of breads and

Fig. 11.4 Seed collection and implementation of living barriers and production areas with native legume and grass species (photo credits: G. Alandia)

cakes, new research is investigating whether quinoa leaf would be a good addition. Tests found that while the leaves improve the antioxidant levels in bread, they may alter the quality of the bread negatively, in particular the protein digestibility (Świeca et al. 2014).

This study is one of very few investigating quinoa leaves as a vegetable crop, but it suggests that they could have great potential in the future.

(a) (b) (c) (d)

Fig. 11.5 Traditional practices, uses and consumption of quinoa in Bolivia: **(a)** Bulk sale of quinoa, **(b)** milling the grain with stone, **(c)** local products selling and **(d)** consumption (photo credits: J.P. Rodriguez)

11.4 Evolution and Perspectives of Quinoa for Food

11.4.1 In Food and Market

Quinoa has been traditionally consumed in the Andean countries, in whole grain preparations including porridges, beverages and soups (Fig. 11.5). The grains are also processed into flours for bread, traditional meals, or toasted flour for snacks and beverages. More recently, industries have started to include quinoa as an ingredient in snacks, for example, extruded quinoa for breakfasts or muesli bars, flours for pasta, cookies, instants soups, pre-cooked quinoa, beer and diverse other products for export (Fig. 11.7). However, the predominant export has been as a raw grain to retailers for processing abroad.

In fact, quinoa development in the Andean countries has been a direct response to demand. The rise of the diet-food market in the 80 s, the organic market in the 90 s, the fair trade market in the 2000s and the promotion of the IYQ in 2013 gave rise to the evolution of this crop in Bolivia and in the Andean region (Fig. 11.6).

Production progress from 1961 to 2015 can be observed in Fig. 11.6. In the 1980s quinoa was established in more areas of Bolivia, Peru and Ecuador, compelled by policies that promoted this grain and the association of producers who started to introduce quinoa to the market. Later in the 1990s the traditional quinoa produced in the Southern highlands of this country fitted well for the organic market and later in the 2000s in the fair trade market. Finally, with the promotion activities motivated by the Bolivian government and the United

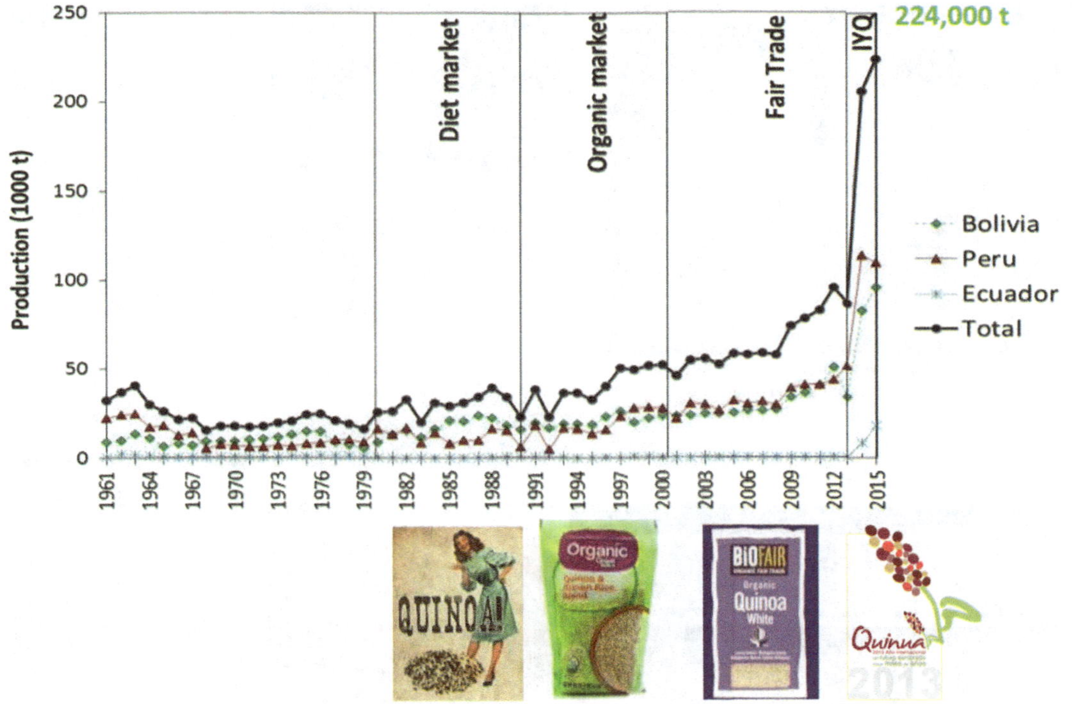

Fig. 11.6 Evolution of quinoa production in the Andean region (*source* elaborated with data from: Alandia et al. 2020)

Nations (UN) through FAO and the IYQ, the demand reached levels that the country was unable to fulfil even with increased areas of production. By 2015, Bolivia produced 60,000 tonnes of quinoa. Similar development happened in Ecuador on a smaller scale and in Peru with higher technology producing a total of 224,000 t in the Andean region, as it is shown in Fig. 11.6.

Nowadays, quinoa is included in many food products; not only in the traditional countries of production but also in several countries where production has begun recently. It started being consumed and purchased as grain, and then progressively the industry started to produce and provide flour, flakes, extruded quinoa for muesli or energy bars. It is now possible to find a diverse range of products with quinoa in different markets globally (Fig. 11.7).

11.4.2 In Healthy Diets and Food Safety

Quinoa has great qualities as a food. It is well known that this grain has a protein content ranging between 12–20% and that it contains all the essential amino acids. Nutritionally it has been compared to dried whole milk (Vega-Gálvez et al. 2010). The body is able to absorb 73% of the protein present in quinoa (a variable known as biological value); this value is similar to the biological value in beef (74%) and higher in relation to other grains (rice, wheat and corn range between 36 and 56%) (Gordillo-Bastidas et al. 2016).

Quinoa is also rich in minerals that are important for body structure and functions, such as calcium for bones and iron for blood. Quinoa also contains zinc, which supports the immune system, as well as magnesium, vitamin E, vitamin B

Pre-cooked quinoa with curry mushrooms and vegetables (CH)
Cost: € 1.62

Multicereal pasta with quinoa (SP) Cost: € 1.69

Cooked quinoa ready to eat (DK)
Cost: € 2.3

Different quinoa products in a supermarket shelve (MX)
Costs: between €2.5 – 10.6

Malta drink with quinoa (BO)
Cost: € 0.9

Multi-fibre cookies with quinoa (AE)
Cost: € 5.32

Fig. 11.7 Different products with quinoa found in different parts of the world. *Note* In parenthesis Alpha-2 ISO country codes (for Chile, Spain, Denmark, Mexico, Bolivia and the United Arab Emirates). All prices correspond to the year 2019. Photo credits: G. Alandia, L. Robles and J.P. Rodriguez

complex, tryptophan amino acid and omega fatty acids, which are relevant for the nervous system. In addition, it is free of gluten, which can cause small intestine inflammation reactions in patients with celiac disease (Gordillo-Bastidas et al. 2016). Finally, quinoa has a low glycemic index (GI) that ranges 35–53 and puts quinoa in the list of food options to reduce the risk of type 2 diabetes.

These functional qualities mean that quinoa is a great addition to diets addressing specific health needs, such as gluten-free and low glycaemic diets. Due to the high bioavailability of its amino acids and the quality of its nutritional components, quinoa represents a good alternative in healthy diets in particular for young children and the elderly.

Economic wealth and education influence personal choices. When these factors are present, people tend to choose more healthy and safe products. These include more natural and less processed products as well as foods enhancing health, such as functional foods. Quinoa comes from traditional production systems and there are sectors of people from growing economies willing to pay for exotic and/or less processed products.

11.4.3 In the Plant-Based Food Sector

The meat alternative industry is growing fast; in fact, financial analysts predict that in ten years, this new sector could replace 10% of the global meat industry (USD 140 billion) targeted by companies such as Beyond Meat, Impossible Foods and Nestle (Franck 2019). Companies such as Danone also project that they will triple their plant-based sector revenues to more than €5 billion by 2025 with products such as plant-based yogurts and ice cream (Gretler 2019). Supermarkets are diversifying their offers in order to meet the demand for meat alternatives. Last year, the UK chain Sainsbury's expanded their range of plant-based products by 35% (Chiorando 2018). The demand for plant-based dairy products is increasing and different products are offered based on soybeans, rice, almonds, oats and in some countries also quinoa. The bakery industry represents an important market for quinoa. Quinoa can be included in different type of products: as food, as an ingredient (grain, flour, flakes) and as protein concentrates, fractions, extracts and isolates.

11.4.4 For Environmentally Responsible Consumption

Countries have promised to concentrate efforts on reducing greenhouse gas emissions and building climate resilience. Agriculture and food constitute important targets to achieve this. Therefore, regions and countries have agreed to build more energy efficient food systems and reduce meat intake in people´s diets to make consumption and production patterns more sustainable. Activities to catalyse the change of diets and food systems are under development to achieve the Sustainable Development Goals (SDGs) for 2030 (United Nations 2015). These include regional and local programmes, projects and funding. The development of all concerned food sectors is taking place gradually.

Diets are changing and responsible consumption is turning people's interests towards flexitarian, vegetarian and vegan diets that reduce or replace meat consumption with alternative-protein food. Quinoa is a good alternative to generate high-quality plant protein food. It has been widely accepted by consumers and in addition, it has carbon and water footprints 30 and 60 times lower than beef (Gordillo-Bastidas et al. 2016).

11.4.5 Perspectives

Consumer demand for quinoa will continue to rise in the coming years, as trends towards healthy foods or climate responsible consumption increase, in particular for plant proteins or diversifying diets as described above. Particularly in the developing world, quinoa provides the possibility to feed more people in fragile areas. In the other hand, markets are specialising and are demanding not only white large seeds, but also other seed colours such as red, or black. For further increase of quinoa production, the facilities to process and transform this grain need to develop in parallel in the new areas where this grain will be implemented.

11.5 Challenges

In order for quinoa production to successfully continue to increase, it has to face a number of challenges, both in the Andean region and in the rest of the world.

Biotic stress resulting from pests and diseases must be addressed to ensure that yields meet their potential. More studies are needed in the areas of plant pathology and entomology in order to identify emerging problems with biotic agents that will keep in rise with climate change. Research continues to advance with downy mildew (*Peronospora variabilis* Gäum), which is the most reported disease in quinoa, but still more can be done to find sources of tolerance to the crop and provide control options applicable to the new countries of production and farming systems. The same situation exists for pests such as aphids (*Aphis* spp), flea beetles (*Phyllotreta* spp) and many other organisms that may appear with the introduction of this crop in different regions. The establishment of monitoring systems at different scales (global, regional, local) to record the emergence of biotic agents will be needed to keep developing this crop. Collaborative work would be a good way to target this to make reality an international research centre for underutilised grains situated in the Andes like it has been done with potato at the International Potato Center (CIP).

There is a need for development of cultivars adapted to different conditions. Quinoa has a huge genetic diversity and has the potential to adapt to many environments and requirements, due to the diverse ecotypes it originates from (Bazile et al. 2016a). In future, breeding efforts can further develop these characteristics to adapt them to the specific needs of different regions. For example, day length, light levels, soils, altitude and heat. In dry areas, the challenge is testing varieties of quinoa that are already tolerant to drought conditions and improving them to adapt to other types of stress such as heat. In Europe and North America, quinoa must be early maturing, in order to adapt to short summers, but this is also helpful in South America in order to address the gradual reduction of the rainy season. Collaborations between breeding programmes can therefore be helpful in achieving the different aims and interests in quinoa globally.

The global impact of quinoa's rise must also be considered in future, in the case of quinoa and with new-emerging crops. Nuñez de Arco (2015) highlights the paradox of the United Nations intention to provide affordable quinoa to the poor, while also lifting many small quinoa farmers out of poverty—a drastic fall in prices means that one is satisfied, but not the other. A recommendation to bodies such as the FAO is that events like the IYQ should integrate the whole process of promoting and increasing demand to ensure a sustainable and stable supply, in order to avoid price collapses that directly affect farmers, as happened in 2014 (Alandia et al. 2020).

Another breeding aim is to further reduce the saponin content of the seed, therefore reducing bitterness. If bitter materials are used, then processing facilities need to be developed to clean the saponins from grains. However, these are expensive and not available in countries starting to produce this grain.

Farming systems adopting quinoa in the future must use adequate technology in order to be able respond to the market demand, but also avoid unsustainable practices. The quinoa boom in the early 2000s raised the question of sustainability due to high demand and the negative effects of extensive and intensive quinoa production on the land of the main exporting countries of South America (Peru and Bolivia). The expansion of the agricultural frontier and unsustainable practices to achieve the demanded production raised social and environmental concerns.

Challenges with genetic material

The ethics of the sharing of genetic materials by Andean countries is also in question. The traditional quinoa-producing countries face tough competition and may lose markets due to the production of quinoa outside the Andean region. However, this is not necessarily with recognition of the origin of the genetic material, the seed property rights, or the equitable distribution of benefits derived from the use of the genetic resources. Currently the development of new cultivars is possible outside of the Andean region without any sharing of benefits or recognition with the region of origin of the genetic material. Quinoa is not included in the system of multilateral exchange, which would protect the biological origin of quinoa taken outside the Andean region (Rojas et al. 2015; Bazile et al. 2016b). The enormously diverse genetic material for quinoa has been developed over thousands of years in the Andes, and for Andean countries to benefit of this, the existing legal framework must be revised and built upon. Intellectual rights protections and recognitions can help to ensure that quinoa farmers also benefit. Better dialogue at all levels of quinoa use is required, in order to better connect farmers, researchers, breeders and politicians, which is necessary for quinoa's future (Alandia et al. 2020).

Furthermore, quinoa breeding outside of the Andean region is limited, as it is dependent on a reduced quinoa biodiversity available in collections established prior to the protective agreements of 1992 and 2004 (the Convention of Biological Biodiversity and the International Treaty on Plant Genetic Resources for Food and Agriculture respectively) (Alandia et al. 20).

11.6 Conclusions and Recommendations

Quinoa is an Andean grain with high nutritional quality and that has recently been qualified as a super-food. This grain, with high biological value, contains high-quality proteins including all the amino acids essential for human nutrition. It has omega acids that benefit health and a glycaemic index (GI) that make it eligible as a great candidate for the new diet trends. Quinoa can also be considered as a super-crop. It can grow under extreme climates and soils where most other crops would fail, and it is particularly useful in areas of drought and salinity. It is due to this combination of benefits that interest in quinoa has risen so drastically over the last few decades. However, such a boom cannot arise without costs. In order to maintain quinoa as a superfood and super-crop in the future, some things must be taken into consideration, in particular the effects in South America.

Firstly, to ensure that future breeding programmes are able to satisfy the demand for new varieties of quinoa which are adapted to different climates and soils, the diverse genetic material originating in South America must be recognised when it is taken out of the area. This could be done within a revision of the existing legal framework, that makes effective and tangible the acknowledgement and compensation to those who have conserved or bred the parental materials. Investment is needed in agricultural research around the world, for example for innovating agricultural techniques, new processing technologies, or to study and ensure the application of positive local/traditional sustainable production practices. Learning from the past and collaborating at different levels (locally, regionally, globally) may be the way forward for a sustainable development of this grain. And finally, we should promote the consumption of quinoa and other plant proteins to reduce the climate footprint, protect our environment and ensure future climate resilient agricultural systems.

References

Alandia, G, Blajos, J, Rojas, W (2011) Chapter 6: economic aspects of quinoa throughout the world. In: Quinoa: an ancient crop to contribute to world food security, FAO—Regional Office for the Americas and the Caribbean, Santiago de Chile, pp 37–42

Alandia G, Rodriguez JP, Jacobsen S.-E, Bazile D, Condori B (2020) Global expansion of quinoa and challenges for the Andean region. Glob Food Secur 26:100429

Bazile D, Pulvento C, Verniau A, Al-Nusairi MS, Ba D, Breidy J, Hassan L, Mohammed MI, Mambetov O, Otambekova M, Sepahvand NA, Shams A, Souici D, Miri K, Padulosi S (2016a) Worldwide evaluations of quinoa: preliminary results from post international year of quinoa FAO projects in nine countries. Front Plant Sci 7

Bazile D, Jacobsen S-E, Verniau A (2016b) The global expansion of quinoa: trends and limits. Front Plant Sci 7:850

Bhargava A, Ohri D (2014) Quinoa in the Indian Subcontinent. In: Bazile D et al (eds) State of the Art Report on Quinoa around the world in 2013, Rome, pp 511–523

Bioversity International (2017) Mainstreaming agrobiodiversity in sustainable food systems: scientific foundations for an agrobiodiversity index. Bioversity International, Italy, Rome

Bonifacio A, Aroni G, Gandarillas A (2013) El compromiso de PROINPA con un sistema sostenible de quinua en el altiplano boliviano. Comercio Exterior 21 (210):21–22 https://ibce.org.bo/images/publicaciones/ce_210_la_quinua_boliviana_traspasa_fronteras.pdf

Bruno M (2006) A Morphological Approach to Documenting the Domestication of Chenopodium in the Andes. In: Zeder Melinda A, Bradley DG, Emshwiller E, Smith BD (eds) Documenting domestication:

new Genetic and archaeological paradigms. University of California Press, Berkeley, pp 32–45

Chiorando M (2018) 'Sainsbury's to launch new 'affordable' vegan meat brand in stores across UK'. Plant Based News, 10 October. https://www.plantbasednews.org/lifestyle/sainsburys-vegan-veggie-range-skyrockets-35-october

Choukr-allah R et al (2016) Quinoa for marginal environments: toward future food and nutritional security in MENA and Central Asia regions. Front Plant Sci 7:346

El-Samad E, Hussin S, El-Naggar A, El-Bordeny Nasr, Eisa Sayed (2018) The potential use of quinoa as a new non-traditional leafy vegetable crop. Biosci Res 15:3387–3403

FAO (2006) Livestock's role in climate change and air pollution. In Livestock's long shadow: environmental issues and opinions, Chap. 3. Food and Agriculture Organisation of the United Nations, Rome

FAO (2019) Quinoa. fao.org/quinoa/en

Franck T (2019) Alternative meat to become $140 billion industry in a decade, Barclays predicts'. CNBC 23 May. https://www.cnbc.com/2019/05/23/alternative-meat-to-become-140-billion-industry-barclays-says.html

Galwey N (1993) The potential of quinoa as a multipurpose crop for agricultural diversification: a review. Ind Crops Prod 1:101–106

Gandarillas A, Wilfredo R, Alejandro B, Norka O (2014) La quinua en Bolivia: Perspectiva de la Fundación PROINPA. In: Bazile D et al (eds) Estado del arte de la quinua en el mundo en 2013. FAO and CIRAD: Santiago de Chile, Montpellier, Francia, pp 410–431

Gandarillas H (1979) Botánica. In: Tapia M, Gandarillas H, Alandia S, Cardozo A, Mujica A, Ortiz R, Otazu V, Rea J, Salas B, Zanabria E (eds) Quinua y la Kañiwa: cultivos andinos. Centro Internacional de Investigaciones para el Desarrollo (CIID), Instituto Interamericano de Ciencias Agrícolas (IICA). Bogotá, Colombia, p 20

Gordillo-Bastidas E, Diaz-Rizzolo DA, Roura E, Massanes T, Gomis R (2016) Quinoa (*Chenopodium quinoa* Wild.), from nutritional value to potential health benefits: an integrative review. J Nutr Food Sci 6:497

Godfray HCJ, Aveyard P, Garnett T, Hall J, Key T et al (2018) Meat consumption, health, and the environment. Science 361:243. https://doi.org/10.1126/science.aam5324

Gretler C (2019) 'Danone CEO says plant-based could become as big as dairy in U.S', eDairy news, 9 May. https://edairynews.com/en/danone-ceo-says-plant-based-could-become-as-big-as-dairy-in-u-s-65237/

IBCE (2015) Bolivia: Productos Alimenticios con Potencial Exportador. Instituto Boliviano de Comercio Exterior 230:4

International Trade Centre - ITC (2019) International Trade Centre List of exporters for the selected product - product: 100850 Quinoa "Chenopodium quinoa" Trade Map: Trade Statistics for International Business Development, International Trade Centre. https://www.trademap.org/Country_SelProduct_TS.aspx?nvpm=1%7c%7c%7c%7c%7c100850%7c%7c%7c6%7c1%7c1%7c2%7c2%7c1%7c2%7c1%7c1

Jacobsen SE (2003) The worldwide potential for quinoa (Chenopodium quinoa Willd.), Food Rev Int 19:1–2, 167–177. https://doi.org/10.1081/fri-120018883

Jarvis DE, Ho YS, Lightfoot DJ, Schmöckel SM, Li B, Borm TJA, Ohyanagi H, Mineta K, Michell CT, Saber N, Kharbatia NM, Rupper RR, Sharp AR, Dally N, Boughton BA, Woo YH, Gao G, Schijlen EGWM, Guo X, Momin AA, Negrão S, Al-Babili S, Gehring C, Roessner U, Jung C, Murphy K, Arold ST, Gojobori T, Linden CGvd, van Loo EN, Jellen EN, Maughan PJ, Tester M (2017) The genome of Chenopodium quinoa. Nat Adv:1–6. https://doi.org/10.1038/nature21370 http://www.nature.com/nature/journal/vaop/ncurrent/abs/nature21370.html#supplementary-information

Kiel University (2018). Growing quinoa in Europe. In: Research news (ed) Plant Breeding Institute University of Kiel. https://www.uni-kiel.de/en/research/details/news/growing-quinoa-in-europe/#

Landbrygsstyrelsen (2018) Statistik over økologiske jordbrugsbedrifter 2015–2018. Ministry of Environment and Food, Denmark. http://www.lbst.dk/

Maliro MFA, Guwela V (2015) Quinoa breeding in Africa: history, goals and progress. In: Murphy KM, Matanguihan JG (eds) Quinoa: improvement and sustainable production. Wiley-Blackwell, Hoboken, NJ, pp 161–172

Núnez de Arco S (2015) Chapter 12: Quinoa's Calling. In: Murphy K, Matanguihan J (eds) Quinoa: improvement and sustainable production. Wiley Blackwell, New Jersery, pp 211–226

Oficina Internacional del Trabajo (2015) Análisis de la cadena de valor en el sector de la quinua en Perú: aprovechando las ganancias de un mercado creciente a favor de los probres: Ginebra

Palmgren M (2019) Raising quinoa: an orphan crop that is healthy and ready for a changing climate. The Carlsberg Foundation. https://www.carlsbergfondet.dk/en/Forskningsaktiviteter/Bevillingsstatistik/Bevillingsoversigt/CF18_1113_Michael-Broberg-Palmgren

Peterson A, Murphy K (2013) Quinoa in the United States of America and Canada. Chapter 6.4.1. In: Bazile D et al (eds) FAO & CIRAD. State of the Art Report of Quinoa in the World in 2013, Rome, p 549

Rodriguez JP, Rahman H, Thushar S, Singh RK (2020) Healthy and resilient cereals and pseudo-cereals for marginal agriculture: molecular advances for improving nutrient bioavailability. Fron Genet 11. https://doi.org/10.3389/fgene.2020.00049

Rojas W, Pinto M, Alanoca C, Gomez Pando L, Leon-Lobos P, Alercia A, Diulgheroff S, Padulosi S, Bazile D (2015). Quinoa genetic resources and ex situ conservation. In: Bazile D, Bertero HD, Nieto C (eds) State of the art report on quinoa around the world in 2013. FAO & CIRAD, Roma, Italy, pp 56–82

Song M, Fung T, Hu F, Willett W, Longo V et al (2016) Association of animal and plant protein intake with all-cause and cause-specific mortality. JAMA Intern Med 176(10):1453–1463. https://doi.org/10.1001/jamainternmed.2016.4182

Swieca M, Seczyk L, Gawlik-Dziki U, Dziki D (2014) Bread enriched with quinoa leaves. the influence of protein–phenolics interactions on the nutritional and antioxidant quality. Food Chem 162:54–62

Tapia M (2014) El largo camino de la quinoa: ¿Quiénes escribieron su historia? In: Bazile D et al (eds) FAO y CIRAD. Estado del arte la quinua en el mundo en 2013, pp 3–10

Tester M (2019) Association Genetics of Quinoa. In: ODEPA (eds) Libro de Resúmenes. VII Congreso Mundial de la Quinua y otros granos Andinos. Ministerio de Agricultura, INDAP, Pontificia Universidad Católica de Chile. Santiago de Chile, p 168

United Nations (2015) Transforming our world: the 2030 agenda for sustainable development. United Nations General Assembly Resolution A/RES/70/1. 70th session

Vega-Gálvez A, Miranda M, Vergara J, Uribe E, Puente L, Martínez EA (2010) Nutrition facts and functional potential of quinoa (Chenopodium quinoa Willd.), an ancient Andean grain: a review. J Sci Food Agric 90:2541–2547

Xiu-shi Y, Pei-you Q, Hui-min G, Gui-xing R (2019) Quinoa industry development in China. Cien Inv Agric 46(2):208–219. https://doi.org/10.7764/rcia.v46i2.2157

The manufacturer's authorised representative in the EU is Springer Nature Customer Service Centre GmbH, Europaplatz 3, 69115 Heidelberg, Germany. If you have any concerns regarding our products, please contact ProductSafety@springernature.com

Printed and bound by CPI Group (UK) Ltd, Croydon, CR0 4YY

30/09/2026

02225693-0009